# Digital Signal Processing Implementations

## Using DSP Microprocessors—with Examples from TMS320C54xx

▷ Avtar Singh
*San Jose State University*

▷ S. Srinivasan
*Indian Institute of Technology, Madras*

Australia • Canada • Mexico • Singapore • Spain
United Kingdom • United States

Publisher/Executive Editor: *Bill Stenquist*
Editorial Assistant *Julie Lynne Ruggiero*
Technology Project Manager: *Burke Taft*
Executive Marketing Manager: *Tom Ziolkowski*
Marketing Assistant: *Jennifer Gee*
Advertising Project Manager: *Vicki Wan*
Project Manager, Editorial Production: *Kelsey McGee*
Print/Media Buyer: *Doreen Suruki*

Permissions Editor: *Sommy Ko*
Production Service: *Matrix Productions*
Copy Editor: *Pamela Rockwell*
Cover Designer: *Roger Knox Design*
Cover Photo: *Courtesy of Texas Instruments*
Composition: *Asco Trade Typesetters*
Cover Printing, Printing and Binding:
*Phoenix Color Corp.*



Printed in the United States
1 2 3 4 5 6 7 07 06 05 04 03

For more information about our products, contact us at:
**Thomson Learning Academic Resource Center**
**1-800-423-0563**

For permission to use material from this text, contact us by:
**Phone:** 1-800-730-2214 **Fax:** 1-800-730-2215
**Web:** http://www.thomsonrights.com

Library of Congress Control Number: 2003107814

ISBN 0-534-39123-0

**Brooks/Cole—Thomson Learning**
**10 Davis Drive**
**Belmont, CA 94002**
**USA**

**Asia**
Thomson Learning
5 Shenton Way #01-01
UIC Building
Singapore 068808

**Australia/New Zealand**
Thomson Learning
102 Dodds Street
Southbank, Victoria 3006
Australia

**Canada**
Nelson
1120 Birchmount Road
Toronto, Ontario M1K 5G4
Canada

**Europe/Middle East/Africa**
Thomson Learning
High Holborn House
50/51 Bedford Row
London WC1R 4LR
United Kingdom

**Latin America**
Thomson Learning
Seneca, 53
Colonia Polanco
11560 Mexico D.F.
Mexico

**Spain/Portugal**
Paraninfo
Calle Magallanes, 25
28015 Madrid, Spain

# Contents

# Preface

Due to advances in VLSI technology, programmable DSP devices are becoming increasingly available and affordable. These devices have, therefore, become popular in the industry for the design of products. Consequently, a large number of undergraduate senior projects and graduate projects are planned and implemented using these devices. Many students attempt these projects based on a first-level course on digital signal processing. The books that are used in these classes do not, however, cover the topics from the implementation point of view. There is generally a wide gap in students' understanding of DSP algorithms and how to use programmable DSP devices to implement them.

This is a DSP implementation-oriented textbook that has been written based on the authors' experience in teaching graduate and undergraduate courses on the subject. The objective of the book is to help the reader to understand the architecture, programming, and interfacing of commercially available programmable DSP devices and to effectively use them in system implementations. The book is intended for senior undergraduate and first-level graduate students in electrical engineering and computer science programs. The book will also be useful to engineers in industry engaged in the design of DSP systems. The background expected from a reader is a course in digital signal processing and a course in microprocessors, both at the undergraduate level.

This book contains 11 chapters and covers the architectural issues of programmable DSP devices and their relationship to the algorithmic requirements, architectures of commercially popular programmable devices, and the use of such devices for software development and system design. These issues are covered using a popular family of DSP devices—TMS320C54xx from Texas Instruments.

Chapter 1 identifies the role of programmable devices in the implementation of DSP-based systems. Chapter 2 reviews the DSP basics so that the reader can correlate the remainder of the book to the theoretical requirements of a DSP system. The aim is not to attempt to teach DSP theory, which is abundantly covered elsewhere, but to highlight the concepts that are relevant from the point of view of implementations. MATLAB is used as a tool in exploring and understanding the basic DSP concepts. Chapter 3 looks at issues that determine the computational accuracy of algorithms when implemented using programmable DSP devices. Although it is desirable to retain as much accuracy as possible when DSP algorithms are implemented in hardware, in a practical implementation, accuracy has to be measured against the speed of operation and hardware complexity. Different number representation schemes are introduced and their effects on precision and dynamic range are discussed. Various sources of errors in a DSP system are described and are quantitatively evaluated in this chapter.

One of the objectives of the book is to give readers sufficient exposure to the architecture of programmable DSP devices so that they can use them effectively and optimally in designing systems. Chapter 4 explains the architectural features of programmable DSP devices based on the operations these devices are required to perform. Various building blocks that constitute a programmable digital signal processor are discussed from the point of view of implementations. Desirable features for each of these blocks are discussed in terms of their hardware realization. Chapter 5 introduces the Texas Instruments' TMS320C54xx family of fixed-point DSP processors and discusses their architecture, software, and hardware features. These devices are used in programming and design examples throughout the book. Chapter 6 introduces the various tools that are available for the development of DSP software on programmable devices. In particular, the use of DSK5416, a system design kit used for program development for the TMS320C54xx, and the development software called Code Composer Studio are described. The DSK5416 is the development board around which all the designs are implemented in subsequent chapters.

In Chapters 7 and 8, programming of the TMS320C54xx devices for several basic DSP algorithms is explained. Examples are constructed to show implementations of FIR filters, IIR filters, decimation filters, interpolation filters, adaptive filters, a PID controller, two-dimensional signal processing, and the FFT algorithms.

Chapters 9 and 10 deal with the signals of a programmable DSP device required for interfacing it to the real world. Interfacing of memory and I/O to the DSP devices are discussed with examples. The system integration topics such as DMA and interrupts are also covered. Programming of a CODEC device interfaced to the DSP on the DSK5416 is covered so as to enable the reader to use its A/D and D/A converters for serial I/O.

Chapter 11 presents several applications of programmable DSP devices. The objective of this chapter is to highlight the suitability of programmable DSP devices for various application areas and motivate readers to design systems around these devices.

The chapters have many end-of-chapter assignment problems and laboratory exercises. The lab exercises require the use of MATLAB as an analysis/design tool and DSK5416 with Code Composer Studio as a hardware/software development tool. The programs in the book are available on the web site. The site also contains additional examples and projects and links to other related information. To access the site requires a password available from the publisher. The programs in the book can be used in many applications with appropriate enhancements. The development tools are inexpensively available from TI. At the end of a course with this book as the text, the student should be comfortable in using both hardware and software for designing with programmable DSP devices.

In conclusion, there is a gap between the algorithm-based DSP courses, generally offered in most universities, and the implementation of these algorithms using commercial devices and tools. The implementation area is becoming increasingly important as it leads to innovative applications for the marketplace. Seeing the importance, many universities have attempted courses in this area, generally without a textbook and mainly relying on the company literature. In our opinion, this book fills this gap between DSP theory and DSP design.

A book of this nature can only be developed with help from both academia and industry. Many of our students at both of our institutions have been the source of motivation for this project and have contributed to its completion. Specifically, we would like to thank our students Ramandeep Kaur Sahi, Ulhas Kotha, Uldarico Muico, and H. Larios of San Jose State University, and Abhishek Tandon, Vineet Jain, Kaushik Raghunath, Gaurav Verma, and Surender Reddy of the Indian Institute of Technology, Madras. Secretarial assistance provided by S. Sreekala and the technical assistance by Narendra S. Sihra are gratefully acknowledged. Chris Petersen and Keith Ogboenyiya of Texas Instruments are specially thanked for arranging a generous donation of the development boards and the software, without which this project could not have been completed.

*Avtar Singh, SJSU*
*S. Srinivasan, IIT, Madras*

# Chapter 1

# Introduction

## 1.1 A Digital Signal-Processing System

Digital signal processing (or DSP) is the technique of performing mathematical operations on signals represented as sequences of samples. These sequences are obtained by converting real-world analog signals by means of analog-to-digital converters. After processing, the digital samples are converted back to analog signals by means of digital-to-analog converters. Although functionally digital signal processing is the heart of a DSP system, the analog front end and the analog back end are equally important, as the system has to be interfaced to the real-world signals, which are mostly analog. Digital processing of signals offers many advantages over analog processing. Some of these are: immunity to environmental noise, predictable and reproducible behavior, programmability, size, and cost. Examples of digital signal-processing systems can be found in speech and audio systems, telecommunication applications such as modems, electronic and biomedical instrumentation, image processing, robotics, control applications, etc.

The block diagram of a typical DSP system is shown in Figure 1.1. It consists of the DSP processor between the analog front end and the analog back end. The analog front end consists of an antialiasing filter, a sample and hold circuit, and an analog-to-digital (A/D) converter feeding into the DSP. The back end consists of a digital-to-analog (D/A) converter to convert the digital output to its analog value followed by a reconstruction filter. The antialiasing filter, an analog lowpass filter, is used to band limit the input analog signal to the required frequency range and prevent frequency components beyond this range from appearing as aliases in the sampled spectrum of the input signal. The sample and hold circuit presents the samples of the input signal at the rate determined by the system design requirements to the input of the analog-to-digital converter. It also holds these samples at constant levels irrespective of the variations in the input signal in the interval between sampling instants. The analog-to-digital converter maps the value of the analog input sample to its equivalent digital representation and feeds it to the DSP.

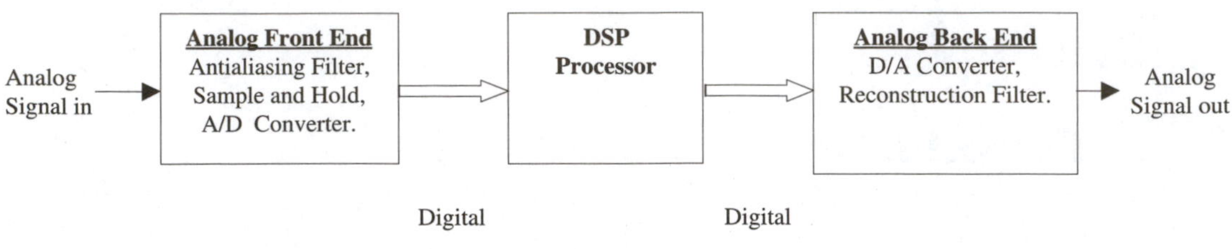


**Figure 1.1** The block diagram of a DSP system

After processing, the digital outputs of the DSP are converted to their equivalent analog values by the digital-to-analog converter. These discrete analog values are converted to a smooth, continuous waveform by the reconstruction filter at the output for use in the real world. Like the antialiasing filter, the reconstruction filter is also an analog lowpass filter.

The following issues are important to be considered in designing and implementing a DSP system.

**Complexity of the algorithm:** The arithmetic operations to be performed and the precision required are decided by the application.

**Sample rate:** The rate at which input samples are received and processed varies with the application, and this rate along with the algorithm complexity determines whether a particular DSP is suitable for a given application.

**Speed:** This depends on the technology. To meet specified throughput requirement with a given sample rate, it must be possible to operate the DSP at a particular clock rate (or speed). If this speed is not achievable in a given technology, a faster technology or other options must be explored.

**Data representation:** The format and the number of bits used for data representation depend on the arithmetic precision and the dynamic range required for the given application.

## 1.2 Programmable Digital Signal Processors

Digital signal processors can be either application-specific or general purpose. Application-specific chips are designed to perform one function more accurately, faster, or more cost-effectively than their general-purpose counterparts. Typical examples are digital filters and fast Fourier Transform chips. Some application-specific chips are programmable, but only within the confines of the chip's function; the coefficients of a filter, for example, can be programmed.

A programmable digital signal processor, on the other hand, is cost-effective. It can be programmed for different applications and has a short design cycle time. Basically, it is a microprocessor whose architecture is optimized to process sampled data at high rates [1]. It performs such operations as accumulating the sum of multiple products much faster than an ordinary microprocessor. Its architecture is designed to exploit the repetitive nature of signal processing by pipelining the data flow and by incorporating parallelism in its operation. These features are designed in the programmable DSP to achieve higher speed and throughput.

For a given application, there is a large number of programmable DSPs to choose from, based on such factors as speed, throughput, arithmetic capability, precision, size, cost, and power consumption. As the technology grows, there are more and more such devices with better and better performance characteristics that are easily incorporated in DSP systems.

## 1.3 Major Features of Programmable Digital Signal Processors

Although there are many unique architectural features implemented in programmable DSP devices [3], following are the ones that are commonly found:

**Multiply-accumulate hardware:** Multiply-accumulate is the most frequently used operation in digital signal processing. In order to implement this efficiently, the DSP has a hardware multiplier, an accumulator with an adequate number of bits to hold the sum of products and an explicit multiply-accumulate instruction.

**Harvard architecture:** In Harvard memory architecture, there are two memory spaces, typically partitioned as program memory and data memory (though there are modified versions that allow some crossover between the two). The processor core connects to these memory spaces by two separate bus sets, allowing two simultaneous accesses to memory. This arrangement doubles the processor's memory bandwidth, and is crucial in keeping the processor core fed with data and instructions. The Harvard architecture is sometimes further extended with additional memory spaces and/or bus sets to achieve even higher memory bandwidths.

**Zero-overhead looping:** One common characteristic of DSP algorithms is that most of the processing time is spent on executing instructions contained within relatively small loops. That is why most DSP processors include specialized hardware for zero-overhead looping. The term *zero-overhead looping* means that the processor can execute loops without consuming cycles to test the value of the loop counter, perform a conditional branch to the top of the loop, and decrement the loop counter.

**Specialized addressing:** DSP processors often support specialized addressing modes that are useful for common signal-processing operations and algorithms. Examples include modulo (circular) addressing, useful for implementing digital-filter delay lines, and bit-reversed addressing, useful for implementing a commonly used DSP algorithm called the *Fast Fourier Transform or FFT*.

## 1.4 The Scope of the Book

Due to advances in VLSI technology, programmable DSP devices are becoming increasingly available and affordable. These devices have, therefore, become popular in the industry for the design of products. Consequently, a large number of undergraduate senior projects and graduate projects are planned and implemented using these devices [2]. This book attempts to bridge the gap between the knowledge of DSP theory and practical implementation of systems using DSP devices.

The scope of this book includes the following:

1. Architectural issues of programmable DSP devices and their relationship to the algorithmic requirements
2. Exposure to commercially popular architectures
3. Use of programmable devices for software development and system design

These topics are covered using a popular family of DSP devices from Texas Instruments (TI), the TMS320C54xx DSP family, similar to the one shown in Figure 1.2. The processors from this family have been used in many digital signal-processing implementations. The processors from other companies, such as Analog Devices and Motorola, can equally be used to implement such systems. In this book, however, we limit our discussion to the TI processors.

The book contains 11 chapters. Chapter 2 reviews the basic DSP concepts. Chapter 3 covers the accuracy in DSP implementations. It discusses the sources of errors in DSP computations. Chapter 4 lists the architectural requirements of digital signal processors for efficient implementation of algorithms. Chapter 5 introduces programmable DSP devices and gives the architectural and programming details of the TMS320C54xx family of devices. Chapter 6 covers the software development tools for programmable DSP devices. Chapters 7 and 8 deal with implementations of DSP algorithms on TMS320C54xx DSP processors. Chapters 9 and 10 discuss interfacing of DSP devices to external peripherals, both serial and parallel. Chapter 11 gives selected examples of applications of programmable DSP devices.

**Figure 1.2** TMS320C54x DSP Microprocessor

(Courtesy of Texas Instruments Inc.)

## References

1. Allen, J., "Computer Architecture for Digital Signal Processing," IEEE Proceedings, Vol. 73, pp. 852–873, May 1985.
2. *Special Issue on Digital Signal Processing in Undergraduate Education*, IEEE Transactions on Education, vol. 39, no. 12, May 1996.
3. Lapsley, P., Bier, J., Shoham, A., and Lee, E. A., *DSP Processor Fundamentals: Architectures and Features*, IEEE Press, Piscataway, NJ, 1997.

# Chapter 2

## Introduction to Digital Signal Processing

### 2.1 Introduction

This chapter reviews the important basic concepts of digital signal processing (DSP). The coverage is brief and is from the viewpoint of implementations of DSP algorithms. The concepts are illustrated with examples using MATLAB's capability to analyze and design algorithms. For comprehensive coverage of DSP algorithms, the reader is advised to consult the references [1, 2] at the end of this chapter. Specifically, the following topics are covered in this chapter:

A digital signal-processing system
The sampling process
Discrete time sequences
Discrete Fourier transform (DFT) and fast Fourier transform (FFT)
Linear time-invariant systems
Digital filters
Decimation and interpolation
Analysis and design tool for DSP systems: MATLAB

### 2.2 A Digital Signal-Processing System

A digital signal-processing (DSP) system uses a computer or a digital processor to process signals. The real-life signals are analog and therefore must be converted to digital signals before they can be processed with a computer. To convert a signal from analog to digital, an analog-to-digital (A/D) converter is used. After processing the signal digitally, it is usually converted to an analog signal using a device called a digital-to-analog (D/A) converter. The block diagram of Figure 2.1 shows the components of a DSP scheme. This

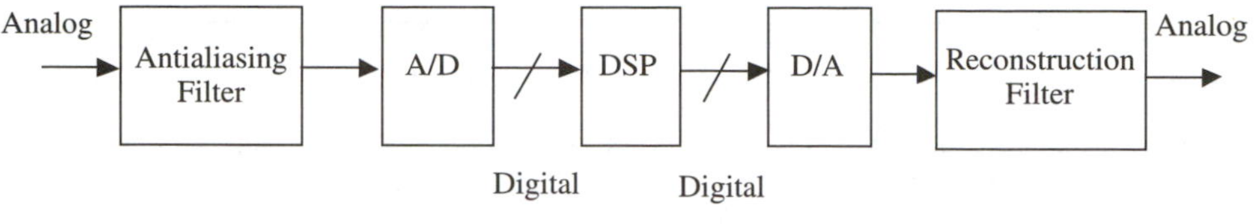


**Figure 2.1** A digital signal-processing system

figure contains two additional blocks, one is the antialiasing filter for filtering the signal before sampling and the second is the reconstruction filter placed after the D/A converter. The antialiasing filter ensures that the signal to be sampled does not contain any frequency higher than half of the sampling frequency. If such a filter is not used, the high-frequency contents sampled with an inadequate sampling rate generate low-frequency aliasing noise. We will discuss the choice of sampling frequency further in the next section. The reconstruction filter removes high-frequency noise due to the "staircase" output of the D/A converter.

The signals that occur in a typical digital signal-processing scheme as shown in Figure 2.2 are: continuous-time or analog signal, sampled signal sampled-data signal, quantized or digital signal, and the D/A output signal.

An analog signal is a continuous-time, continuous-amplitude signal that occurs in real systems. Such a signal is defined for any time and can have any amplitude within a given range. The sampling process generates a sampled signal. A sampled signal value is held by a hold circuit to allow an A/D converter to change it to the corresponding digital or quantized signal. The signal at the A/D converter input is called a *sampled-data* signal and at the output is the digital signal. The processed digital signal, as obtained from the digital signal processor (DSP), is the input to the D/A converter. The analog output of a D/A converter has "staircase" amplitude due to the conversion process used in such a device. The signal, as obtained from the D/A, can be passed through a reconstruction lowpass filter to remove its high-frequency contents and hence smoothen it.

## 2.3 The Sampling Process

The process of converting an analog signal to a digital signal involves sampling the signal, holding it for conversion, and converting it to the corresponding digital value. The sampling frequency must be high enough so as to avoid aliasing. Aliasing is a phenomenon due to which a high-frequency signal when sampled using a low (inadequate) sampling rate becomes a low-frequency signal that may interfere with the signal of interest. To avoid

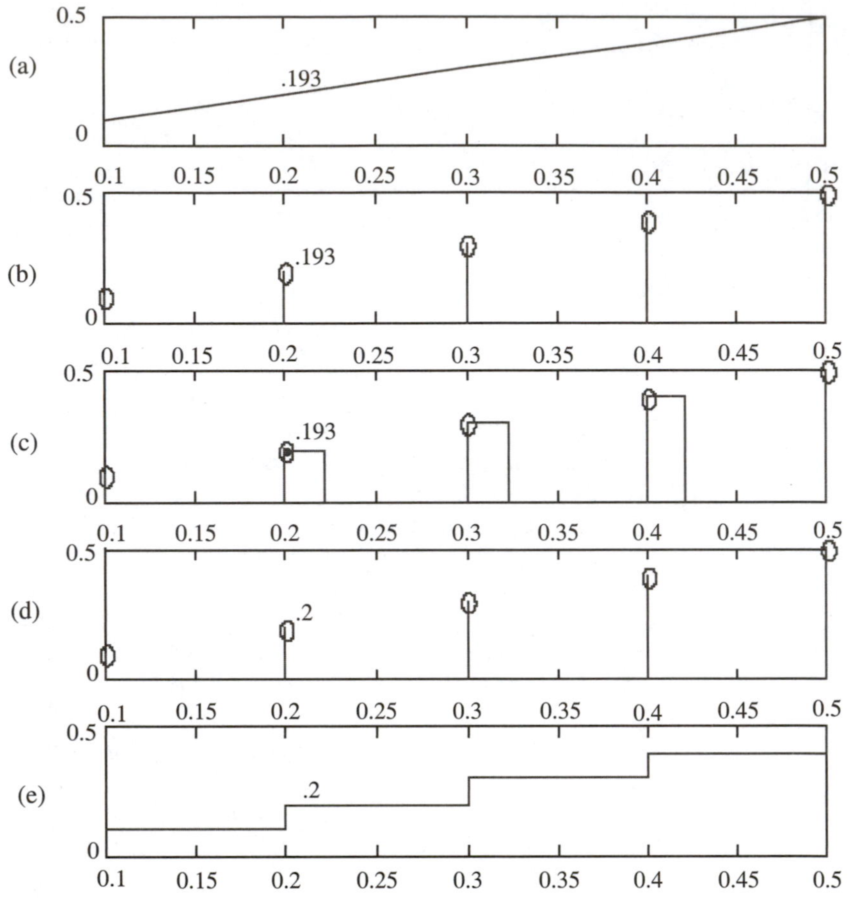


**Figure 2.2** Typical signals in a DSP scheme: (a) continuous-time signal, (b) sampled signal, (c) sampled-data signal, (d) quantized (digital) signal, (e) digital-to-analog converter ouput signal

aliasing, the sampling theorem states that the following requirement must be satisfied:

$$f_s = 1/T > 2f_{max} \tag{2.1}$$

where

$f_s$ is the sampling frequency in Hz,
$T$ is the sampling interval in seconds, and
$f_{max}$ is the highest frequency contents of the analog signal

For instance, if we sample a signal with its highest frequency content as 10 KHz, it must be sampled using a sampling rate of more than 20 KHz. In order to satisfy this requirement, an antialiasing filter is used. This filter limits the frequency contents of the signal to satisfy the sampling theorem. One has to sacrifice (unimportant, one hopes) frequency contents to avoid violation of the sampling theorem, or else the sampling rate must be increased. The actual sampling frequency must be higher than this theoretical limit to avoid tight constraints for the implementation of the antialiasing filter.

## 2.4 Discrete Time Sequences

The result of sampling an analog signal is a sequence representing the signal samples. The sequence that results depends upon the signal that is sampled. For instance, when

$$x(t) = A \cos 2\pi ft$$

is sampled using $T$ as the sampling interval, it yields the samples as

$$x(nT) = A \cos 2\pi fnT, \quad \text{where } n = 0, 1, 2, \ldots, \text{etc.}$$

For simplicity, the sequence $x(nT)$ is denoted as $x(n)$. Thus,

$$x(n) = A \cos 2\pi fnT$$

Since the sampling frequency $f_s = 1/T$, and substituting $\theta$ for $2\pi fT$, we obtain

$$x(n) = A \cos 2\pi fnT = A \cos 2\pi fn/f_s = A \cos \theta n$$

The quantity, denoted by $\theta$, is called the **digital frequency.** Note that the units for the digital frequency are radians. The general equation that relates the digital frequency to analog frequency is

$$\theta = 2\pi fT = 2\pi f/f_s \tag{2.2}$$

Note that the digital frequency range, for a properly sampled signal ($f_s > 2f_{max}$) as obtained from Eq. 2.2, is from 0 to $\pi$.

The above $x(n)$ sequence, called the *sinusoidal sequence*, occurs frequently in DSP systems. Another important sequence that arises in DSP schemes is the complex exponential sequence given by

$$p(n) = e^{j2\pi n/N}, \quad n = \ldots -1, 0, 1, 2, \ldots, \text{etc.}$$

where $N$ is an integer.

A sequence that repeats is called a **periodic sequence**. Periodic sequences result from sampling periodic signals and satisfy the following relation:

$$x(n) = x(n + N), \quad n = \ldots -1, 0, 1, 2, \ldots \tag{2.3}$$

where $N$ is called the *sequence period.* It is easy to show that the sinusoidal sequence $x(n)$ above has a period $f_s/f$, and the exponential sequence $p(n)$ has a period equal to $N$ samples.

The **frequency response** associated with a time domain $N$-point sequence $x(n)$ can be determined from

$$X(e^{j\theta}) = \sum_{n=0}^{N-1} x(n)e^{-jn\theta} \tag{2.4}$$

where $\theta$ is the digital frequency, which ranges from 0 to $2\pi$ radians corresponding to the analog frequency from 0 to $f_s$ Hz. Note that the frequency response is a complex continuous function of $\theta$ and provides both the magnitude response and the phase response.

## 2.5 Discrete Fourier Transform (DFT) and Fast Fourier Transform (FFT)

The discrete Fourier transform, or DFT, is used to transform a time domain $x(n)$ sequence to a frequency domain $X(k)$ sequence. To transform $X(k)$ to $x(n)$, the inverse discrete Fourier transform, or IDFT, is used. Algorithms for fast computation of DFT and IDFT are known as FFT algorithms.

### 2.5.1 The DFT Pair

The two equations that relate the time domain $x(n)$ and the frequency domain $X(k)$ sequences are called the *DFT pair* and are given as

$$X(k) = \sum_{n=0}^{N-1} x(n)e^{-j2\pi nk/N}, \quad k = 0, 1, 2, \ldots (N-1) \tag{2.5}$$

$$x(n) = 1/N \sum_{k=0}^{N-1} X(k)e^{j2\pi nk/N}, \quad n = 0, 1, 2, \ldots (N-1) \tag{2.6}$$

The first equation is called the DFT and the second is called the IDFT. The $N$ in the DFT pair denotes the number of elements in the $x(n)$ or $X(k)$ sequence.

## 2.5.2 The Relationship between DFT and Frequency Response

The frequency response of a sequence (Eq. 2.4) and its DFT (Eq. 2.5) are related as follows:

$$X(k) = X(e^{j\theta})|_{\theta=2\pi k/N}, \quad k = 0, 1, 2, \ldots (N-1) \tag{2.7}$$

The elements of $X(k)$ as obtained from this equation are spaced at a digital frequency of $2\pi/N$ radian. The equation allows us to use DFT to compute points on the frequency response of the $x(n)$ sequence. The corresponding analog frequency spacing $\Delta f$, between elements of the $X(k)$ sequence, using Eq. 2.2, can be shown to be

$$\Delta f = f_s/N = 1/NT = 1/T_0 \tag{2.8}$$

where $T_0$ is called the *signal record length*. From the above relation, it is easy to conclude that the larger the signal record, the smaller (or better) is the frequency spacing.

The significance of this result lies in the fact that it describes the trade-off between the sampling rate ($f_s$), number of sequence points ($N$), and the frequency spacing ($\Delta f$). To decrease the frequency spacing, $N$ can be increased by simply appending zeros to the $x(n)$ sequence before computing $X(k)$.

## 2.5.3 The Fast Fourier Transform (FFT)

The direct computation of DFT and IDFT requires a large number of complex multiplies. A number of algorithms have been developed to efficiently compute DFT and IDFT. These algorithms use power of 2 points and exploit the periodic nature of the complex exponential $e^{j2\pi nk/N}$ occurring in the DFT and IDFT equations. Table 2.1 compares the complex multiplies needed to compute DFT directly by using an FFT algorithm called the *radix-2 algorithm*. The radix-2 algorithm uses $N$ that is an integer power of 2, such as 2, 4, 8, 16, etc.

It is possible to show that the DFT requires $N^2$ complex multiplies and the radix-2 FFT algorithm requires $\frac{N}{2}\log_2 N$. This produces computational savings for larger values of $N$.

An application of FFT can be to use it to compute signal power spectral density (PSD) or simply the signal spectrum. The FFT result $X(k)$ can be used to compute the spectrum as follows:

$$S(k) = (1/N)|X(k)|^2 = (1/N)X(k)X^*(k), \quad k = 0, 1, 2, \ldots N-1 \tag{2.9}$$

The plot of $S(k)$ provides power density associated with various frequencies and is used to characterize the signal in the frequency domain.

**Table 2.1** Complex Multiplies for Direct DFT and FFT-based DFT Computations

| N | Direct DFT Computation | FFT Based Computation | DFT Multiplies/FFT Multiplies |
|---|---|---|---|
| 2 | 4 | 1 | 4.0 |
| 4 | 16 | 4 | 4.0 |
| 16 | 256 | 32 | 8.0 |
| 64 | 4096 | 192 | 21.3 |
| 256 | 65536 | 1024 | 64.0 |
| 512 | $512^2$ | $512/2 \log_2 512$ | $2 \times 512 \div \log_2 512$ |
| . | . | . | . |
| . | . | . | . |
| . | . | . | . |

## 2.6 Linear Time-Invariant Systems

To represent the input/output relation of a discrete system, the block diagram of Figure 2.3 can be used. A system to which the superposition theorem can be applied is known as a linear system. A system that is described by the same input/output relation at all times is called a *time-invariant system*. A system that is both a linear as well as time-invariant is called *linear time-invariant*, or LTI, *system*.

The LTI systems can be represented in the time domain using linear constant coefficient difference equations. A unit sample (or impulse) response is used to characterize an LTI system. Time domain convolution can be used to determine the response of an LTI system.

In the frequency domain, the system transfer function is used to represent such a system. We now briefly discuss these concepts.

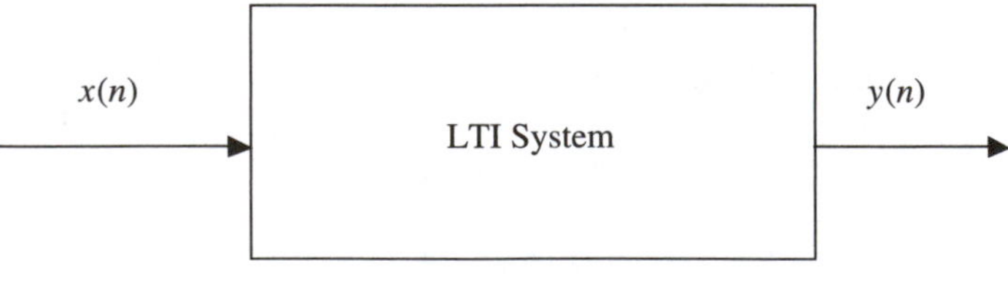


**Figure 2.3** Representation of a linear time-invariant system

### 2.6.1 Convolution

Convolution is an operation that relates the input/output of an LTI system to its unit sample response. It is given by the equation

$$y(n) = \sum_{m=-\infty}^{\infty} h(n)x(n-m) = \sum_{m=-\infty}^{\infty} x(n)h(n-m) = h(n) * x(n) \quad (2.10)$$

where $x(n)$ represents the input, $y(n)$ the output, and $h(n)$ the unit sample response of the system. The $*$ in Eq. 2.10 is used to represent the convolution operation. This result can be derived using the impulse response definition as applied to the sampled $x(n)$ sequence. This equation is used to compute the time-domain response of a system to an arbitrary input sequence.

### 2.6.2 *Z*-Transform

We have seen in Section 2.4 that the frequency response associated with the $N$-point sequence $x(n)$ is given as

$$X(e^{j\theta}) = \sum_{n=0}^{N-1} x(n)e^{-jn\theta} \quad (2.11)$$

Using the substitution

$$z = e^{j\theta} \quad (2.12)$$

in the above equation yields

$$X(z) = \sum_{n=0}^{N-1} x(n)z^{-n} \quad (2.13)$$

where, $X(z)$ is called the $Z$-transform of $x(n)$. Since the parameter $z$ is related to the digital frequency, $X(z)$ represents the frequency response in terms of $z$.

### 2.6.3 The System Function

The ratio of $Z$-transform of $y(n)$ to that of $x(n)$

$$H(z) = Y(z)/X(z) \quad (2.14)$$

is called the *system function* or the *transfer function* of the LTI system. The system function characterizes the system in the frequency domain. $H(z)$ is a complex function of $z$. The magnitude and the phase of $H(z)$ describe the frequency response of the system.

A way to characterize an LTI system is to specify its poles and zeros. The poles are the roots of the denominator of the transfer function and the zeros are the roots of its numerator. The locations of poles and zeros lead one to determine the frequency response of the system [1].

## 2.7 Digital Filters

A filter is a sequence $h(n)$ that operates on an input sequence $x(n)$ to generate a filtered output sequence $y(n)$. A filter sequence may be represented in terms of other sequences (coefficient sequences, such as $a_k$ and $b_k$). The general difference equation for an $N$th order filter is given as

$$y(n) = \sum_{k=1}^{N} a_k y(n-k) + \sum_{k=0}^{L} b_k x(n-k) \tag{2.15}$$

Figure 2.4 represents the above general difference equation in a block diagram form. Notice that the structure employs feedback, as the current output depends on past outputs in addition to current and past inputs. Any kind of filter can be realized by appropriately selecting the coefficient sequences. For instance, the standard filter types, such as lowpass, highpass, bandpass, and bandstop, can be implemented using appropriate $a_k$s and $b_k$s in the general filter equation given above. Thus, a filter design involves determining these coefficients to obtain the desired frequency response.

### 2.7.1 Finite Impulse Response (FIR) Filter

A simpler version of the general filter difference equation is

$$y(n) = \sum_{k=0}^{L} b_k x(n-k) \tag{2.16}$$

This equation defines a finite impulse response (FIR) filter. The unit sample response of the FIR filter lasts for a finite time dependent on the number of filter coefficients represented by $b_k$. It is for this reason that such a filter is called a FIR filter.

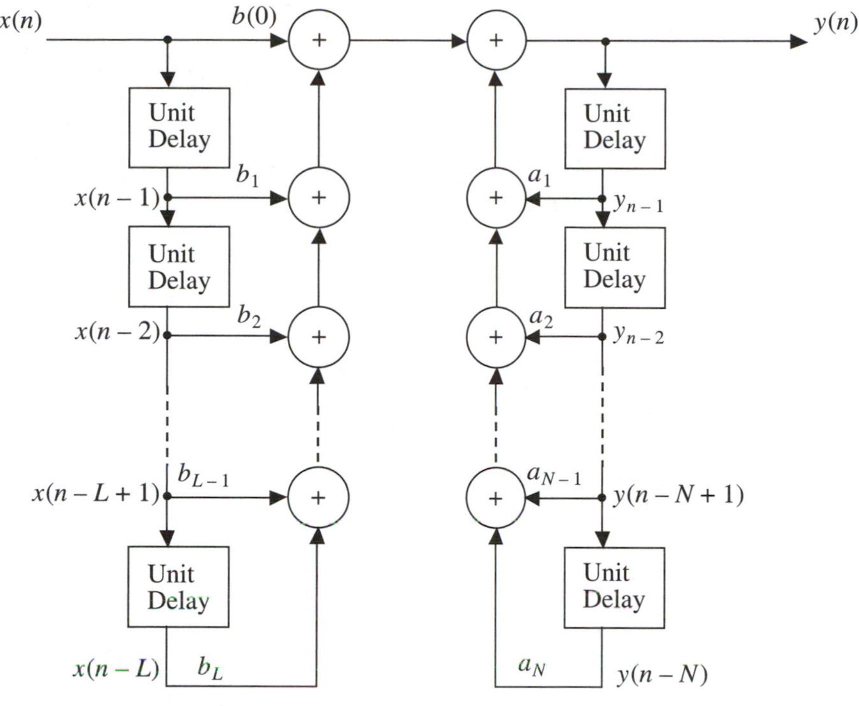


**Figure 2.4** Block diagram representation of a digital filter

The frequency response of a FIR filter can be determined from

$$H(e^{j\theta}) = \sum_{k=0}^{L} b_k e^{-jk\theta} \tag{2.17}$$

In terms of the $Z$-transform it can be expressed as

$$H(z) = \sum_{k=0}^{L} b_k z^{-k} \tag{2.18}$$

Since a FIR filter has no feedback in its structure, it is always a stable filter. A symmetric coefficient FIR filter provides linear phase or constant group delay. This property makes this filter suitable for applications that cannot tolerate phase distortion. The disadvantage of a FIR filter is that to obtain any desired frequency response the number of coefficients is generally quite large. Larger numbers of coefficients require larger computation time. Larger computation time limits the sampling rate and hence the bandwidth of the signal that can be processed in a real-time signal-processing scheme.

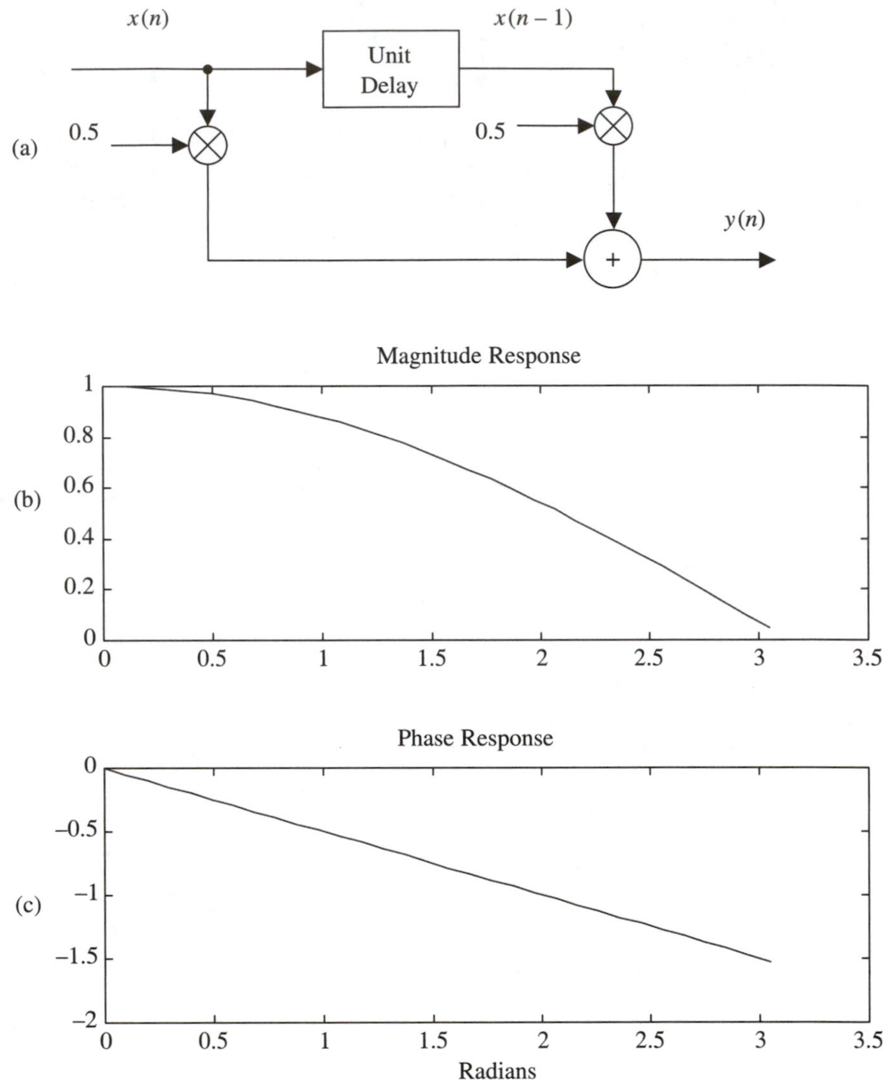


**Figure 2.5** FIR filter in Example 2.1: (a) block diagram, (b) magnitude frequency response, (c) phase frequency response

▷ **Example 2.1 A FIR Filter**

The equation

$$y(n) = 0.5x(n) + 0.5x(n-1)$$

describes a simple FIR filter whose output is the average of the current input $x(n)$ and the past input $x(n-1)$.

The unit sample response of this filter is obtained by substituting $\delta(n)$ for $x(n)$. Thus, we have

$$h(n) = 0.5\delta(n) + 0.5\delta(n-1)$$
$$= [0.5 \;\; 0.5] \quad \text{as a sequence.}$$

The frequency response, using Eq. 2.17, is obtained as

$$H(e^{j\theta}) = 0.5 + 0.5e^{-j\theta} = e^{-j\theta/2} \cos \theta/2$$

or

$$H(z) = 0.5 + .5z^{-1}$$

The magnitude response is given as

$$|H(e^{j\theta})| = M(\theta) = \cos \theta/2$$

and the phase response is given as

$$\angle H(e^{j\theta}) = P(\theta) = -\theta/2 + \angle \cos \theta/2$$

The group delay, which represents the delay to various signal frequencies, can be obtained by differentiating and negating the phase response function. For this example case it is obtained as

$$\text{Group delay} = \tfrac{1}{2}$$

Figure 2.5 describes this filter with its magnitude and phase responses. Implementing this filter requires a unit delay, two multiplies, and an addition.

---

## 2.7.2 Infinite Impulse Response (IIR) Filters

The general difference Eq. 2.15 for an LTI system defines an infinite impulse response (IIR) filter. The corresponding transfer function for this filter can be shown to be

$$H(z) = \frac{b_0 + b_1 z^{-1} + b_2 z^{-2} + b_3 z^{-3} + \cdots + b_L z^{-L}}{1 - a_1 z^{-1} - a_2 z^{-2} - a_3 z^{-3} - \cdots - a_N z^{-N}} \tag{2.19}$$

Since an IIR filter has feedback in its structure, its stability depends upon the number and values of coefficients. In general, an IIR filter has nonlinear phase response and does not provide constant group delay. This property makes this filter unsuitable for applications that cannot tolerate phase distortion. The advantage of an IIR filter is its smaller number of coefficients to realize a desired frequency response relative to an FIR filter. Fewer coefficients require shorter computation time, providing capability to handle a larger bandwidth for a signal-processing scheme.

▷ **Example 2.2 An IIR Filter**

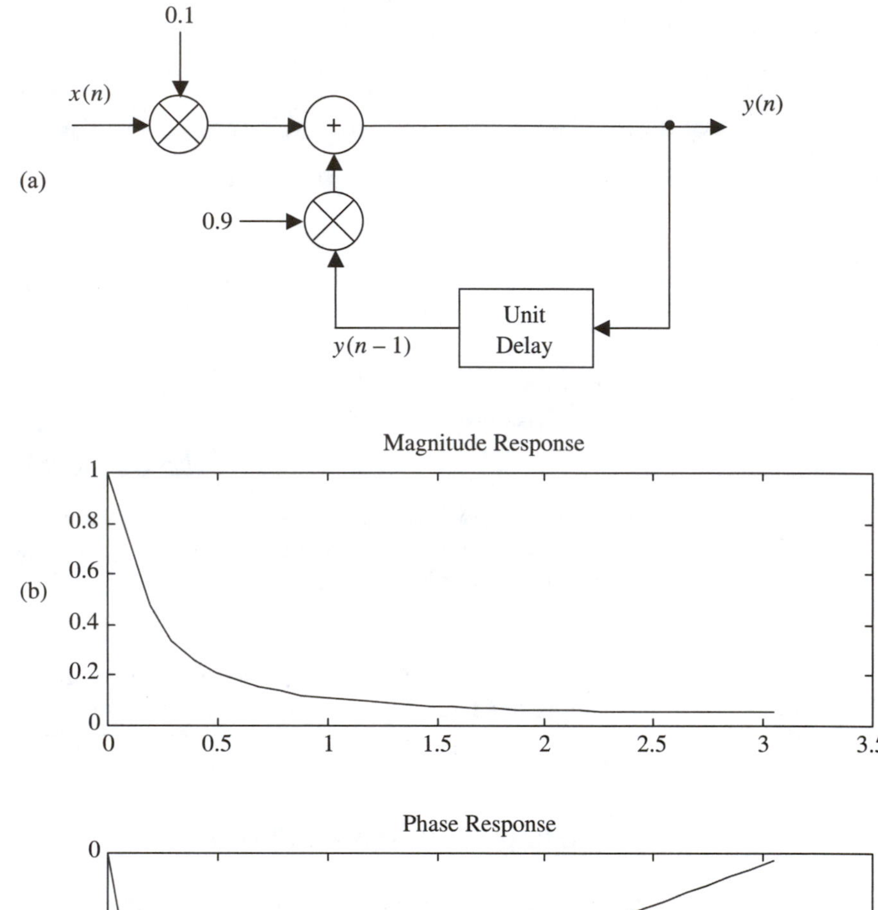

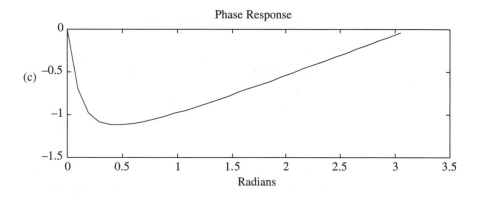


**Figure 2.6** The IIR filter in Example 2.2: (a) block diagram, (b) magnitude frequency response, (c) phase frequency response

The difference equation

$$y(n) = 0.9y(n-1) + 0.1x(n)$$

defines an IIR filter whose output is computed by taking 90% of past output $y(n-1)$ and 10% of the current input $x(n)$.

The transfer function of this filter is obtained as

$$H(z) = \frac{0.1}{(1 - 0.9z^{-1})} = \frac{0.1z}{(z - 0.9)}$$

or

$$H(e^{j\theta}) = \frac{0.1e^{j\theta}}{(e^{j\theta} - 0.9)}$$

Figure 2.6 describes this filter and its magnitude and phase frequency responses. The magnitude and phase frequency responses can be computed by substituting values for the digital frequency $\theta$ in the equation above and finding the absolute value for the magnitude and angle for the phase. To implement this filter requires a unit delay, two multiplies, and an addition.

---

## 2.7.3 FIR Filter Design

We have seen that a FIR filter's frequency response can be obtained from Eq. 2.17. Solving the equation for $b_k$ for a desired frequency response $H(e^{j\theta})$ yields the design equation for the FIR filter. The solution involves integration and is given as

$$b_k = 1/2\pi \int_{\pi}^{\pi} H(e^{j\theta})e^{-jk\theta}\, d\theta \tag{2.20}$$

where $k$ is an integer from $-\infty$ to $+\infty$. An algebraic closed-form solution of the above equation may not be possible for an arbitrary frequency function $H(e^{j\theta})$. In such a case, a computer-based solution can be obtained.

The impulse response $b_k$ as obtained by solving the above equation may be extremely long and may have to be truncated. The truncation results in a distortion called *Gibb's phenomenon* that introduces ripple in the passband of a filter's frequency response. To control the Gibb's phenomenon, special truncation windows are used. These windows, in general, provide smooth truncation to control the ripple in the passband of the filter. Window-based FIR filter design methods are covered in many DSP books, including the references at the end of this chapter.

### Parks–McClellan FIR Filter Design

This is a computer method for the design of FIR filters. It is based on the Remez exchange algorithm and Chebyshev approximation theory and involves minimization of the maximum error between the actual and the desired frequency responses. It allows arbitrary frequency response specification and designs an equiripple FIR filter. This technique has been implemented in many filter design packages and is available in the MATLAB program. The technique will be used to design FIR filters for the examples in this book.

## 2.7.4 IIR Filter Design

Two approaches are used to design IIR filters. One is based on analog filter design techniques and the other, called *direct design*, is based on a least-squares fit to achieve the desired frequency response.

### IIR Filter Design Based on Analog Filter Design Techniques

Digital IIR filters are designed using techniques that are based on analog filter design methods such as Butterworth filter design, Chebyshev1 filter design, Chebyshev2 filter design, and elliptic filter design. These methods are covered in many DSP books, including the references at the end of this chapter.

The approach consists of designing an analog filter to satisfy the filter specifications and then converting it to the equivalent digital filter using an appropriate transformation. The filter specifications consist of: passband ripple (dB), stopband attenuation (dB), and the transition width (ws – wp). For a lowpass filter the specifications are illustrated in Figure 2.7. These design methods are available in the MATLAB program and are used for examples in this book.

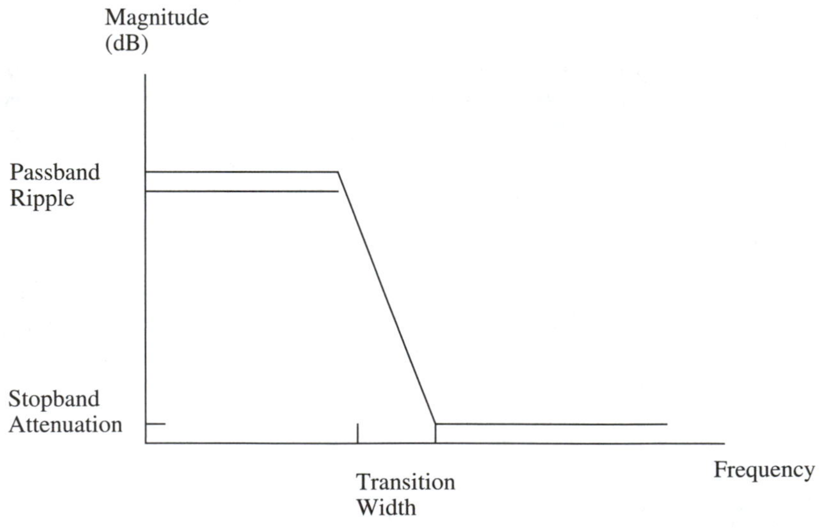


**Figure 2.7** Lowpass filter design specifications

### Direct IIR Filter Design

The direct IIR filter design methods are based on least-squares fit to a desired frequency response. These methods allow arbitrary frequency response specifications. One such method is called *Yulewalk design*, which is available in the MATLAB program and will be illustrated with an example in Section 2.10.

## 2.8 Decimation and Interpolation

Decimation and interpolation are the DSP operations that are used to change the sampling rate of a sequence. Decimation is used to decrease the sampling rate and interpolation to increase it. The decimation involves dropping samples without violating the sampling theorem. The interpolation involves inserting samples with appropriate consideration to the samples around the point of insertion.

▷ **Example 2.3** **The Decimation Process**

Let $x(n) = [3\ \ 2\ \ 2\ \ 4\ \ 1\ \ 0\ \ {-3}\ \ {-2}\ \ {-1}\ \ 0\ \ 2\ \ 3]$ be decimated by a factor of 2.

After filtering with an appropriate lowpass filter to satisfy the sampling theorem, let the filtered sequence be given as

$$w(n) = [2.1\ \ 2\ \ 3.9\ \ 1.5\ \ .1\ \ {-2.9}\ \ {-2}\ \ {-1.1}\ \ .1\ \ 1.9\ \ 2.9]$$

The decimated sequence is obtained by dropping every other sample. This gives the decimated sequence as

$$y(m) = [2\ \ 1.5\ \ {-2.9}\ \ {-1.1}\ \ 1.9]$$

---

The factor by which the signal is decimated is called the *decimation factor*. The input–output relation, for decimation of the signal by an integer factor $M$, is given as

$$y(m) = w(mM) = \sum_{k=-\infty}^{\infty} b_k x(mM - k) \tag{2.21}$$

where

$$w(n) = \sum_{k=-\infty}^{\infty} b_k x(n - k) \tag{2.22}$$

This is a case in which the sampling rate of $x(n)$ is divided by $M$ to obtain the sampling rate for $y(m)$. To prevent sampling theorem violation, the signal is

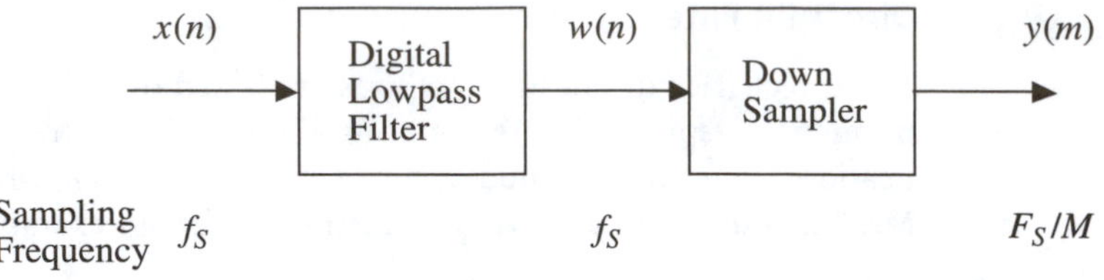


**Figure 2.8** Decimation process using a decimation factor $M$

first bandlimited using the filter with $b_k$ coefficients. Equation 2.22 allows computation of all samples. However, Eq. 2.21 specifies computation of only those samples that need to be kept, i.e., every $M$th sample. The decimation process based on the two equations is shown in the block diagram of Figure 2.8.

The input–output relation for the interpolation, where the sampling rate is increased by a factor $L$, is given as

$$y(m) = \sum_{k=-\infty}^{\infty} b_k w(m-k) \tag{2.23}$$

where

$$\begin{aligned} w(m) &= x(m/L), \quad m = 0, \pm L, \pm 2L, \ldots \\ &= 0 \quad \text{for other values of } m. \end{aligned} \tag{2.24}$$

The process of interpolation, as implemented by the above equations, requires first generating the sequence $w(m)$ by inserting $(L-1)$ zeros and then applying the filter with $b_k$ coefficients. The filter computes the interpolated samples using the samples of the original signal. The order of the filter equals the number of original signal samples used in computing the interpolated signal sample. The interpolating filter is a lowpass filter used to filter the image frequencies generated by increasing the sampling rate. The interpolation process is shown in the block diagram of Figure 2.9.

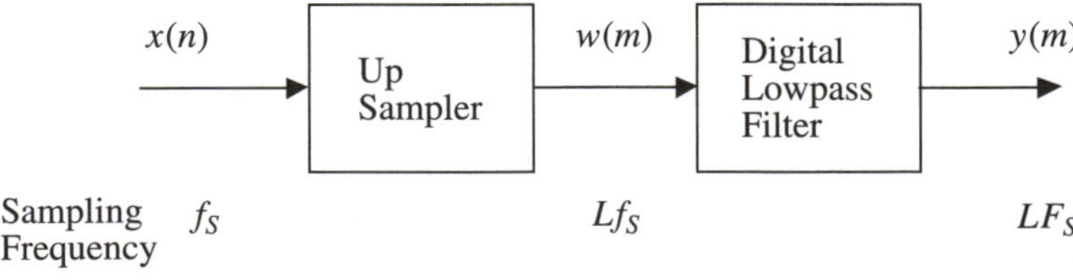


**Figure 2.9** Interpolation process using an interpolation factor $L$

▷ **Example 2.4 The Interpolation Process**

Let $x(n) = [0\ \ 3\ \ 6\ \ 9\ \ 12]$ be interpolated using $L = 3$. After inserting zeros to increase the sampling rate, we get

$$w(m) = [0\ \ 0\ \ 0\ \ 3\ \ 0\ \ 0\ \ 6\ \ 0\ \ 0\ \ 9\ \ 0\ \ 0\ \ 12]$$

Using the lowpass filter given by $b_k = [1/3\ \ 2/3\ \ 1\ \ 2/3\ \ 1/3]$, we get the interpolated sequence as

$$y(m) = [0\ \ 1\ \ 2\ \ 3\ \ 4\ \ 5\ \ 6\ \ 7\ \ 8\ \ 9\ \ 10\ \ 11\ \ 12]$$

This is an example of linear interpolation, as the filter used computes linearly the interpolated samples from the original samples.

## 2.9 Analysis and Design Tool for DSP Systems: MATLAB

A tool for DSP analysis and design must provide functions for carrying out the following basic operations:

1. Signal data generation and presentation
2. Convolution
3. Frequency response
4. Discrete Fourier transform (DFT)
5. Filtering
6. Spectrum estimation
7. FIR filter design, and
8. IIR filter design

MATLAB [3, 4] is a program that provides the above functions to process signals in addition to many more. The program is based on manipulation of data represented as vectors. The data can be one-dimensional, such as speech, or two-dimensional, such as an image.

Signal data input to MATLAB is by way of data files or direct keyboard entries for matrix elements. For signal processing, program files incorporating the DSP functions can be used. These files are called M-files. MATLAB also provides the capability to use command mode execution. In the command mode, the commands can be entered directly to process signals.

MATLAB provides an extensive list of commands or statements usable for signal-processing analysis and design. The signals can be presented and viewed using its extensive data presentation capability, including various types of plots.

MATLAB is supported with Help and Demo facilities that can be used to learn the program. It also provides an editor to create program and data files. This is the program we use in this book to design and analyze the DSP algorithms.

## 2.10 Digital Signal Processing Using MATLAB

In this section, we present MATLAB examples to illustrate the basic digital signal-processing operations covered in this chapter. Each program is followed by the results it produces when executed. The reader is advised to become familiar with the commands used in the following programs by using MATLAB's extensive Help and Demo facility.

▷ **Example 2.5 Convolution of Two Sequences [Figure 2.10]**

```
% Convolution of sequence x and sequence h to generate sequence y
x = [1 2 3 4];
h = [3 2 1];
y = conv(x,h)
```

```
y=
  3  8  14  20  11  4
```

**Figure 2.10** Result of convolution of sequence [1 2 3 4] and sequence [3 2 1]

▷ **Example 2.6 Frequency Response of an FIR Filter [Figure 2.11]**

```
% Frequency response of a digital differentiator (FIR Filter):
% y(n) = x(n) - x(n - 1)

% Filter definition
b = [1 -1];
a = 1;

% Frequency response computation
[h,th] = freqz(b,a,32);

% Frequency response plot
clf
figure(1)
subplot(211), plot(th,abs(h)), title('Magnitude Response'),
subplot(212), plot(th,angle(h)), title('Phase Response'),
xlabel('Radians')
```

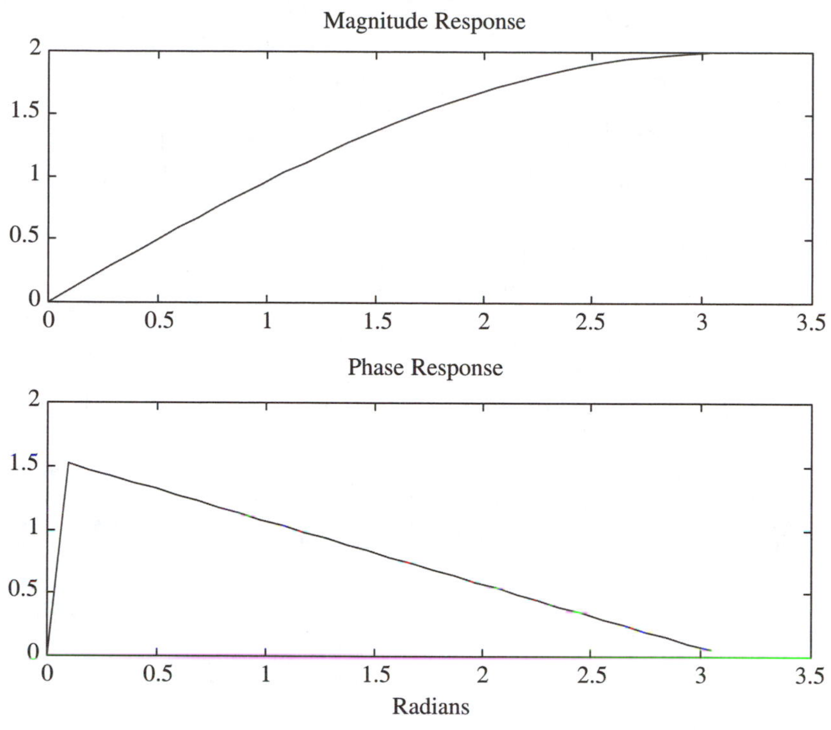


**Figure 2.11** Frequency response of the FIR filter, $y(n) = x(n) - x(n-1)$

▷ **Example 2.7** **Spectrum of a Noisy Sinusoidal Sequence [Figure 2.12]**

```
% Generate a 5 Hz signal of 1 sec duration sampled at 100 Hz.
t = 0:.01:1;
x = sin(2*pi*5*t);
clf
figure(1)
plot(t,x), title('Original Signal'), xlabel('Time in sec.')

% Add random noise with a standard deviation of 1 to produce a noisy
% signal y
y = x + 1*randn(1,101);
figure(2)
plot(t,y), title('Noisy Signal'), xlabel('Time in sec.')

% Compute the DFT and power spectral density of the noisy signal y
using 128 point FFT
Y = fft(y,128);
Pyy = Y.*conj(Y)/128;
```

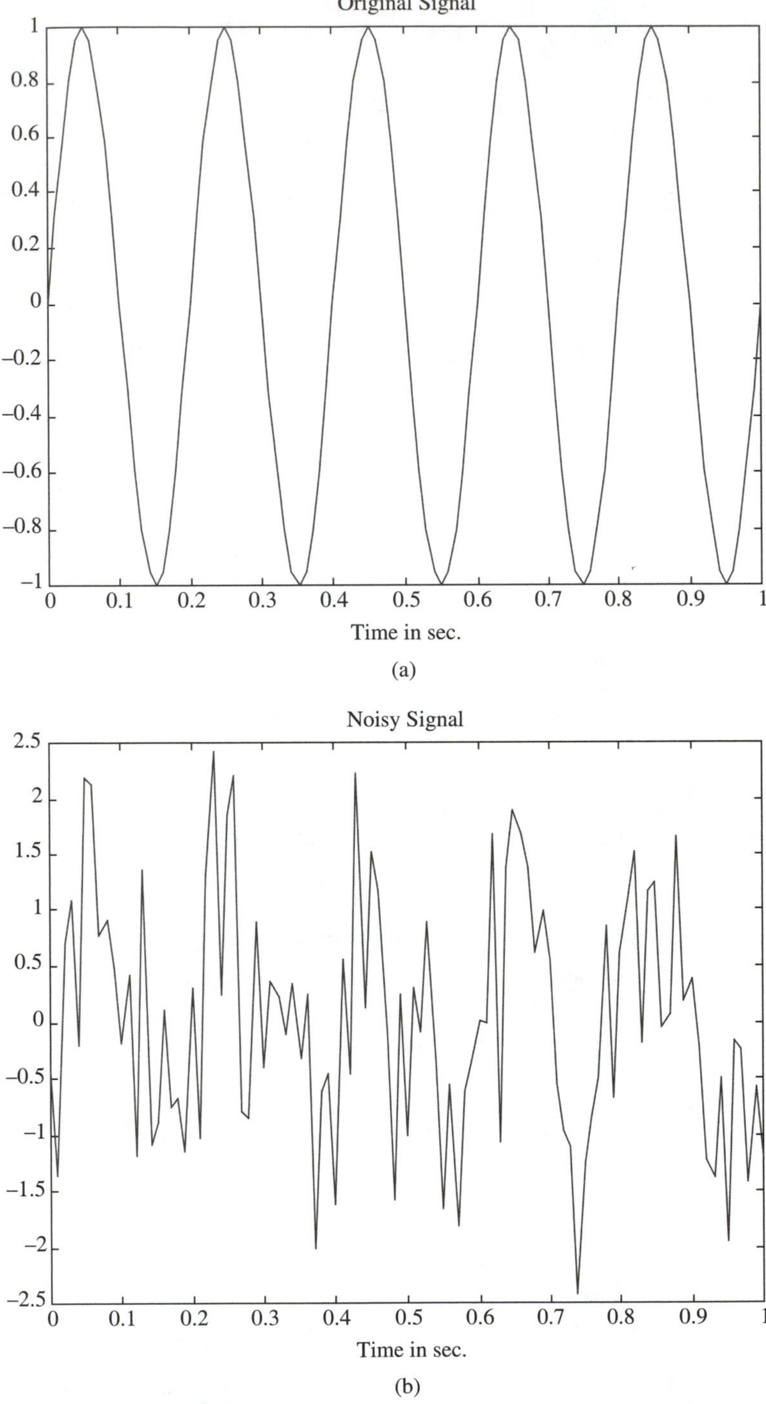


**Figure 2.12** Power spectral density of a noisy sinusoidal signal: (a) original sinusoidal signal, (b) noisy sinusoidal signal, (c) power spectral density of the noisy sinusoidal signal *(continued)*

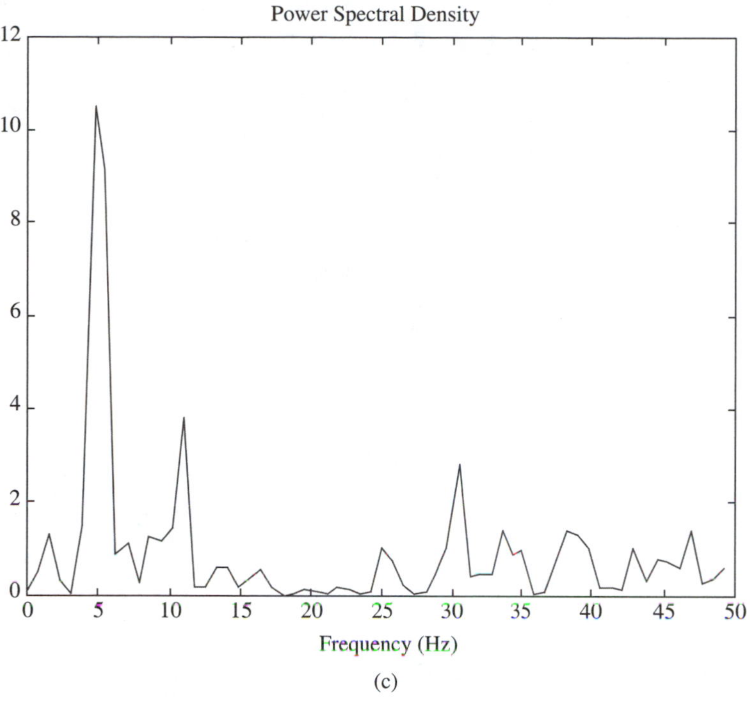


(c)

**Figure 2.12** Continued

```
% Change the horizontal axis to represent analog frequency in the
frequency response plot
f = 100/128*(0:63);
figure(3)
plot(f,Pyy(1:64)), title('Power Spectral Density'),
xlabel('Frequency (Hz)')
```

▷ **Example 2.8 FIR Filter Analysis [Figure 2.13]**

```
% Example 2.8: FIR Filter Analysis

% Filter definition (a 5-point averager)
b = [.2 .2 .2 .2 .2];
a = [1. .0 .0 .0 .0];

% Frequency response calculations and plots
[h,th] = freqz(b,a,32);
figure(1)
plot(th,abs(h)), title('Magnitude Response'), xlabel('Radians');
```

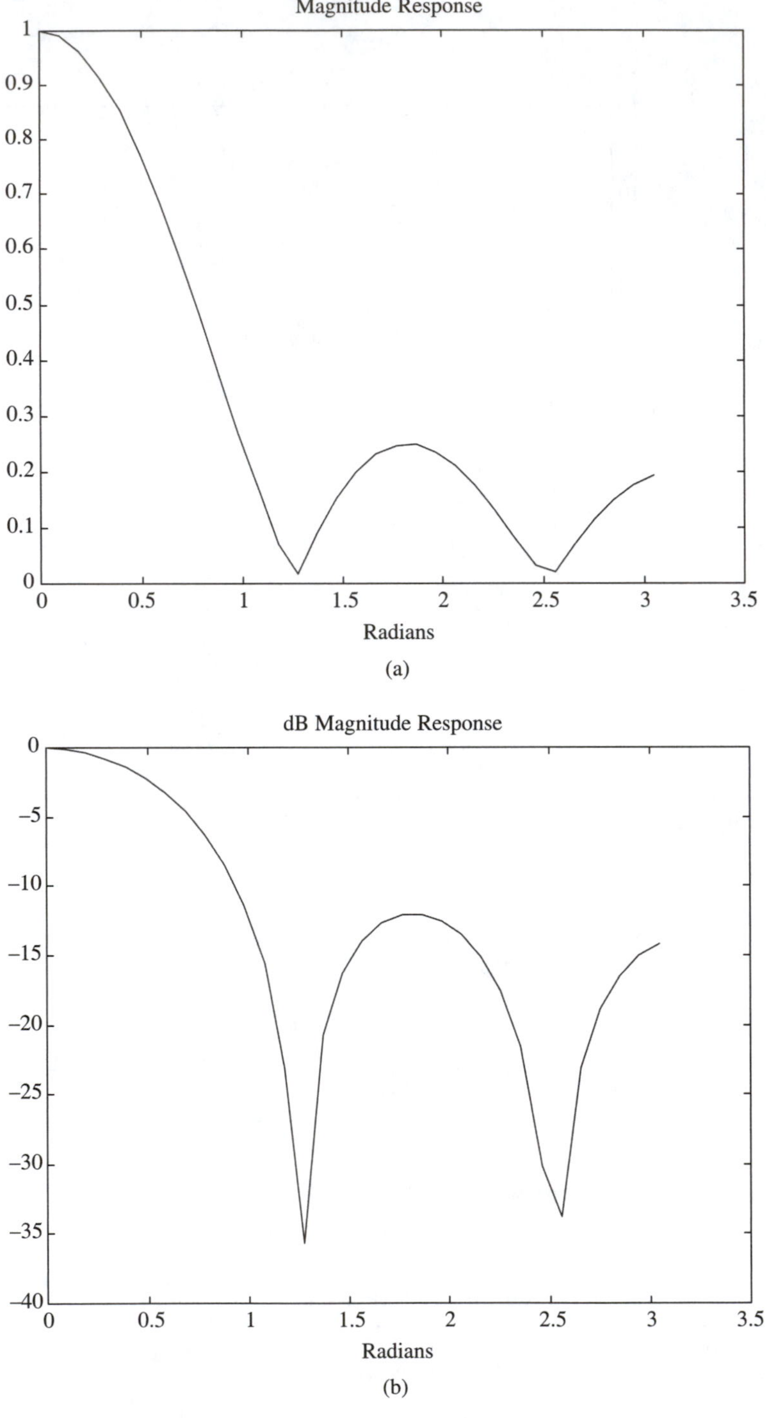


**Figure 2.13** Analysis of a FIR filter: (a) magnitude response, (b) dB magnitude response, (c) phase and group delay responses, (d) impulse response, (e) pole-zero plot

*(continued)*

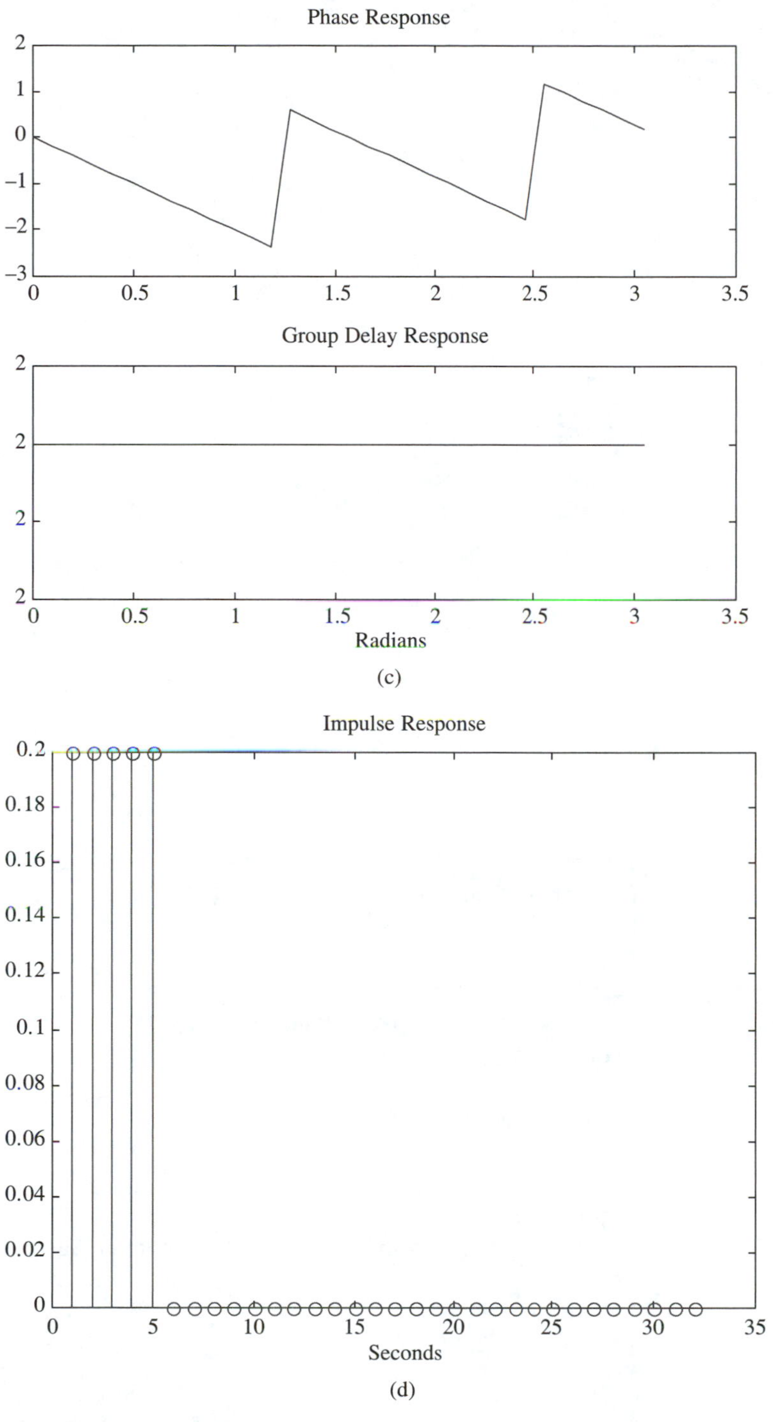


**Figure 2.13** Continued

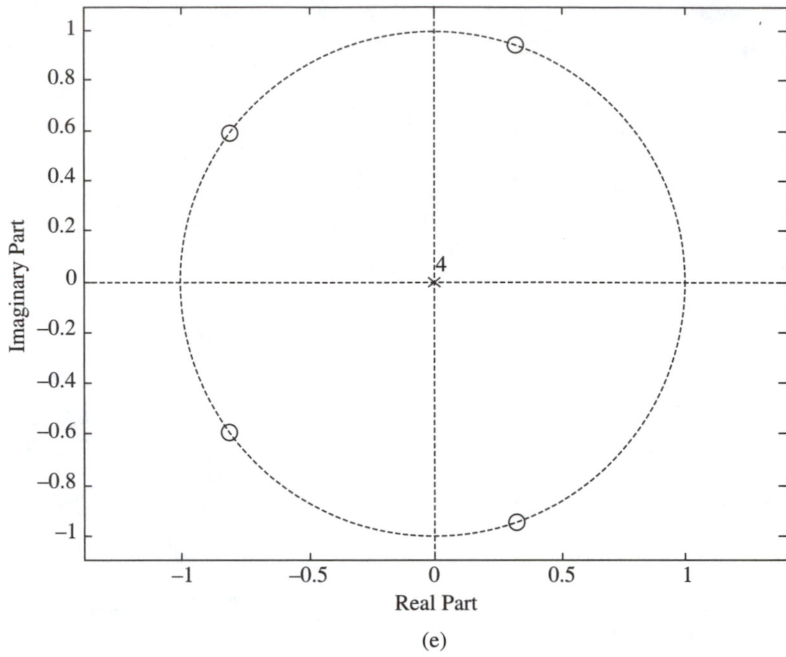


**Figure 2.13** Continued

```
figure(2)
plot(th,20*log10(abs(h))), xlabel('Radians'), title('dB Magnitude
Response');
figure(3)
subplot(211), plot(th, angle(h)), title('Phase Response')
subplot(212), plot(th, grpdelay(b,a,32)), xlabel('Radians'),
title('Groupdelay Response');

% Impulse response calculations and plot
x = [1 zeros(1,31)];
y = filter(b,a,x);
figure(4)
stem(y), title('Impulse Response'), xlabel('Seconds');

% Pole-Zero Plot
[z,p,k] = tf2zp(b,a);
figure(5)
zplane(z,p)
```

▷ **Example 2.9 IIR Filter Analysis [Figure 2.14]**

```
% Example 2.9: IIR Filter Analysis [Figure 2.14]
% Filter definition
b = [.0013 .0064 .0128 .0128 .0064 .0013];
a = [1.0 -2.9754 3.8060 -2.5453 0.8811 -0.1254];

% Frequency response
[h,th] = freqz(b,a,128);
clf
figure(1)
plot(th,abs(h)), title('Magnitude Response'), xlabel('Radians')
figure(2)
subplot(211), plot(th,angle(h)), title('Phase Response'),
ylabel('Radians');
subplot(212), plot(th,grpdelay(b,a,128)), title('Groupdelay
Response'), xlabel('Radians'), ylabel('Seconds');
```

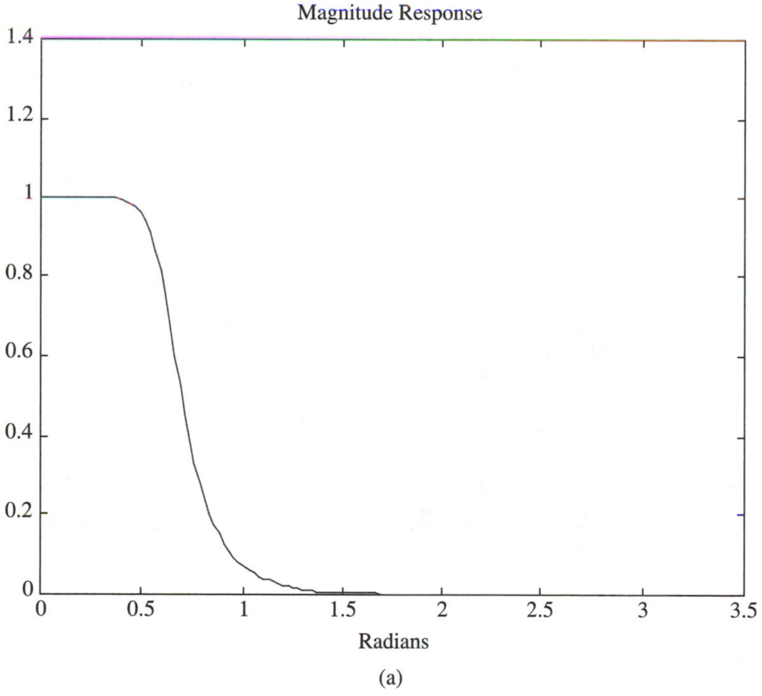


**Figure 2.14** Analysis of an IIR filter: (a) magnitude response, (b) phase and group delay responses, (c) impulse response, (d) pole-zero plot *(continued)*

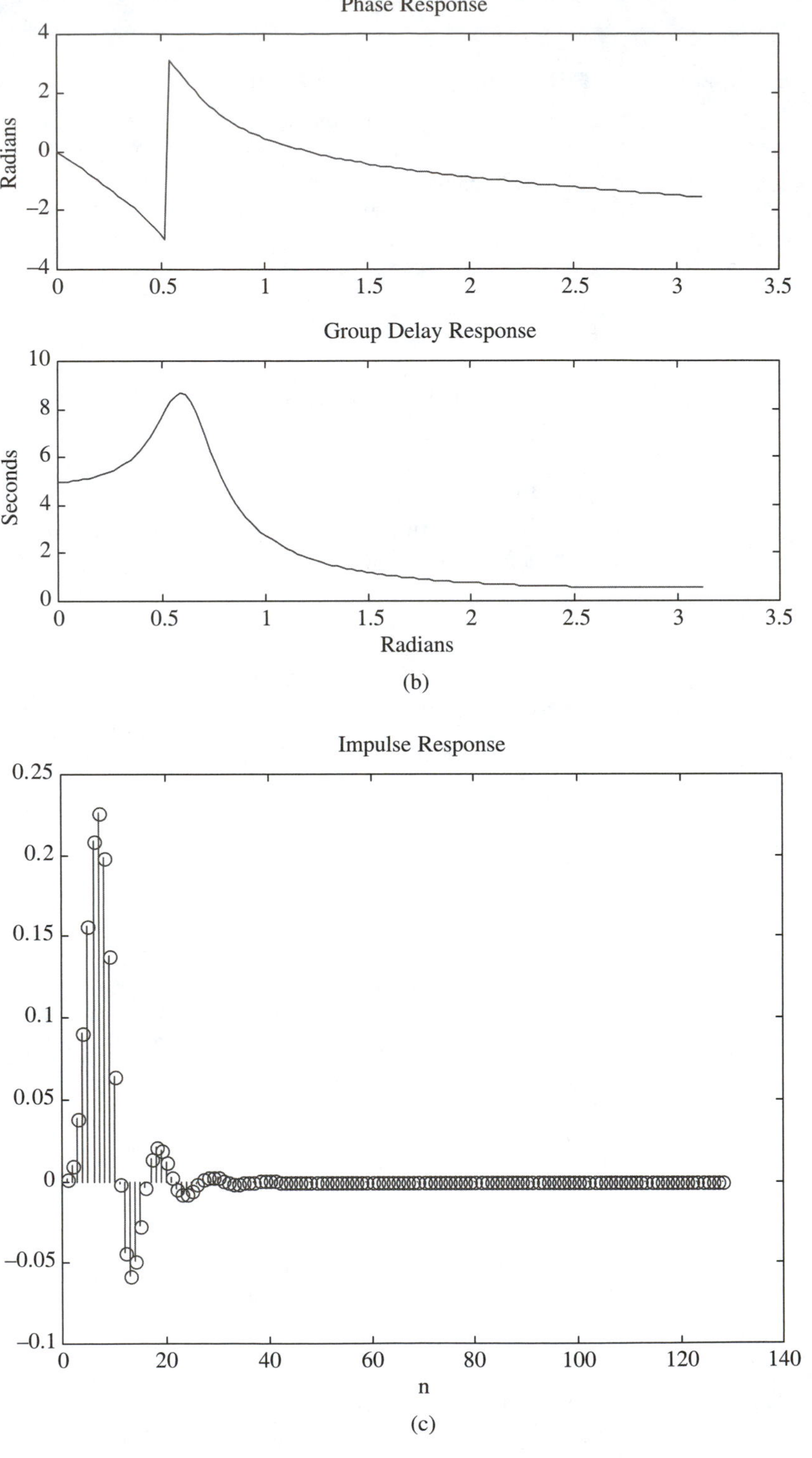


**Figure 2.14** Continued

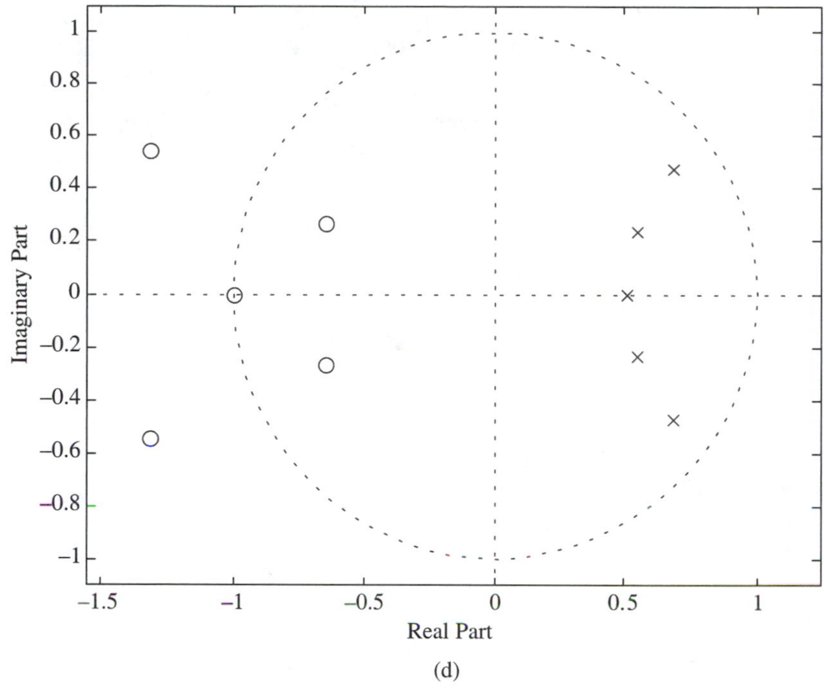


**Figure 2.14** Continued

```
% Impulse Response
x = [1 zeros(1,127)];
y = filter(b,a,x);
figure(3)
stem(y), title('Impulse Response'), xlabel('n')

% Pole-Zero Plot
[z,p,k] = tf2zp(b,a);
figure(4)
zplane(z,p)
```

▷ **Example 2.10** **Butterworth Lowpass IIR Filter Design [Figure 2.15]**

```
% Filter specifications
N = 5; % Filter order
fs = 200; % Sampling frequency
fc = 30; % Cut-off frequency

% Filter design
[b,a] = butter(N, 2*fc/fs)
```

```
b=
  0.0069  0.0347  0.0693  0.0693  0.0347  0.0069

a=
  1.0000  −1.9759  2.0135  −1.1026  0.3276  −0.0407
```

(a)

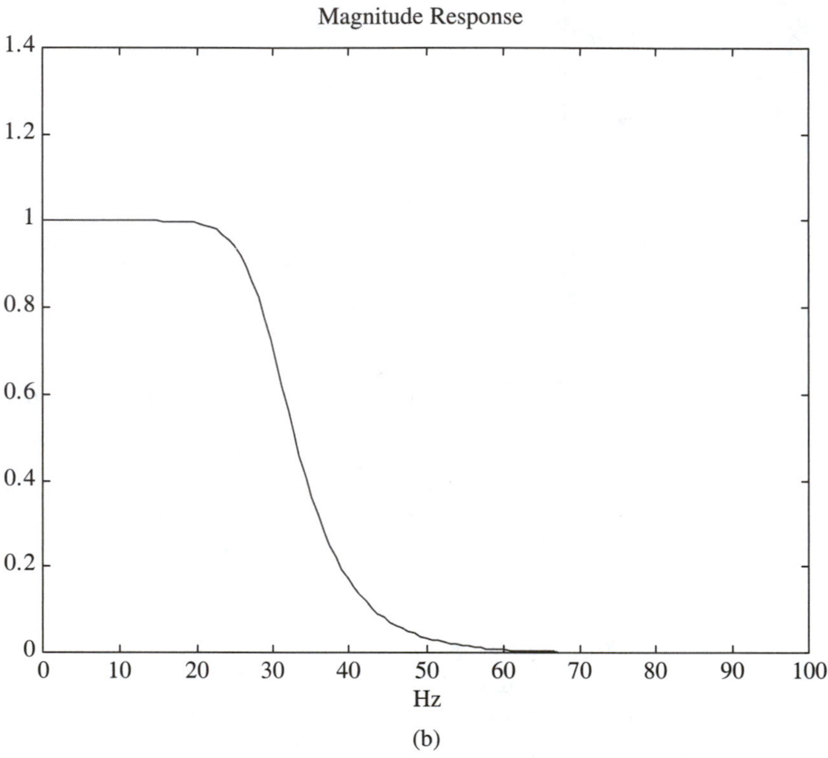


(b)

**Figure 2.15** Lowpass IIR filter design using the Butterworth technique: (a) designed filter coefficients, (b) designed filter magnitude response, (c) designed filter phase and group delay responses *(continued)*

```
% Designed filter frequency response
[h,th] = freqz(b,a,128);
f = (th/pi)*(fs/2);
clf
figure(1)
plot(f,abs(h)), title('Magnitude Response'), xlabel('Hz')
figure(2)
subplot(211), plot(f,angle(h)), title('Phase Response'),
ylabel('Hertz')
```

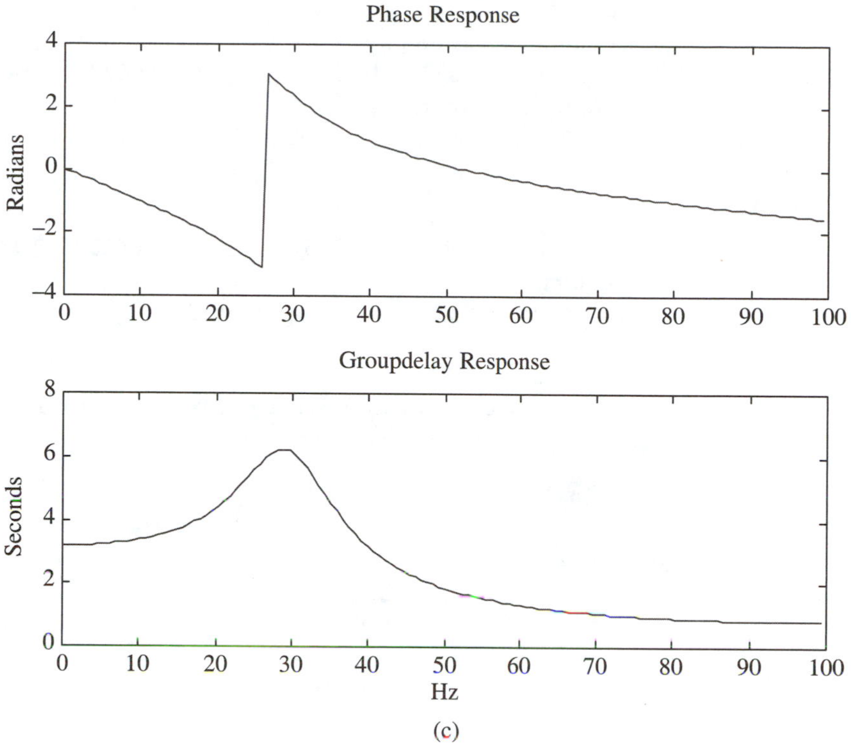


(c)

**Figure 2.15** Continued

```
subplot(212), plot(f,grpdelay(b,a,128)), title('Groupdelay
Response'), xlabel('Hz'), ylabel('Seconds')
```

▷ **Example 2.11 Yulewalk IIR Filter Design [Figure 2.16]**

```
% Filter specifications (Bandpass filter)
f = [0 .1 .2 .3 .4 .6 .7 .8 .9 1];
m = [0 0 1 1 1 1 0 0 0 0];

% Filter design
N = 10; % Filter order
[b,a] = yulewalk(N,f,m)

% Designed filter frequency response
[h,th] = freqz(b,a,128);

% Specified(solid curve) and designed(x curve)filter frequency
responses comparison
```

```
b=
  Columns 1 through 7
    0.1467   0.1368   –0.1699   –0.3064   0.0072   0.2344   0.0883
  Columns 8 through 11
    –0.1106   –0.0771   0.0366   0.0664

a=
  Columns 1 through 7
    1.0000   –0.9551   1.2125   –1.5030   1.6430   –0.9850   0.8491
  Columns 8 through 11
    –0.5510   0.2769   –0.0668   0.0462
```

(a)

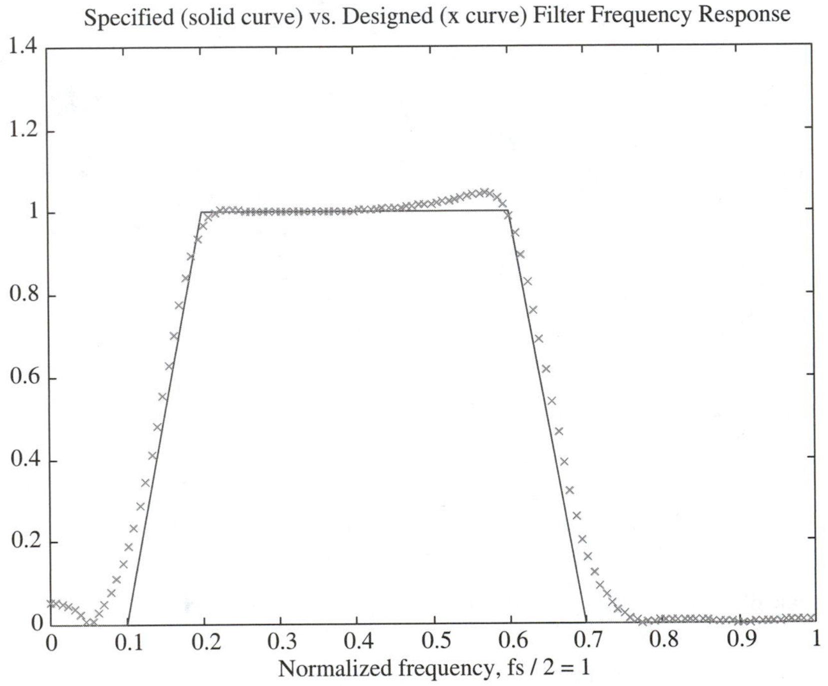

(b)

**Figure 2.16** Bandpass IIR filter design using the Yulewalk technique: (a) designed filter, (b) designed vs. specified filter magnitude response

```
figure(1)
plot(f,m,th/pi,abs(h),'x'), title('Specified (solid curve) vs
Designed (x curve)Filter Frequency Response'), xlabel('Normalized
frequency, fs/2 = 1')
```

▷ **Example 2.12** **Parks–McClellen FIR Filter Design [Figure 2.17]**

```
% Filter specifications
f = [0 .1 .2 .3 .4 .6 .7 .8 .9 1];
m = [0 0 1 1 1 1 0 0 0 0];

% Filter design
N = 20; % Filter order;
b = remez(N,f,m)
```

*b*=

Columns 1 through 7

0.0520 0.0101 −0.0001 0.0398 −0.0339 −0.0822 0.0000

Columns 8 through 14

−0.1181 −0.2571 0.1348 0.5000 0.1348 −0.2571 −0.1181

Columns 15 through 21

0.0000 −0.0822 −0.0339 0.0398 −0.0001 0.0101 0.0520

(a)

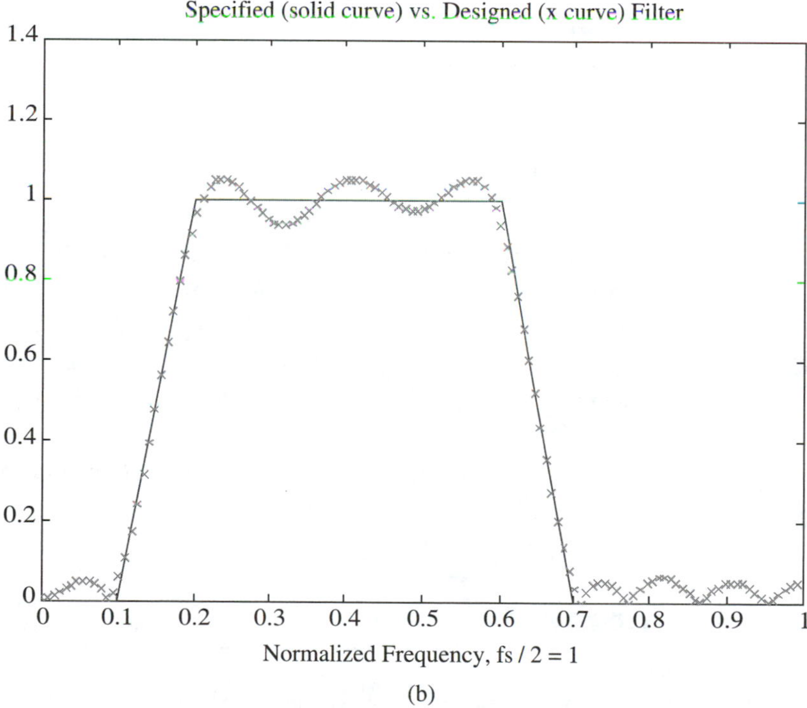


(b)

**Figure 2.17** Filter design using the Parks–McCullen technique: (a) designed filter, (b) designed vs. specified filter magnitude response

```
% Frequency response
[h,th] = freqz(b,1,128);

% Specified vs designed frequency response
figure(1)
plot(f,m,th/pi,abs(h),'x')
title('Specified (solid curve) vs Designed (x curve) Filter'),
xlabel'Normalized Frequency, fs/2 = 1'
```

## 2.11 Summary

This chapter is a brief review of digital signal-processing fundamentals. The basic DSP concepts are discussed from the implementation point of view. The topics that are covered consist of: a digital signal-processing system, sampling process and the sampling theorem, digital signal sequences, DFT and FFT, linear time-invariant systems, the convolution theorem, digital filters, FIR and IIR filters, and filter design techniques. Thus most of the basic techniques of DSP analysis and design have been introduced. The techniques are illustrated with MATLAB examples.

## References

1. Strum, R. D., and Kirk, D. E., *First Principles of Discrete Systems and Digital Signal Processing*, Addison-Wesley, 1988.
2. Ifeacho, E. C., and Jervis, B. W., *Digital Signal Processing: A Practical Approach*, Addison-Wesley, 1993.
3. Mitra, S. K., *Digital Signal Processing Laboratory using MATLAB*, McGraw-Hill, 1999.
4. The Math Works, *Student Edition of MATLAB* and various Toolboxes, *http://www.mathworks.com/products/education/*, 2003.

## Assignments

**2.1** A signal whose spectrum is shown in Figure P2.1 is to be sampled so that no aliasing results. Determine the minimum sampling rate that can be used to sample the signal. If the sampling rate must be 8 KHz, determine the type and the cutoff frequency of the antialiasing filter.

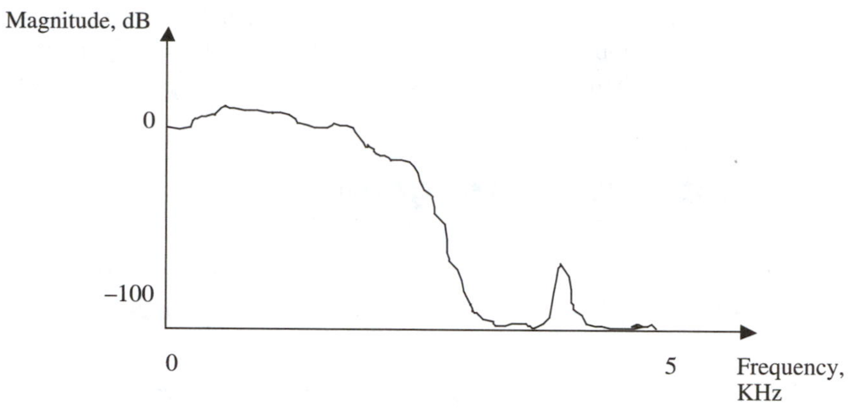


**Figure P2.1** Magnitude spectrum for the signal in Problem 2.1

**2.2** Redraw the frequency spectrum for the signal in 2.1 using the digital frequency as the horizontal axis. Let the sampling frequency be 8 KHz. Determine the analog frequencies for the digital frequencies 0, $\pi/4$, $\pi/2$, $3\pi/4$, and $\pi$ radian.

**2.3** Determine the periods for the periodic sequences: (a) $e^{-jn\pi/8}$, (b) $e^{-jn3\pi/8}$.

**2.4** The signal in 2.1 is filtered and sampled using the sampling rate of 8 KHz. If 512 samples of this signal are used to compute the Fourier transform $X(k)$, determine the frequency spacing between adjacent $X(k)$ elements. What is the analog frequency corresponding to $k = 64$, 128, and 200. Repeat this problem using 1024 samples and an 8 KHz sampling rate.

**2.5** Assuming $X(k)$ as a complex sequence, determine the number of complex and real multiplies for computing IDFT using direct and radix-2 FFT algorithms.

**2.6** For the FIR filter

$$y(n) = (x(n) + x(n-1) + x(n-2))/3$$

determine the (a) system function, (b) magnitude response function, (c) phase response function, (d) impulse response, (e) step response, and (f) poles and zeros.

**2.7** For the IIR filter

$$H(z) = \frac{(z-1)}{(z-0.25)(z-0.5)}$$

determine the (a) magnitude response function, (b) phase response function, (c) impulse response, (d) step response, and (e) poles and zeros.

**2.8** Determine the lowpass filter cutoff frequency that must be used to decimate to reduce the sampling rate from 8 KHz to 4 KHz.

**2.9** The signal sequence $x(n) = [0\ \ 2\ \ 4\ \ 6\ \ 8]$ is interpolated using the interpolation filter sequence $b_k = [.5\ \ 1\ \ .5]$ and the interpolation factor is 2. Determine the interpolated sequence $y(m)$.

## Laboratory Assignment

Use the MATLAB program to do the following laboratory assignments:

**L2.1** Generate and plot each of the following sequences:

a. $x(n) = [3\ \ 2\ \ -2\ \ 0\ \ 7]$, $n = 0, 1, 2, 3, 4$

b. a ramp of length 64 with minimum value 0 and maximum value 1

c. a triangular waveform of length 64, period 16, minimum value 0, and maximum value 1

d. $x(n) = 1.5\sin(\pi n/10 + \pi/4)$, $n = 0, 1, \ldots, 63$.

**L2.2** Generate $x(n) = 2\sin(0.1\pi n + 0.1) + w(n)$, $n = 0, 1, \ldots, 255$, where $w(n)$ is Gaussian noise with zero mean and unit variance.

**L2.3** Given the sequences

$$xm(n) = \sin 2\pi n/100, \quad n = 0, 1, \ldots, 255$$

and

$$xc(n) = \sin 2\pi n/10, \quad n = 0, 1, \ldots, 255$$

use the given sequences to generate the following sequences:

a. $xam(n) = [1 + .7xm(n)]xc(n)$, $n = 0, 1, \ldots, 255$

b. $xsc(n) = xm(n)xc(n)$, $n = 0, 1, \ldots, 255$

**L2.4** For the 12-point sequence

$$\begin{aligned} x(n) &= 1, \quad n = 0, 1, \ldots, 5 \\ &= 0, \quad n = 6, 7, \ldots, 11 \end{aligned}$$

use 64-point FFT to compute the following sequences:

a. $|X(k)|$, $k = 0, 1, \ldots, 63$

b. $\angle X(k)$, $k = 0, 1, \ldots, 63$

c. $\text{Real}(X(k))$, $k = 0, 1, \ldots, 63$

d. $\text{Imag}(X(k))$, $k = 0, 1, \ldots, 63$

Also plot all the above sequences. Determine the frequency resolution of the FFT. How can the resolution be improved and at what cost?

**L2.5** Given the sequences

$$x1(n) = [3\ \ 4.2\ \ 11\ \ 0\ \ 7\ \ -1\ \ 0\ \ 2], \quad n = 0, \ldots, 7$$

$$x2(n) = [1.2\ \ 3\ \ 0\ \ -.52], \quad n = 0, \ldots, 4$$

compute and plot the sequence $x1(n) * x2(n)$. Determine the length of the computed sequence.

**L2.6** For the sequence in Problem L2.5, find the sequences, $X1(k)$ and $X2(k)$ using 8-point FFT. Next, multiply the two sequences to generate the sequence $Y(k) = X1(k).X2(k)$. Now use 8-point IFFT to compute $y(n)$. Repeat using 16-point FFT and IFFT. Compare these results to the one obtained in Problem L2.5 and explain any discrepancy in the two approaches.

**L2.7** Find and plot the (a) impulse, (b) unit step, (c) magnitude, (d) phase, and (e) group delay responses for the system with transfer function

$$H(z) = \frac{(z-1)}{(z-0.25)(z-0.5)}$$

**L2.8** Given a three-tap averaging filter

$$y(n) = [(x(n) + x(n-1) + x(n-2)]/3$$

obtain and plot the (a) magnitude, (b) dB magnitude, (c) phase, and (d) group delay frequency response for the filter. Comment on the lowpass filtering nature of the filter.

**L2.9** Repeat Problem L2.8 for the filter

$$y(n) = [-3x(n) + 12x(n-1) + 17x(n-2) + 12x(n-3) - 3x(n-4)]/35$$

**L2.10** Design a 31-tap bandpass FIR filter with cutoff frequencies of 25 and 75 Hz and sampling frequency of 200 Hz. Calculate the passband ripple and the stopband attenuation for the designed filter.

Use this filter to filter the noisy signal

$$x(t) = 2\sin(100\pi t) + w(t)$$

where $w(t)$ is a uniformly distributed noise with amplitude range from $-.25$ to $+.25$. Evaluate the performance using FFT.

**L2.11** For the filter of Problem L2.10, determine the transition widths, when gain drops from 90% to 10%, around the cutoff frequencies. How will you reduce the transition to obtain a sharper response? Demonstrate with an example.

**L2.12** Design a second-order Butterworth IIR lowpass filter with a cutoff frequency of 50 Hz for a signal sampled at 250 Hz. Determine its dc gain, poles, and zeros.

**L2.13** Design an elliptic IIR lowpass filter with cutoff frequency of 50 Hz for a signal sampled at 250 Hz. The filter order should be such that the passband ripple is less than .2 dB and the stopband attenuation is more than 20 dB.

# Chapter 3

# Computational Accuracy in DSP Implementations

## 3.1 Introduction

In this chapter, we shall study the issues related to computational accuracy of algorithms when implemented using programmable digital signal processors. We shall first study the various formats of number representation and their effect on the dynamic range and precision of signals represented using these formats. We shall also study the various sources of errors in the implementation of DSP algorithms and how to control these errors while designing DSP systems. Specifically, we discuss the following topics in this chapter:

Number formats for signals and coefficients in DSP systems
Dynamic range and precision
Sources of error in DSP implementations
A/D conversion errors
DSP computational errors
D/A conversion errors

## 3.2 Number Formats for Signals and Coefficients in DSP Systems

In a digital signal processor, as in any other digital system, signals are represented as numbers right from the input, through different stages of processing, to the output. The DSP structures, such as filters, also require numbers to specify coefficients [1]. There are various ways of representing these numbers [4], depending on the range and precision of signals and coefficients to be represented, hardware complexity, and speed requirements. In this section, we look at the typical formats used for numbers to represent signals and coefficients in DSP systems.

### 3.2.1 Fixed-Point Format

The simplest scheme of number representation is the format in which the number is represented as an integer or fraction using a fixed number of bits. An $n$-bit fixed-point signed integer shown in Figure 3.1(a) specifies the value $x$ given as

$$x = -s.2^{n-1} + b_{n-2}.2^{n-2} + b_{n-3}.2^{n-3} + \cdots + b_1.2^1 + b_0.2^0 \tag{3.1}$$

where $s$ represents the sign of the number: $s = 0$ for positive numbers and $s = -1$ for negative numbers. The range of signed integer values that can be represented with this format is $-2^{n-1}$ to $+(2^{n-1} - 1)$.

Similarly, a fraction can also be represented using a fixed number of bits with an implied binary point after the most significant sign bit. An $n$-bit fixed-point signed fraction representation shown in Figure 3.1(b) specifies the value given as

$$x = -s.2^0 + b_{-1}.2^{-1} + b_{-2}.2^{-2} + \cdots + b_{-(n-2)}.2^{-(n-2)} + b_{-(n-1)}.2^{-(n-1)} \tag{3.2}$$

The range of signed fractions that can be represented with this format is $-1$ to $+(1 - 2^{-(n-1)})$.

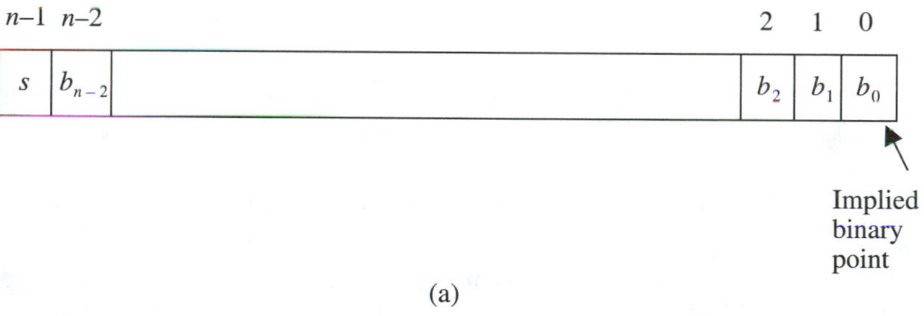


**Figure 3.1(a)** Fixed-point format to represent signed integers

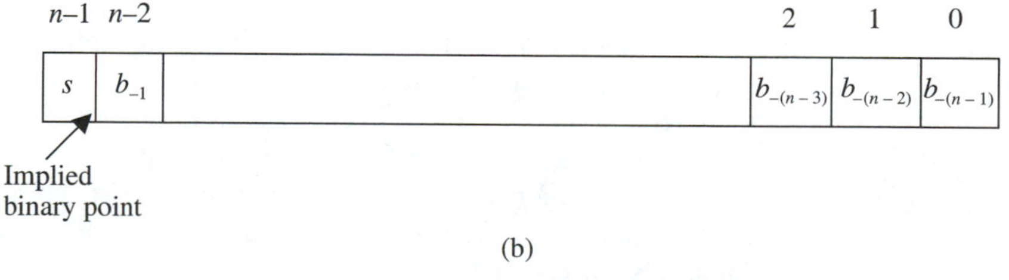


**Figure 3.1(b)** Fixed-point format to represent signed fractions

▷ **Example 3.1** What is the range of numbers that can be represented in a fixed-point format using 16 bits if the numbers are treated as (a) signed integers, (b) signed fractions?

Solution

a. Using 16 bits, the range of integers that can be represented is determined by substituting $n = 16$ in Eq. 3.1 and is given as

$$-2^{15} \text{ to } +2^{15} - 1$$

i.e., −32,768 to +32,767.

b. The range of fractions, as determined from Eq. 3.2 using $n = 16$, is given as

$$-1 \text{ to } +(1 - 2^{-15})$$

i.e., −1 to +.999969482.

---

In DSP implementations, multiplication of integers produces numbers that may require more bits to represent, and in the event of a fixed number of available bits, it may create wraparound. The wraparound generates the most negative number after the most positive number, and vice versa. The problem can be tackled by using fractional representation. When two fractions are multiplied, the result is still a fraction. The resulting fraction may use the same number of bits as the original fractions by discarding the less significant bits.

### 3.2.2 Double-Precision Fixed-Point Format

To increase the range of numbers that can be represented in fixed-point format, one obvious approach is to increase its size. If the size is doubled, the range of numbers increases substantially. Simply doubling the size and still using the fixed-point format creates what is known as the *double-precision fixed-point format.* However, one should remember that such a format requires double the storage for the same data and may need double the number of accesses for the same size of data bus of the DSP device.

### 3.2.3 Floating-Point Format

For DSP applications, if an algorithm involves summation of a large number of products, it requires a large number of bits to represent the signal to allow for adequate signal growth over the summation. However, since a processor architecture will not allow for an unlimited number of bits, some processors choose a *floating-point format* for signal-processing computations. A floating-point number is made up of a mantissa $M_x$ and an exponent $E_x$ such that its value $x$ is represented as

$$x = M_x 2^{Ex} \tag{3.3}$$

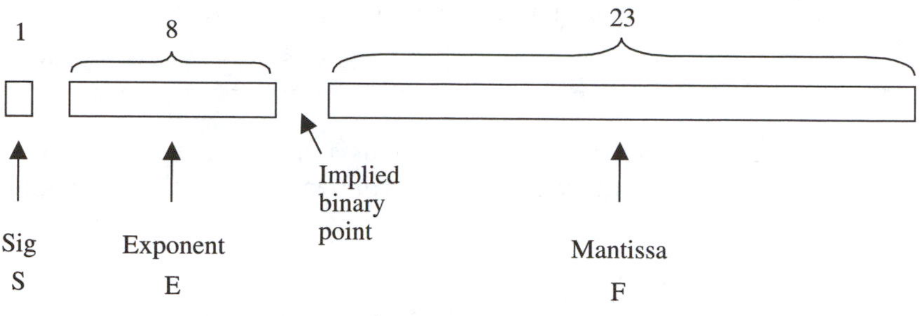


**Figure 3.2** IEEE-754 format for floating-point numbers

If two floating-point numbers $x$ and $y$ are multiplied, the product $xy$ is given by

$$xy = M_x M_y 2^{Ex+Ey} \tag{3.4}$$

Implementation of a floating-point multiplier must contain a multiplier for the mantissa and an adder for the exponent. An addition of floating-point numbers requires normalization of the numbers to be added so that they have the same exponents.

A commonly used single-precision floating-point representation (IEEE-754 format) is shown in Figure 3.2.

The value represented by the data format in Figure 3.2 is given as

$$x = (-1)^S \times 2^{(E\text{-bias})} \times 1.F \tag{3.5}$$

$F$ represents the magnitude fraction of the mantissa, and the exponent $E$ is an integer. Further, in determining the mantissa, an implied 1 is placed immediately before the binary point of the fraction. The sign bit provides the sign of the fractional part of the number. That is to say, with $n$ bits for $F$, the range of fractional numbers that can be represented in the mantissa is $-(2 - 2^{-n})$ to $+(2 - 2^{-n})$. The bias depends upon the bits reserved for the exponent. In Figure 3.2, the bias is 127, the largest positive number represented by 8 bits. The value of $E$ can be from 0 to 255. In double-precision representation, the exponent uses 11 bits, making the bias value as 1023.

▷ **Example 3.2** Find the decimal equivalent of the floating-point binary number 1011000011100. Assume a format similar to IEEE-754 in which the MSB is the sign bit followed by 4 exponent bits followed by 8 bits for the fractional part.

**Solution** The number is negative, as the sign bit is 1.

$$F = 2^{-4} + 2^{-5} + 2^{-6} = .109375$$

$$E = 2^1 + 2^2 = 6$$

Thus the value of the number is

$$x = -1.109375 \times 2^{(6-7)} = -0.5546875.$$

---

▷ **Example 3.3** Using 16 bits for the mantissa and 8 bits for the exponent, what is the range of numbers that can be represented using the floating-point format similar to IEEE-754?

Solution The most negative number will have as its mantissa $-2 + 2^{-16}$ and as its exponent $(255 - 127) = 128$. The most negative number is, therefore,

$$-1.999984741 \times 2^{128}$$

Similarly, the most positive number is

$$+1.999984741 \times 2^{128}$$

---

Floating-point format, used to increase the range of numbers that can be represented, suffers from the problem of speed reduction for DSP computation. More steps are required to complete a floating-point computation compared to a fixed-point computation. For instance, a floating-point multiplication requires addition of exponents in addition to the multiplication of mantissas. Floating-point additions, on the other hand, require the exponents to be normalized before the addition of the mantissas. For these reasons, a floating-point processor requires a more complex hardware compared to a fixed-point processor and requires more time to do computations.

### 3.2.4 Block Floating-Point Format

An approach to increase the range and precision of the fixed-point format is to use the *block floating-point* format [3]. In this approach, a group or block of fixed-point numbers are represented as though they were floating-point numbers with the same exponent value and different mantissa values. Mantissas are stored and handled similar to fixed-point numbers. The common exponent of the block is stored separately and is used to multiply the numbers as they are read off the memory. The exponent is decided by the smallest number of leading zeros in the fixed-point representation of the given block of numbers. The numbers are then shifted by this value to accommodate the maximum number of nonzero bits using the given fixed-point format.

The block floating-point format increases the range and precision of a given fixed-point format by retaining as many lower-order bits as is possible. The scheme does not require any additional hardware resources except an extra memory location to store the block exponent. However, programming overhead is needed to find the block exponent and to normalize and denormalize the given numbers using this exponent.

▷ **Example 3.4** The following 12-bit binary fractions are to be stored in an 8-bit memory. Show how they can be represented in block floating-point format so as to improve accuracy.

000001110011

000011110000

000000111111

000010101010

Solution If these fractions are represented using an 8-bit fixed-point format, they will be represented as

00000111

00001111

00000011

00001010

The last 4 bits of the numbers would have been discarded, thereby losing the precision corresponding to those 4 bits.

However, since all four numbers have at least four leading zeros, they can be rewritten as

$01110011 \times 2^{-4}$

$11110000 \times 2^{-4}$

$00111111 \times 2^{-4}$

$10101010 \times 2^{-4}$

Eight bits of each number can be stored without discarding any bit. The block exponent is −4 and will have to be stored separately. When the numbers are read from the memory for any computation, they have to be shifted by four bit positions to the right to bring them to their original values.

---

Similar operation can also be performed on a block of integers if there are zeros to the right.

## 3.3 Dynamic Range and Precision

The *dynamic range* of a signal is the ratio of the maximum value to the minimum value that the signal can take in the given number representation scheme. The dynamic range of a signal is proportional to the number of bits used to represent it and increases by 6 dB for every additional bit used for the representation. The number of bits used to represent a signal also determines the resolution or the precision with which the signal can be represented. However, the time taken for certain operations such as the A/D conversion increases with the increase in the number of bits.

*Resolution* is the minimum value that can be represented using a number representation format. For instance, if $N$ bits are used to represent a number from 0 to 1, the smallest value it can take is the resolution and is given as

$$\text{Resolution} = 1/2^N \quad \text{for large } N \tag{3.5}$$

Resolution of a number representation format is normally expressed as number of bits used in the representation. At times, it is also expressed as a percentage.

*Precision* is an issue related to the speed of DSP implementation. In general, techniques to improve the precision of an implementation reduce its speed. Larger word size improves the precision but may pose a problem with the speed of the processor, especially if its bus width is limited. For example, if the 32-bit product of a $16 \times 16$ multiplication has to be preserved without loss of precision, two memory accesses are required to store and recall this product using a 16-bit bus. Another example is the rounding off, as against the truncation, used to limit the word size in the fixed-point representation of numbers. The former is slightly more accurate than the latter, but requires more time to carry out computations.

When the floating-point number representation is used, the exponent determines the dynamic range. Since the exponent in the floating-point representation is a power, the dynamic range of a floating-point number is very large. The resolution or precision of a floating-point number is determined by its mantissa. Since the mantissa uses fewer bits compared to fixed-point representation, the precision of floating-point number representation is smaller than a comparable fixed-point representation.

It is important to be aware of the speed implications when adopting schemes to improve precision or the dynamic range and not just choose higher precision or larger dynamic range than what is required for a given application.

▷ **Example 3.5** Calculate the dynamic range and precision of each of the following number representation formats.

a. 24-bit, single-precision, fixed-point format

b. 48-bit, double-precision, fixed-point format

c. a floating-point format with a 16-bit mantissa and an 8-bit exponent

Solution

a. Since each bit gives a dynamic range of 6 dB, the total dynamic range is $24 \times 6 = 144$ dB. Percentage resolution is $(1/2^{24}) \times 100 = 6 \times 10^{-6}$.

b. Since each bit gives a dynamic range of 6 dB, the total dynamic range is $48 \times 6 = 288$ dB. Percentage resolution is $(1/2^{48}) \times 100 = 4 \times 10^{-13}$.

c. For floating-point representation, the dynamic range is determined by the number of bits in the exponent. Since there are 8 exponent bits, the dynamic range is $(2^8 - 1) \times 6 = 255 \times 6 = 1530$ dB.

The percentage resolution depends on the number of bits in the mantissa. Since there are 16 bits in the mantissa, the resolution is

$$(1/2^{16}) \times 100 = 1.5 \times 10^{-3}\%$$

These results are summarized in Table 3.1.

**Table 3.1** Dynamic Range and Precision for Various Number Representations

| Format of Representation | Number of Bits Used | Dynamic Range | Percentage Resolution (Precision) |
|---|---|---|---|
| Fixed-point | 24 bits | 144 dB | $6 \times 10^{-6}$ |
| Double-precision | 48 bits | 288 dB | $4 \times 10^{-13}$ |
| Floating-point | 24 bits (16-bit mantissa, 8-bit exponent) | 1530 dB | $1.5 \times 10^{-3}$ |

## 3.4 Sources of Error in DSP Implementations

A DSP system consists of an A/D converter, a DSP device, and a D/A converter. The accuracy of a DSP implementation depends upon a number of factors contributed by the A/D and D/A conversions and how the calculations are performed in the DSP device. The error in the A/D and D/A in the representation of analog signals by a limited number of bits is called the *quantization error* [2]. The quantization error decreases with the increase in the number of bits used to represent signals in A/D and D/A converters.

The errors in the DSP calculations are due to the limited word length used. These errors depend upon how the algorithm is implemented in a given DSP architecture. This error can be reduced by using a larger word length for data and by using rounding, instead of truncation, in calculations.

In the following sections, we consider the quantization and rounding errors in A/D converters, DSP computations, and D/A converters.

## 3.5 A/D Conversion Errors

Consider an A/D converter, shown in Figure 3.3(a), with $b$ bits used to represent an unsigned signal value. Its digital representation is of the form

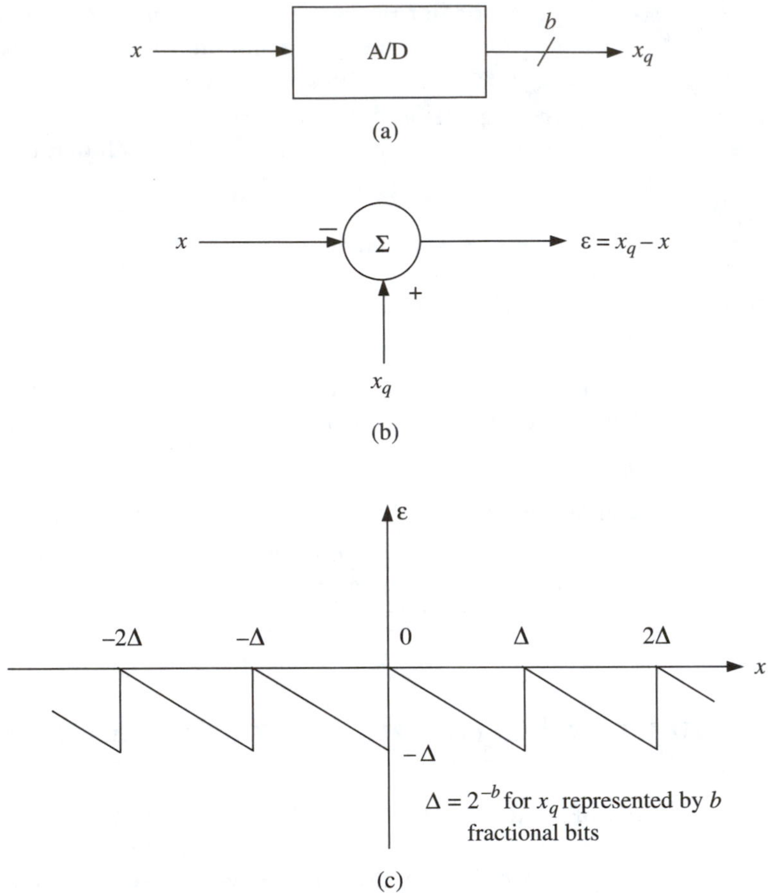


**Figure 3.3** (a) An A/D converter with *b* bits for signal representation, (b) quantization model for the A/D converter, (c) quantization error in truncation A/D converter, (d) quantization error in rounding A/D converter, (e) probability density function for truncation error, (f) probability density function for rounding error

*(continued)*

*.xxx ... x*, where there are *b* bits after the assumed binary point. In this kind of binary representation, the value of the least significant bit is given by

$$\Delta = 2^{-b} \tag{3.6}$$

The maximum error due to quantization depends on *b*. The quantization error for a given conversion as shown in the model of Figure 3.3(b) is given by

$$\varepsilon = x_q - x \tag{3.7}$$

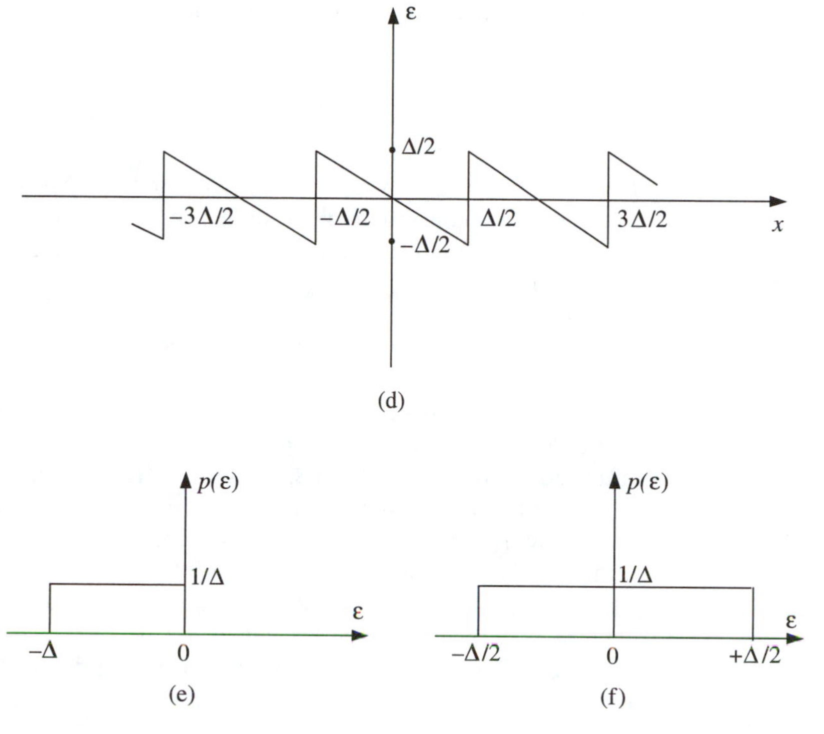


**Figure 3.3** Continued

where $x$ is the input and $x_q$ is the quantized output. This error is called the *truncation error* if the signal value above the largest integral multiple of $\Delta$ is simply dropped. It is called the *rounding error* if the value is rounded to the nearest integral multiple of $\Delta$. This way the rounding limits the error to $\pm\Delta/2$. Figures 3.3(c) and (d) show these two types of errors. The statistical interpretation of these errors can be used to evaluate their effect on DSP implementations. Assuming that the truncation and rounding errors in the A/D converter are uniformly distributed, their probability density functions are given in Figures 3.3(e) and (f), respectively. Analysis of Figure 3.3(e) for the mean and the variance of the error yields

$$m_\varepsilon = -\Delta/2 = -2^{(-b-1)} \tag{3.8}$$

$$\begin{aligned} {\sigma_\varepsilon}^2 &= \int_{-\Delta}^{0} (\varepsilon - (-\Delta/2))^2 p(\varepsilon)\, d\varepsilon \\ &= \int_{-\Delta}^{0} (\varepsilon + \Delta/2)^2 1/\Delta\, d\varepsilon \\ &= -\Delta^2/12 = 2^{-2b}/12 \end{aligned} \tag{3.9}$$

Similarly, the analysis of Figure 3.3(f) yields

$$m_\varepsilon = 0 \tag{3.10}$$

$$\sigma_\varepsilon{}^2 = 2^{-2b}/12 \tag{3.11}$$

That is, the variance of error is the same in both cases; the mean is zero in rounding and nonzero in truncation. The *signal-to-noise ratio* (SNR) is a measure that is used to evaluate the performance of the A/D converter. It can be calculated from

$$\text{SNR} = 10\log(\sigma_x{}^2/\sigma_\varepsilon{}^2) \tag{3.12}$$

where $\sigma_x{}^2$ is the signal power and $\sigma_\varepsilon{}^2$ is the noise variance.

The SNR cannot be calculated unless an assumption about the input signal amplitude is made. Practically speaking, too little a signal amplitude will result in a poor SNR, yet assuming the maximum signal amplitude in Eq. 3.12 will show only the best SNR. For the signal representation considered here (value from 0 to 1), it is customary to assume the root mean square (rms) value of the signal ($\sigma_x$) as 1/4 for SNR calculations. This leaves enough bits for the maximum possible value of the signal, yet it yields a more realistic SNR for evaluation of an A/D converter. With this assumption and substituting for $\sigma_x{}^2$ and $\sigma_\varepsilon{}^2$ in Eq. 3.12, we get

$$\text{SNR} = 10\log(1/16)/(2^{-2b}/12) = 10\log((3/4)(2^{2b})) \tag{3.13}$$

It is clear from Eq. 3.13 that using an A/D converter with a larger word length gives a larger SNR. As an example, if $b = 14$, the SNR is given as

$$\text{SNR} = 10\log((3/4)(2^{2\times 14})) = 83.04\text{ dB}.$$

## 3.6 DSP Computational Errors

The DSP computations involve using the digitized signal values and DSP structures represented by coefficients. These numbers are typically represented in the signed fractional 2's complement form. The computations almost always involve multiplications or *multiply and accumulate* (MAC) operations. In this section, we discuss the error in the multiplication carried out using the fixed word length arithmetic logic unit. Consider a specific DSP device that provides a 16 × 16 multiplier with a 32-bit result interfaced to a 16-bit A/D and a 16-bit D/A converter. The error in the computation will be due to discarding the 16 least significant bits of the 32-bit multiplication product. Assuming that the signal and the coefficients use *s.xxx ... x* format representation for signed numbers, and the multiplier used is also a signed binary multiplier, the multiplier result will be of the form *ss.xx. … x*. Before truncating (or rounding), this result can be shifted left by 1 bit (to discard the extra sign bit) to generate $s.b_{-1}b_{-2} \ldots b_{-30}0$ and then the 16 least significant bits can be dropped. The error in this computation is then given by

$$\varepsilon = 0 + 2^{-30}.b_{-30} + 2^{-29}.b_{-29} + 2^{-28}.b_{-28} + \cdots + 2^{-16}.b_{-16}. \tag{3.14}$$

Maximum error occurs when all the discarded bits are 1s. That is,

$$\varepsilon_{\text{max}}\Delta = 2^{-30} + 2^{-29} + \cdots + 2^{-16} = (2^{-15} - 2^{-30})$$

and the minimum error is when all the discarded bits are 0s. That is,

$$\varepsilon_{\text{min}}\Delta = 0$$

Assuming that $\varepsilon$ is uniformly distributed, we can compute mean as

$$m_\varepsilon = -\Delta/2 = (2^{-15} - 2^{-30})/2 = (2^{-16} - 2^{-31}) \approx 2^{-16} \tag{3.15}$$

and the variance as

$$\sigma_\varepsilon{}^2 = \Delta^2/12 = (2^{-15} - 2^{-30})^2/12 \approx 2^{-30}/12 \tag{3.16}$$

Using the argument of the last section, we can assume that the multiplier result has the rms value $\sigma_x$ of 1/4. Using this assumption leads to the following SNR:

$$\begin{aligned} \text{SNR} &= 10\log(\sigma_x{}^2/\sigma_\varepsilon{}^2) \\ &= 10\log(1/16)/(2^{-30}/12) \\ &= 10\log((3/4)(2^{30})) \\ &= 89.06\text{ dB} \end{aligned} \tag{3.17}$$

In a multiply and accumulate process using a fractional signed multiplier and a 32-bit accumulator, assuming no overflow condition, the SNR will be even better due to the averaging effect of the accumulator. It can be shown that in such a case the error variance is given as

$$\sigma_\varepsilon{}^2 = (1/N)(2^{-30}/12) \tag{3.18}$$

for $N$ accumulations. As is obvious, in most cases an individual DSP operation is not the dominant factor in error calculations. The overall calculation error depends upon the DSP algorithm that is being implemented.

Another type of computational error in DSP implementations is the overflow error.

If the result of a computation cannot be held in the accumulator register, an overflow condition occurs. If nothing is done to avoid or correct the overflow condition, the arithmetic wraparound occurs, in which case after the most positive number an overflow generates the most negative number, and vice versa. In a signal, it amounts to presence of a glitch with serious consequences.

A solution to the overflow problem is to provide extra bits called *guard bits* in the accumulator to accommodate the overflow bits. For instance, a provision of 4 extra bits ensures that there will not be any overflow for up to 16 accumulations.

If enough guard bits cannot be provided, there is need to implement saturation logic to at least keep the overflow under control and not let it produce a glitch in the signal. This is done by replacing the overflowed result with the most positive number, in the case of overflow from the most positive number to a negative number. For the case where the wraparound occurs from the most negative to a positive number, the result is replaced with the most negative number. This implementation ensures a glitch-free signal, although it still has calculation error, the amount of which depends upon the amount of the overflow.

## 3.7 D/A Conversion Errors

A source of error in a D/A converter is due to the fact that, typically, a D/A converter uses fewer bits in conversion than the number of bits required by the computed result, produced by the DSP device. This is equivalent to the truncation or the rounding off error in the A/D converter and can be handled in the same way as the computational error described in the previous section.

Another and more serious error occurs in the D/A converter due to the fact that the D/A converter output is not ideally reconstructed. Typically, the output samples from the DSP are applied to the input of a reconstruction filter through a zero-order hold, which maintains the input to the filter constant during the periods between successive samples. This is equivalent to saying that the input to the reconstruction filter is the convolution of the DSP output samples with a unit pulse of width equal to the sampling interval. The effect of this convolution is a reduction in the amplitude of the analog output. A compensating filter can compensate for this reduction in the amplitude. The frequency response of the compensating filter should be the inverse of the frequency response of the convolving pulse.

The source of error explained above can be illustrated by means of Figure 3.4. Consider the sequence of output samples of a DSP as shown in Figure 3.4(a). These samples are passed through a D/A converter with a zero-order

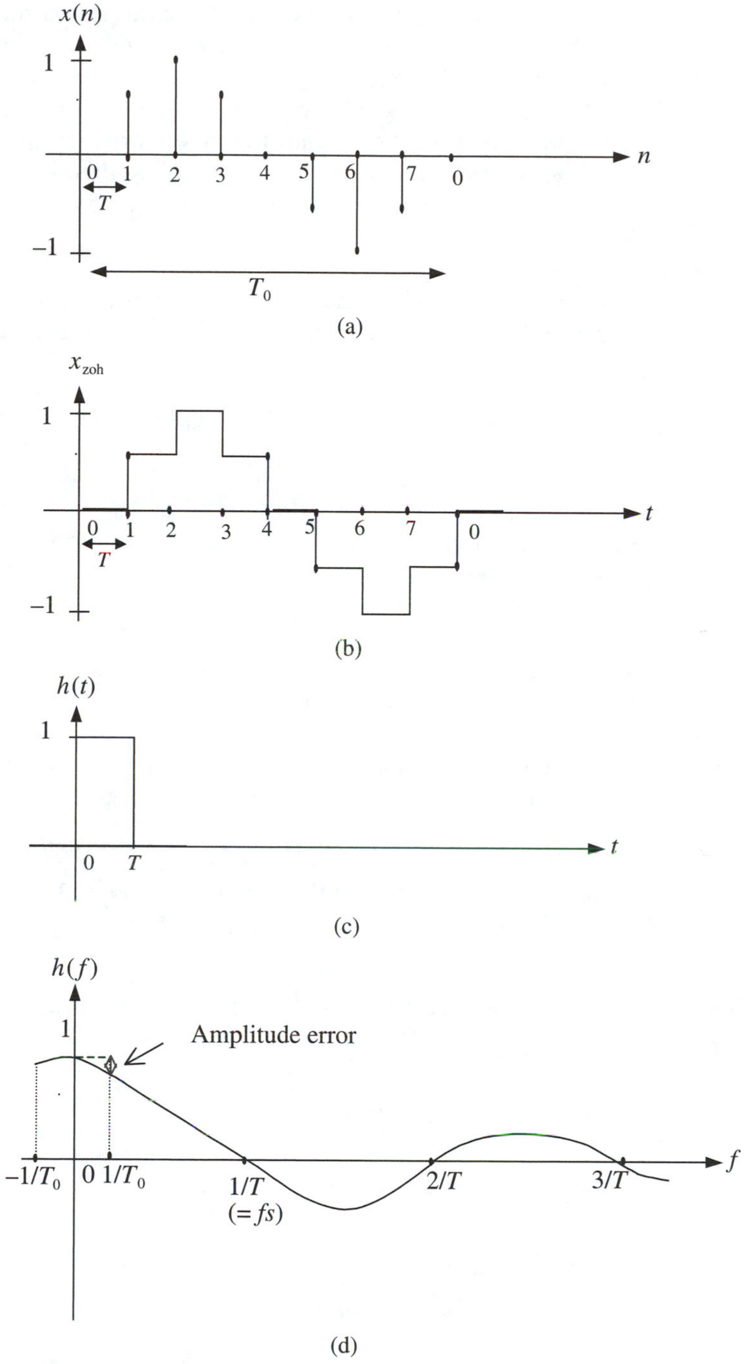

**Figure 3.4** An example showing the D/A converter error due to the zero-order hold at its output: (a) DSP output, (b) D/A output, (c) the convolving pulse that generates (b) from (a), (d) frequency contents of the convolving pulse in (c)

hold at its output. The output of the D/A converter is shown in Figure 3.4(b). Figure 3.4(c) shows the shape of the convolving pulse that generates the output of Figure 3.4(b) from the DSP output of Figure 3.4(a). Figure 3.4(d) shows the frequency contents of the convolving pulse as well as the degradation (amplitude error) of the output of the reconstruction filter from an ideal output. The compensating filter to restore the required output of the reconstruction filter should have a frequency response, which is the inverse of Figure 3.4(d).

▷ **Example 3.6** Find the degradation in amplitude gain when a sine wave of unit amplitude and 50 Hz frequency, sampled at 400 Hz, is reconstructed using a zero-order hold.

Solution The amplitude of the sine wave at a sampling instant is given by

$$x(n) = \sin 2\pi f n / f_s \tag{3.18}$$

where $f$ is the frequency of the sine wave and $f_s$ is the sampling frequency. In this example, $f = 50$ Hz and $f_s = 400$ Hz. Substituting these values in Eq. 3.18 yields

$$x(n) = \sin 2\pi n / 8 \tag{3.19}$$

The values of the amplitude computed using Eq. 3.19 are valid only for the ideal case. In order to compute the degradation in the amplitude due to the zero-order hold, these values have to be modified by the frequency response of the convolution pulse shown in Figure 3.4(d). In the frequency domain, the amplitude or the gain is a *sinc* function and is given by

$$\text{Gain} = H(f) = (\sin \pi f / f_s)/(\pi f / f_s) \tag{3.20}$$

Table 3.2 gives the values of the gain given by Eq. 3.20 for different frequencies expressed as a fraction of $f_s$. The gain at 50 Hz ($f_s/8$) is 0.9745 instead of 1.

**Table 3.2** Amplitude Degradation of D/A Output Due to the Zero-Order Hold.

| Frequency | Gain | 1/Gain |
|---|---|---|
| 0 | 1 | 1 |
| $f_s/32$ | 0.9984 | 1.0016 |
| $f_s/16$ | 0.9936 | 1.0064 |
| $f_s/8$ | 0.9745 | 1.0261 |
| $f_s/4$ | 0.9003 | 1.1107 |
| $f_s/3$ | 0.8270 | 1.2092 |
| $f_s/2.5$ | 0.7568 | 1.3213 |
| $f_s/2$ | 0.6366 | 1.5708 |

### 3.7.1 Compensating Filter

One can design a filter with a frequency response, which is the inverse of the gain $H(f)$ as shown in Table 3.2, and place it at the output of the D/A converter to compensate for the amplitude degradation of the D/A output due to the zero-order hold. Such a filter can be an IIR filter that can be designed using the techniques discussed in Chapter 2.

▷ **Example 3.7** Design a first-order IIR compensating filter having the frequency response depicted in Table 3.2.

**Solution** A first-order IIR filter can be designed using the program in Figure 3.5(a). Notice that the program uses the direct design method called the Yulewalk technique. As shown in Figure 3.5(b), the design produces the following coefficients for the filter:

$$b = [1.1752 \;\; 0.0110]$$

$$a = [1.0000 \;\; 0.1495]$$

which corresponds to the difference equation

$$y(n) = -0.1495y(n-1) + 1.1752x(n) + 0.0110x(n-1) \tag{3.21}$$

```
% Compensating filter specifications
f = [0 1/32 1/16 1/8 1/4 1/3 1/2.5 1/2]*2;
m = [1 1.0016 1.0064 1.055 1.1107 1.2092 1.3213 1.5708];

% Filter design
[b,a] = yulewalk(1,f,m)

% Designed filter frequency response
[h,th] = freqz(b,a,128);
plot(th/pi,abs(h)), title('Designed Compensating Filter Frequency
Response'), xlabel('f*2/fs'), ylabel('Magnitude')
```

(a)

```
b=
  1.1752   0.0110

a=
  1.0   0.1495
```

(b)

**Figure 3.5** Design of the compensating filter of Example 3.7: (a) a MATLAB program, (b) designed filter coefficients, (c) designed filter frequency response

*(continued)*

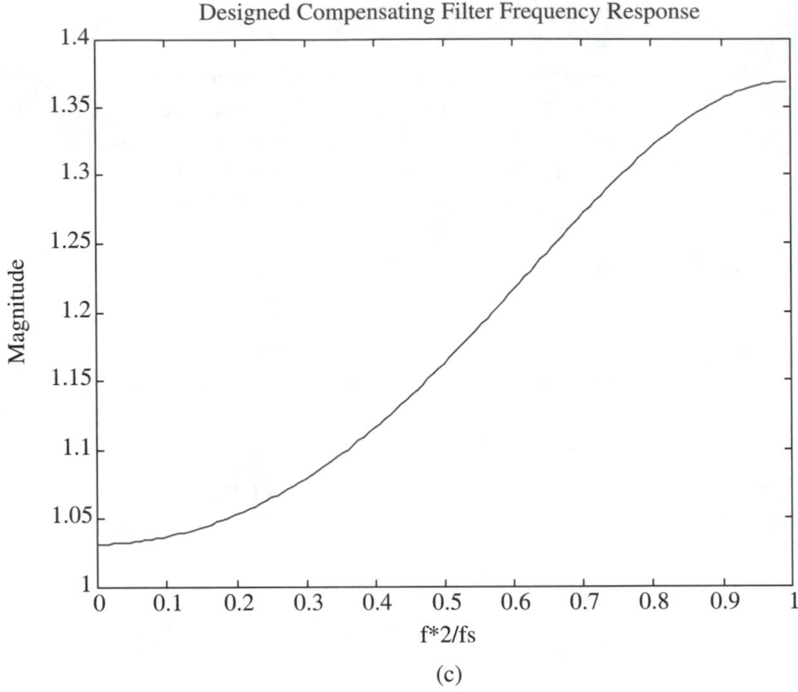


**Figure 3.5** Continued

The transfer function of the filter in the *z*-domain is given by

$$H(z) = \frac{(1.1752 + 0.0110z^{-1})}{(1 + 0.1495z^{-1})} \tag{3.22}$$

The frequency response of this compensating filter is shown in Figure 3.5(c). Since the compensating filter is a digital filter, it can be merged with the DSP calculations. That is, the input to the D/A converter is first passed through the filter before it is applied to the D/A converter, thus eliminating the need for a filter to be placed after the D/A converter. In general, making the compensating filter a part of the DSP eliminates additional computations, since the filter computations can be merged with the DSP computations.

---

The analysis presented above can be extended to correct degradation more accurately. However, it should be noted that the compensating filter in such a case will be more complex and will be of orders higher than 2.

## 3.8 Summary

In this chapter, we studied various number formats for representing signals and coefficients, consisting of the fixed-point format, floating-point format, double-precision format, and block floating-point format. We also studied the dynamic range and precision of signals represented by each of these formats. We identified the sources of errors in DSP implementations, such as A/D conversion errors, DSP computational errors, and D/A conversion errors. For each category, we have estimated the errors and have suggested ways to minimize them in the implementation of DSP systems.

## References

1. Ifeacho, E. C., and Jervis, B. W., *Digital Signal Processing: A Practical Approach*, Redding, MA, Addison-Wesley, 1993.
2. Bateman, A., and Yates, W. *Digital Signal Processing Design*, Los Alamitos, CA, Computer Science Press, 1989.
3. Higgin, R. J. *Digital Signal Processing in VLSI*, Englewood Cliffs, Prentice Hall, 1990.
4. Lapsley, P., Bier, J., Shoham, A., and Lee, E. A. *DSP Processor Fundamentals: Architectures and Features*, Piscataway, NJ, IEEE Press, 1997.

## Assignments

**3.1** Determine (a) the most positive, (b) the least positive, (c) least negative, and (d) the most negative values for the following number representation formats.

a. 32-bit 2's complement integer format

b. 32-bit floating-point format given as:

| s | eee...e | fff....f | |
|---|---|---|---|
| 1 | 8 | 23 | (bits) |
| s | exp (unsigned) | frac (unsigned) | |

where the value of the number is computed as 1.frac $\times$ $2^{exp}$ if $s = 0$, and $-1$.frac $\times$ $2^{exp}$ if $s = 1$.

**3.2** Determine the maximum truncation error for both positive and negative numbers for the two formats in Problem 3.1.

**3.3** Show that the dynamic range of a signal increases by 6 dB for each additional bit used to represent its value.

**3.4** Compute the dynamic range and percentage resolution of a signal that uses

a. 16-point fixed-point format

b. 32-point floating-point format with 24 bits for the mantissa and 8 bits for the exponent.

**3.5** Compute the dynamic range and the percentage resolution for a block floating-point format with a 4-bit exponent used in a 16-bit fixed-point processor.

**3.6** For the DSP system shown in the block diagram of Figure P3.6, the analog input is a 50 Hz sinusoidal signal with 2 V peak value. Both the A/D and D/A converters are 0–5 V devices. Determine (a) the SNR of A/D, (b) the SNR of DSP, and (c) the peak output of the D/A converter. Assume a sampling rate of 400 samples/sec. State other assumptions that are needed for calculations.

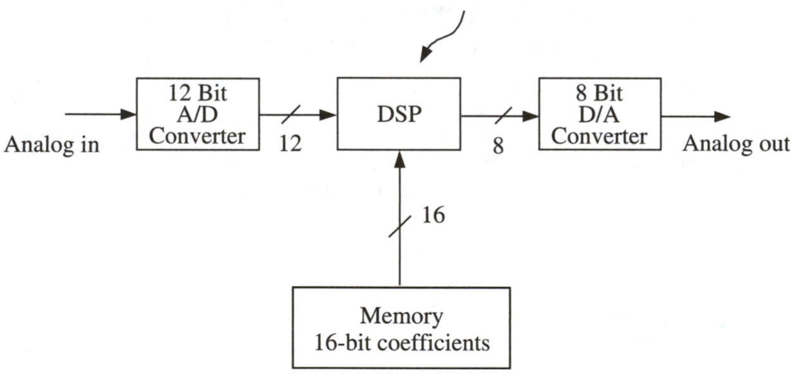


**Figure P3.6** A DSP system block diagam

**3.7** One can use the filter of Eq. 3.22 to compensate for the D/A converter error in Problem 3.6. This filter, however, does not compensate the D/A error completely. There remains some error at different frequencies. Prepare a table to show the error that remains uncompensated.

**3.8** Determine the frequency response for the filter

$$H(z) = \frac{1.125}{(1 + 0.1807z^{-1})}$$

Compare its frequency response to the one in Table 3.2 and discuss its suitability as a zero order hold D/A compensating filter.

# Chapter 4

## Architectures for Programmable Digital Signal-Processing Devices

## 4.1 Introduction

In this chapter, architectural features of programmable DSP devices are described based on the DSP operations these devices are generally required to perform. The features are examined from the points of view of functional needs, programmability, speed, and interfacing requirements of these devices. Commonly used hardware implementations are also described for various functional units. Following are the topics covered in this chapter:

Basic architectural features

DSP computational building blocks

Bus architecture and memory

Data addressing capabilities

Address generation unit

Programmability and program execution

Speed issues

Features for external interfacing

## 4.2 Basic Architectural Features

A programmable DSP device should provide instructions similar to a microprocessor. These instructions can then be used to design programs for implementing DSP algorithms. The basic computational capabilities provided by way of instructions should include the following [1–3, 11]:

- Arithmetic operations such as add, subtract, and multiply.
- Logic operations such as AND, OR, XOR, and NOT.

- Multiply and accumulate (MAC) operation.
- Signal scaling operations for scaling the signal before and/or after digital signal processing.

It is important that dedicated high-speed hardware be provided to carry out these operations. For instance, multiply operation can be done much faster on a hardware multiplier than on a microcoded multiplier realized using the shift and add technique, as is often done in microprocessors.

In addition to the computational units, support architecture should include the following hardware features [10]:

- On-chip registers for storage of intermediate results.
- On-chip memories for signal samples (RAM).
- On-chip program memory for programs and fixed data such as filter coefficients (ROM).

▷ **Example 4.1** Investigate the basic features that should be provided in the DSP architecture to be used to implement the following $N^{\text{th}}$-order FIR filter:

$$y(n) = \sum_{i=0}^{N-1} h(i)x(n-i); \quad n = 0, 1, 2, \ldots \tag{4.1}$$

where $x(n)$ denotes the input sample; $y(n)$, the output sample; and $h(i)$, the $i$th filter coefficient. $x(n-i)$ is the input sample $i$ samples earlier than $x(n)$.

Solution The FIR filter requires the following basic features for implementing Eq. 4.1:

1. Memory for storage of signal samples $x(n), x(n-1), \ldots$, etc. (RAM).
2. Memory for storage of filter coefficients: $h(0), h(1), \ldots$, etc. (ROM).
3. A hardware multiplier and an adder to carry out the multiply and accumulate (MAC) operation.
4. A register to keep track of accumulation (accumulator).
5. A register to point to the current signal sample being used (signal pointer).
6. A register to point to the current filter coefficient being used (coefficient pointer).
7. A register to keep count of the MAC operations that remain to be done (counter).
8. Capability to scale the signal value $x(n)$ as it is read from the memory and the computed signal $y(n)$ as it is stored in the memory (shifters at input and output).

---

Computational units such as the multiplier, the arithmetic logic unit (ALU), shifters, etc. will be described in the next section. Subsequent sections will examine the other functional units such as the memory, the addressing unit and the program execution unit.

## 4.3 DSP Computational Building Blocks

In this section, we learn about the hardware building blocks that carry out the basic DSP computations. While choosing these computational building blocks, we keep in mind the requirements of speed and accuracy, which are the two key issues in the design of DSP systems. At the same time, we should ensure that such building blocks could be configured to implement many different applications. That is, while each building block should be optimized for functionality and speed, the design should be sufficiently general so that it can be easily integrated with other blocks to implement overall DSP systems.

Following are the basic building blocks that are essential to carry out DSP computations [5–9]:

- Multiplier
- Shifter
- Multiply and accumulate (MAC) unit
- Arithmetic logic unit

In the following subsections, we shall discuss each of these blocks in detail.

### 4.3.1 Multiplier

The advent of single-chip multipliers and their integration into the microprocessor architecture are the most important reasons for the availability of commercial VLSI chips capable of implementing DSP functions. These multipliers, called *parallel* or *array multipliers*, implement complete multiplication of two binary numbers to generate the product in a single processor cycle. Earlier multiplication schemes relied either on software such as the shift and add algorithm or on microcoded controllers, which implement the same algorithm in hardware. Both these options require several processor cycles to complete the multiplication. The advances made in VLSI technology, both in terms of speed and size, have made possible the hardware implementation of parallel multipliers.

From earlier chapters, it is apparent that multiplication is one of the key operations in implementing DSP functions. However, before we design an actual multiplier, we should be clear about its specifications such as speed, accuracy, and dynamic range. The number of bits used to represent the multiplication operands and whether they are represented in fixed-point or floating-point format decide the accuracy and dynamic range of the multiplier. The speed, on the other hand, is decided by the architecture employed. For a given technology, there are several architectures for parallel multipliers, which trade off speed for reductions in circuit complexity and power dissipation. The choice of the architecture depends on the application.

| | | | | $A_3$ | $A_2$ | $A_1$ | $A_0$ |
|---|---|---|---|---|---|---|---|
| | | | | $B_3$ | $B_2$ | $B_1$ | $B_0$ |
| | | | | $A_3B_0$ | $A_2B_0$ | $A_1B_0$ | $A_0B_0$ |
| | | | $A_3B_1$ | $A_2B_1$ | $A_1B_1$ | $A_0B_1$ | |
| | | $A_3B_2$ | $A_2B_2$ | $A_1B_2$ | $A_0B_2$ | | |
| | $A_3B_3$ | $A_2B_3$ | $A_1B_3$ | $A_0B_3$ | | | |
| $P_7$ | $P_6$ | $P_5$ | $P_4$ | $P_3$ | $P_2$ | $P_1$ | $P_0$ |

(a)

**Figure 4.1(a)** The 4 × 4 binary multiplication

### Parallel Multiplier

Let us consider the multiplication of two unsigned numbers A and B. Let the number A be represented using $m$ bits ($A_{m-1}A_{m-2}\ldots A_0$) and the number B, using $n$ bits ($B_{n-1}B_{n-2}\ldots B_0$). The multiplicand A, the multiplier B, and the product P are given by [4–6]

$$A = \sum_{i=0}^{m-1} A_i 2^i \tag{4.2}$$

$$B = \sum_{j=0}^{n-1} B_j 2^j \tag{4.3}$$

$$P = \sum_{j=0}^{n-1} \left[ \sum_{i=0}^{m-1} A_i B_j 2^{i+j} \right] \tag{4.4}$$

and can have a maximum of $(m+n)$ bits. Each bit of the product P is obtained by a summation of bits $A_iB_j$ using an array of single-bit adders. The bits $A_iB_j$, where the index $i$ takes on values from 0 to $m-1$, and the index $j$ from 0 to $n-1$, are formed using AND gates. Figure 4.1(a) shows the multiplication operation using 4 bits for both A and B ($A = A_3A_2A_1A_0$ and $B = B_3B_2B_1B_0$). Figure 4.1(b) shows the hardware structure of the multiplier for this example. The structure is regular and requires twelve 3 input, 2 output adders. It can be shown that for an $n \times n$ multiplier, the number of adders required is $n(n-1)$.

### Multiplier for Signed Numbers

The multiplier shown in Figure 4.1(b) is known as *Braun multiplier* [7] and is the basis for most of today's commercial implementations. Several improve-

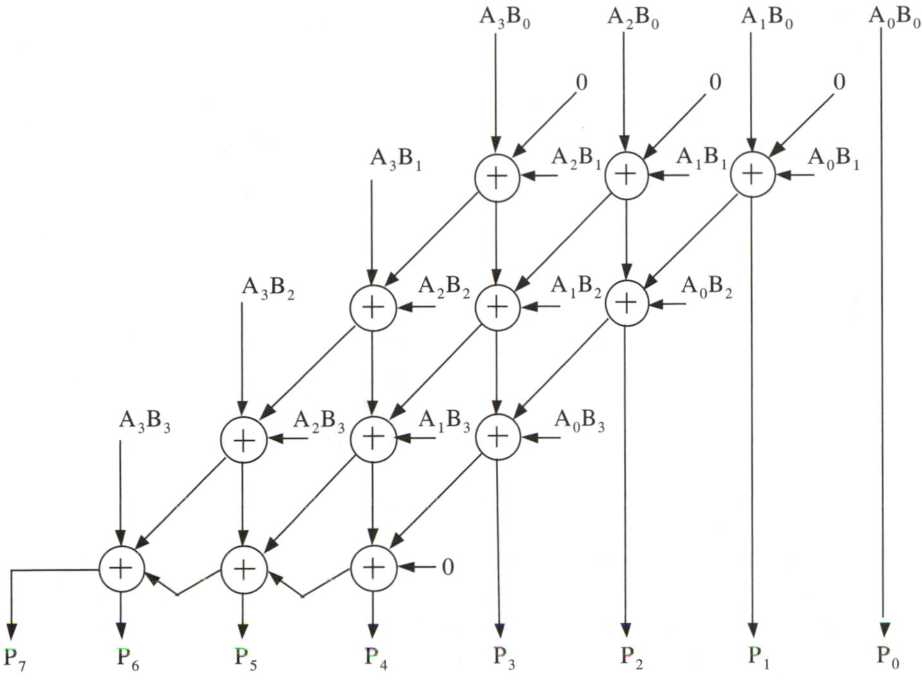


**Figure 4.1(b)** The structure of a 4 × 4 Braun multiplier

ments on this basic structure are possible and have been used to increase the speed and reduce the hardware complexity and power dissipation. We will not be dealing with these variations here. However, we will consider one modification of the Braun structure, which is essential to carry out multiplication of signed numbers.

Braun's multiplier does not take into account the signs of the numbers that are being multiplied. Additional hardware is required before and after the multiplication when signed numbers, represented in 2's complement form, are used. It would be desirable to have a structure that can directly operate on 2's complement numbers.

Consider two numbers A and B represented in 2's complement format. Let A have $m$ bits and B, $n$ bits. A and B can be written as follows:

$$A = -A_{m-1}2^{m-1} + \sum_{i=0}^{m-2} A_i 2^i \tag{4.5}$$

$$B = -B_{n-1}2^{n-1} + \sum_{j=0}^{n-2} B_j 2^j \tag{4.6}$$

The product $P = P_{m+n-1} \dots P_1 P_0$ can be written as

$$P = A_{m-1}B_{n-1}2^{m+n-2} + \sum_{i=0}^{m-2}\sum_{j=0}^{n-2} A_i B_j 2^{i+j} - \sum_{i=0}^{m-2} A_i B_{n-1}.2^{n-1+i} - \sum_{j=0}^{n-2} A_{m-1} B_j 2^{m-1+j} \tag{4.7}$$

The two subtractions in Eq. 4.7 can be expressed as additions of 2's complement numbers. In doing so, Eq. 4.7 gets modified to an expression with only additions and no subtractions and can then be implemented through a structure similar to the Braun multiplier using only adders. The modified structure for handling signed numbers is called the *Baugh–Wooley multiplier* [8].

### Speed

The shift and add technique of multiplication normally used in microprocessors requires $n$ processor cycles to carry out an $n \times n$ multiplication. The cycle time is the time to access the operands, perform add and shift, and store the result in the product register. The parallel multiplier, on the other hand, is a fully combinational implementation, and once the operands are made available to the multiplier, the multiplication time is only the longest path delay time through the gates and adders.

Normally, one would want to achieve the highest possible speed of operation for a given DSP function. This would mean a multiplication time comparable to the processing times of other computational units as well as the access times of memories holding the program and data. As memory technology advances, lower and lower access times are achieved. In order to make the best use of such speeds in a DSP implementation, it would be highly desirable to design multipliers operating at the highest possible speeds. This is possible only with a fully parallel implementation.

### Bus Widths

Consider a multiplier with inputs X and Y and the product Z. If X and Y are represented with $n$ bits each, Z can have a maximum of $2n$ bits. Let us assume that both X and Y are in the memory and the product Z has also to be written back to the memory. A single-cycle execution of the multiplication will then require two buses of width $n$ bits each (for X and Y) and a third bus of width $2n$ bits (for Z). This type of bus architecture is expensive to implement. A number of practical considerations, however, make it possible to realize the multiplication with a less extensive bus architecture. First, the program bus can be used to transfer one of the operands (say, Y) after the multiplication instruction has been fetched from the program memory. This does not cause an additional overhead when repeated multiplications are carried out, as is generally the case with many DSP algorithms. This is because, the instruction, once fetched, usually resides in an on-chip cache. Second, a separate bus for the product Z can be dispensed with, since one of the buses (say, that of X) can be used to transfer the product to the memory as the operand X would have been latched long before the product Z is made available. To handle the $2n$ bits of Z, there are two available alternatives:

a. Use the X bus ($n$ bits) and save Z at two successive memory locations using two memory accesses.

b. Discard the lower $n$ bits of Z and save only the higher $n$ bits. This is the option most often used since one of the two operands X and Y (usually Y) is normalized to one before multiplication so that the $n$ bits discarded from Z are the less significant fractional bits. However, if the product Z is to be further processed (e.g., added to the previous result as is the case in a multiply and accumulate operation), all $2n$ bits of the product Z are retained and passed on to the next stage to retain the accuracy of the product. The decision on discarding lower-order bits or saving the entire word is made after the accumulation process is completed.

For applications in which speed is not the main issue, buffers and latches may be provided at inputs and the output, as shown in Figure 4.2. A single bus can then be used to preload the operands in the input latches before the multiplication and transfer the result from the output latches/buffers to the memory or the next stage, if necessary in two cycles after the multiplication.

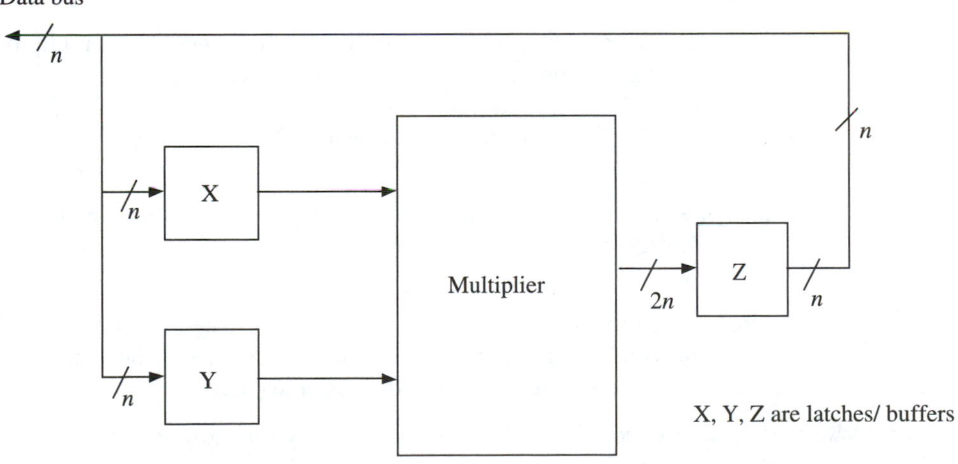


**Figure 4.2** A multiplier with input and output latches/buffers

### 4.3.2 Shifter

Shifter is an essential component of any DSP architecture. Shifters are required to scale down or scale up operands and results to avoid errors resulting from overflows and underflows during computations. Let us consider the following cases:

a. It is required to compute the sum of $N$ numbers, each represented by $n$ bits. As the accumulated sum grows, the number of bits required representing it increases. The maximum number of bits to which the sum can grow is $(n + \log_2 N)$ bits. However, if each of the $N$ numbers is scaled down by $\log_2 N$ bits prior to the addition, the loss of the result due to overflow can be avoided. The accumulator will then hold the sum scaled down by $\log_2 N$ bits. Although the accuracy of the sum is reduced because of the loss of $\log_2 N$ lower-order bits, the summation would be completed without the occurrence of the overflow error. The actual sum can be obtained by scaling up the result by $\log_2 N$ bits, when required.

b. When two numbers, each represented by $n$ bits, are multiplied, the product can have a maximum of $2n$ bits. When this product is saved in memory, which is also $n$ bits wide, the lower-order $n$ bits are generally discarded, resulting in loss of accuracy. However, in the case of multiplication of two signed numbers, the accuracy can be slightly improved by shifting the product by one bit position to the left before saving the $n$ higher-order bits. This is because the $2n$-bit product will have two sign bits, and even after discarding one of them (by a single-bit left shift), the sign of the product is still preserved. The accuracy improves because, instead of discarding all the $n$ lower-order bits, we now discard only $(n - 1)$ bits.

c. When carrying out floating-point additions, the operands should be normalized to have the same exponent. This is accomplished by shifting one of the operands by the required number of bit positions so that it has the same exponent as the other operand.

The cases illustrated above are examples of situations that require shifting of data while implementing DSP operations.

▷ **Example 4.2** It is required to find the sum of 64 numbers each represented by 16 bits. How many bits should the accumulator have so that the sum can be computed without the occurrence of overflow error or loss of accuracy?

Solution When 64 numbers are added, the sum can grow by a maximum of $\log_2 64 = 6$ bits. To avoid overflow, the total number of bits the accumulator should have is $16 + 6 = 22$.

---

▷ **Example 4.3** If, for the problem of Example 4.2, it is decided to have an accumulator with only 16 bits but shift the numbers before the addition to prevent overflow, by how many bits should each number be shifted?

Solution Since the sum can grow by 6 bits, in order to prevent overflow, each number should be shifted by 6 bits to the right before the addition.

---

▷ **Example 4.4** If all the numbers in the problem of Example 4.3 are fixed-point integers, what is the actual sum of the numbers?

Solution Since each number has been shifted to the right by 6 bits, the sum should be shifted left by 6 positions to get the actual value.

$$\text{The actual sum} = (\text{content of the accumulator}) \times 2^6$$

---

▷ **Example 4.5** What is the error in the computation of the sum in the problem of Example 4.4?

Solution Since the six lowest significant bits have been lost in the process of summation, the sum could be off by as much as $2^6 - 1 = 63$.

---

### Barrel Shifter

In conventional microprocessors shifting is normally implemented by an operation similar to the one performed in a shift register. The operation takes one clock cycle for every single bit shift. Such a scheme requires unduly large amounts of time to implement multibit shifts, which are generally required in DSP computations. In DSPs, on the other hand, in order to preserve the computational speed of single-cycle instruction execution, shifts by several bits should be accomplished in a single cycle. This is possible by a combinational circuit known as the *barrel shifter*. The barrel shifter connects the input lines representing a word to a group of output lines with the required shift determined by its control inputs, as shown in Figure 4.3(a). Control input also determines the direction of the shift (left or right). If the input word has $n$ bits, and shifts from 0 to $n - 1$ bit positions to the right or left are to be implemented, the control input requires $\log_2 n$ lines to determine the number of bits to be shifted. Further, an additional line is also required for the control input to indicate the direction of the shift. In practice, however, the direction of shift is usually fixed, with the result that only $\log_2 n$ lines are required for the control input. Bits shifted out of the input word are discarded and the new bit positions are filled with zeros in the case of left shift. In the case of right shift, the new bit positions are replicated with the most significant bit to maintain the sign of the shifted result.

Figure 4.3(b) shows an implementation of a barrel shifter with four input bits, $(A_3A_2A_1A_0)$ and four output bits $(B_3B_2B_1B_0)$. Using this shifter, it is

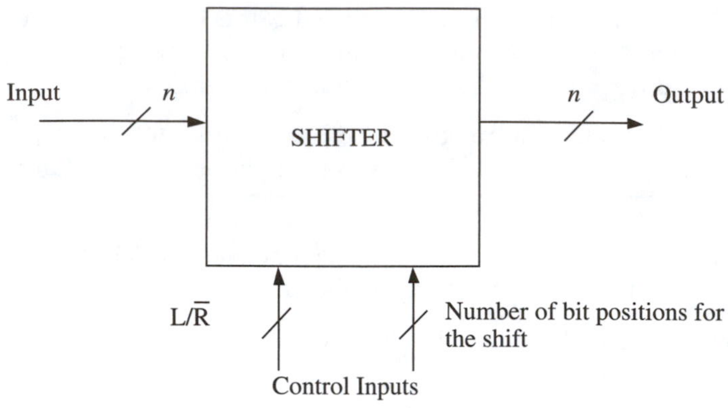


**Figure 4.3(a)** Block diagram of a barrel shifter

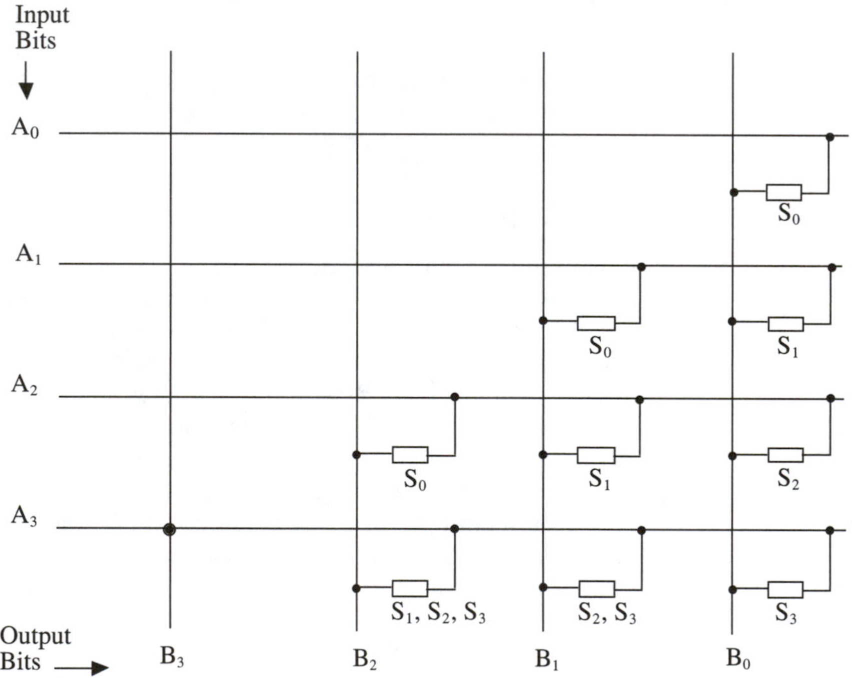


| Input | Shift (Switch) | Output ($B_3B_2B_1B_0$) |
|---|---|---|
| $A_3A_2A_1A_0$ | 0 ($S_0$) | $A_3A_2A_1A_0$ |
| $A_3A_2A_1A_0$ | 1 ($S_1$) | $A_3A_3A_2A_1$ |
| $A_3A_2A_1A_0$ | 2 ($S_2$) | $A_3A_3A_3A_2$ |
| $A_3A_2A_1A_0$ | 3 ($S_3$) | $A_3A_3A_3A_3$ |

**Figure 4.3(b)** Implementation of a 4-bit, shift-right barrel shifter

possible to realize right shift by 0, 1, 2, or 3 bit positions by setting the control inputs ($S_0$, $S_1$, $S_2$, or $S_3$) high, respectively. Only one of the control inputs can be high at any time and this input closes all the switches controlled by it and enables the appropriate paths between the inputs and the outputs.

Since the circuit for a barrel shifter is a combinational logic circuit, the time taken to implement the shift is the total combinational delay involved in decoding the control lines and setting up the path from the input lines to the output lines. This delay is only a fraction of a clock cycle. In fact, in practical DSPs, shifting is combined with data transfer. Both operations are executed in a single clock cycle.

▷ **Example 4.6** A barrel shifter is to be designed with 16 inputs for left shifts from 0 to 15 bits. How many control lines are required to implement the shifter?

Solution The number of control lines required is four, since 4 bits are needed to code any number between 0 and 15, the range over which the shift is required to be accomplished.

---

### 4.3.3 Multiply and Accumulate (MAC) Unit

Most DSP applications such as filters require the accumulation of the products of a series of successive multiplications. In order to implement this accumulation, we need an add/subtract unit and an additional register called the *accumulator* at the output of the multiplier. The configuration of such a multiply and accumulate unit, commonly known as the MAC unit, is shown in Figure 4.4.

The MAC unit consists of a multiplier that multiplies two $n$-bit numbers X and Y and gives a product $2n$ bits wide. This is added to or subtracted from the contents of the accumulator in the add/sub unit. The result is saved in the accumulator. The MAC unit can thus be used to implement functions of the type $A + BC$. If the accumulator is cleared at the start of a series of multiplications, it will contain the accumulated sum of the products on completion of all the multiplications.

Although multiplication and accumulation are two distinct operations, each normally requiring a separate instruction execution cycle, the two can work in parallel. At a time when the multiplier is computing a product, the accumulator accumulates the product of the previous multiplication. If $N$ products are to be accumulated, $N - 1$ multiplies can overlap with accumulations. During the very first multiply, the accumulator is idle since there is nothing to accumulate. Likewise, during the very last accumulation, the multiplier is idle since all the $N$ products have been computed. Thus it takes a total of $N + 1$ instruction execution cycles to compute the sum of products of $N$ multiplications. If $N$ is large, this works out to a speed of nearly one multiply and accumulate (MAC) operation per instruction execution cycle. This pipelined

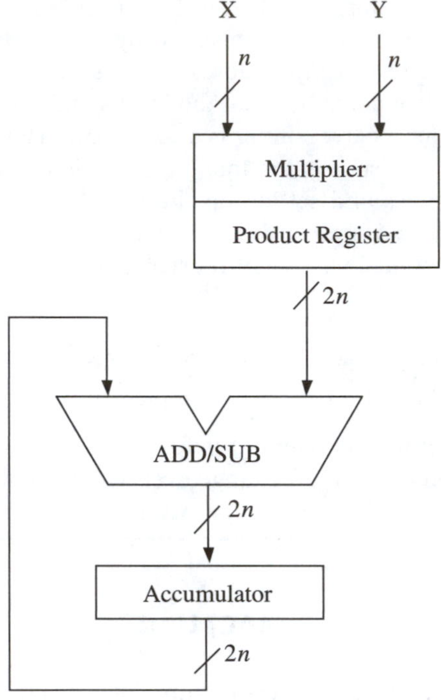


**Figure 4.4** A MAC unit

operation of a multiplier and an accumulator working in parallel to effectively execute a MAC operation per cycle is a standard feature of many commercial DSP devices.

▷ **Example 4.7** If a sum of 256 products is to be computed using a pipelined MAC unit, and if the MAC execution time of the unit is 100 nsec, what will be the total time required to complete the operation?

Solution To carry out 256 MAC operations, 257 execution cycles are required.

$$\text{The total time required} = 257 \times 100 \times 10^{-9} \text{ sec} = 25.7 \ \mu\text{sec}.$$

### Overflow and Underflow

When designing a MAC unit, one has to pay attention to the word sizes encountered at the input of the multiplier and the sizes of the add/subtract unit and the accumulator, as overflow and underflow conditions may be encountered otherwise. Provision of barrel shifters at the inputs and the output of the MAC unit, provision of guard bits in the accumulator, and provision of saturation logic are the frequently used techniques to prevent overflow and underflow conditions from occurring in the MAC unit. Now let us consider each of these provisions in detail.

### Shifters

Shifters are normally provided at the inputs and the output of the MAC unit. The input shifters help to normalize data samples and/or filter coefficients as they are fed into the multiplier, to avoid overflow of the accumulated result at the output. Likewise, the shifter at the output is used to denormalize the result after the sum of products computation, before being saved in the memory. In addition, the output shifter may also be used to discard the redundant sign bit in 2's complement product or to shift the output by the required number of positions before saving to preserve the maximum possible accuracy. This is done when the number to be saved is preceded by several leading 0s or 1s. As shifters provided in the MAC unit are typically barrel shifters, they do not require additional clock cycles to implement the shifts.

### Guard Bits

Sometimes, in order to preserve accuracy, the inputs to the multiplier are not normalized. In such a case, when repetitive MAC operations are performed, the accumulated sum grows with each MAC operation. This increases the number of bits required to represent the result without loss of accuracy. One way to handle this growth is to provide extra bits in the accumulator. These extra bits, called *guard bits* or *extension bits*, allow for the growth of the accumulated sum as more and more product terms are added up. When the computation of the required sum of products is completed, the extension bits may be saved as a separate word, if required. Alternatively, the sum along with the guard bits may be shifted by the required amount and saved as a single word. When guard bits are provided in the accumulator, the size of the add/subtract unit also increases correspondingly.

▷ **Example 4.8** Consider a MAC units whose inputs are 16-bit numbers. If 256 products are to be summed up in this MAC, how many guard bits should be provided for the accumulator to prevent overflow condition from occurring?

Solution In general, the product of a $16 \times 16$ multiplication has 32 bits. Since 256 such products are to be summed, the sum can grow by a maximum of $\log_2 256 = 8$ bits. Therefore, the number of guard bits required to prevent the occurrence of overflow is 8.

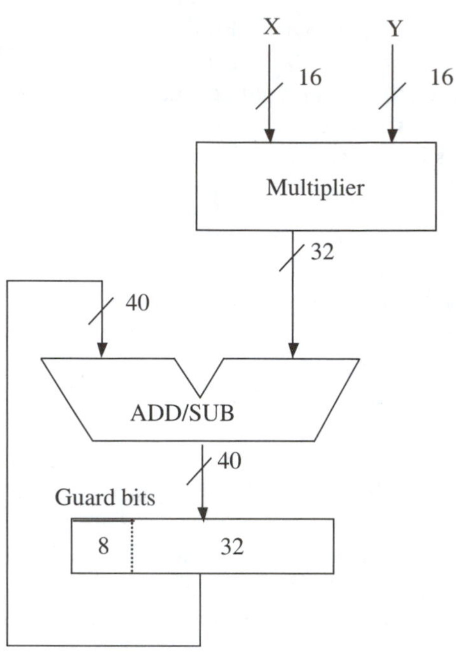


**Figure 4.5** A MAC unit with accumulator guard bits

Figure 4.5 shows a block diagram of the MAC unit with guard bits for this example.

## Saturation Logic

With or without guard bits, an overflow condition occurs when the accumulated result becomes larger than the largest number it can hold. Likewise, when handling a negative number, an underflow will occur if the contents of the accumulator become smaller than the smallest number it can hold. In such situations, it may be better to limit the accumulator contents to the most positive (or the most negative) value to avoid an error known as the wraparound error.

Limiting the accumulator contents to its saturation limits is achieved with a simple logic circuit called the *saturation logic*. The circuit, shown in Figure 4.6, detects the overflow and underflow condition and accordingly loads the accumulator with the most positive or the most negative value, overriding the value computed by the MAC unit. The overflow/underflow condition is detected by monitoring the carry into the MSB and the carry out of the MSB. If carry-in is not equal to carry-out, the overflow/underflow condition occurs. The selection between the most negative and the most positive numbers is made based on the sign bit of the number.

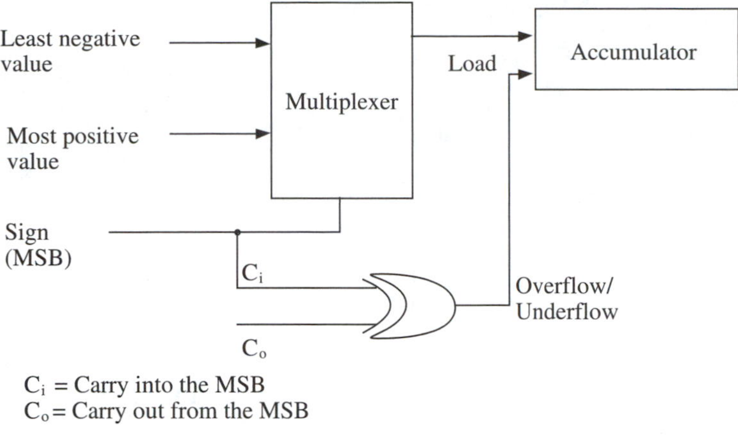


**Figure 4.6** A schematic diagram of the saturation logic

## 4.3.4 Arithmetic and Logic Unit

In addition to shift, multiply, and multiply-and-accumulate (MAC) operations, a DSP is required to carry out several arithmetic and logic operations. These are the operations, such as add, subtract, increment, decrement, negate, AND, OR, NOT, EXOR, and compare, that are also implemented in a conventional microprocessor. This means that the ALU of a DSP is similar to the ALU of a microprocessor but with additional features such as shift and multiply discussed in the earlier sections. Figure 4.7 shows the block diagram of the ALU of a typical DSP device.

Apart from providing arithmetic and logic functions, the design of an ALU for a DSP incorporates several other features borrowed from a general-purpose microprocessor. Three of these features are discussed next.

### Status Flags

It is important to know the status of the accumulator after arithmetic or a logic operation. This information is used for program sequencing and scaling. The ALU includes circuitry to generate status flags after arithmetic and logic operations. These flags include sign, zero, carry, and overflow. For instance, if the execution of an instruction results in overflow, the overflow flag is set; otherwise it is reset.

### Overflow Management

Features similar to those explained in the previous section on MAC are also required in the ALU for overflow management. These features are generally

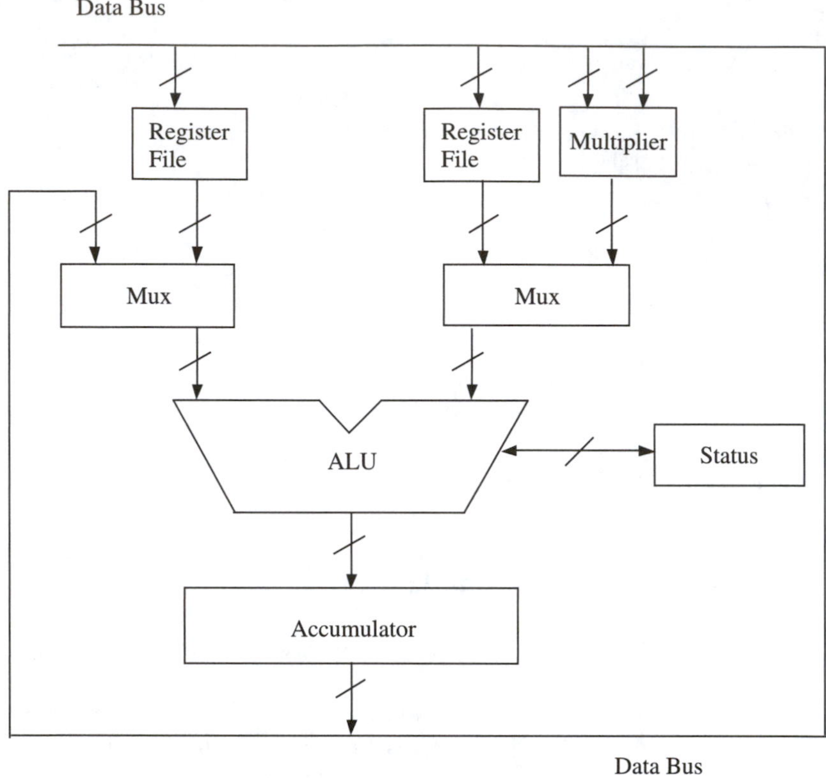


**Figure 4.7** Block diagram of an arithmetic logic unit

combined with the status flags. For example, depending on the status of the overflow and the sign flags, the saturation logic can come into effect to limit the accumulator contents to its most positive or the most negative value.

### Register File

A feature that improves the efficiency of an ALU is the implementation of a large general-purpose register file. Instead of moving data in and out of the ALU to memory during the course of an arithmetic computation, it may be faster to have intermediate results of arithmetic computations stored in the ALU until the computation is complete and the result is ready to be saved. This is possible by providing a file of general-purpose registers in addition to the accumulator as part of the ALU architecture.

## 4.4 Bus Architecture and Memory

In conventional microprocessors, the von Neumann architecture is used, wherein the program and the data reside in the same memory and a single bus (Address + Data) is used to access both, as is shown in Figure 4.8(a). This slows down the program execution considerably as the processor has to wait for the data even after the instruction is made available to it. In order to avoid this waiting and to speed up the program execution, it is desirable to have the program and data reside in two separate memories and have two buses for the processor to access the two memories. This modification, which is called

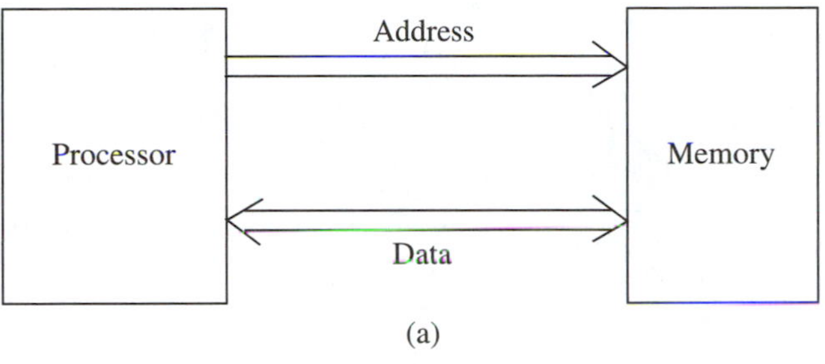


**Figure 4.8(a)** The bus structure of von Neumann architecture

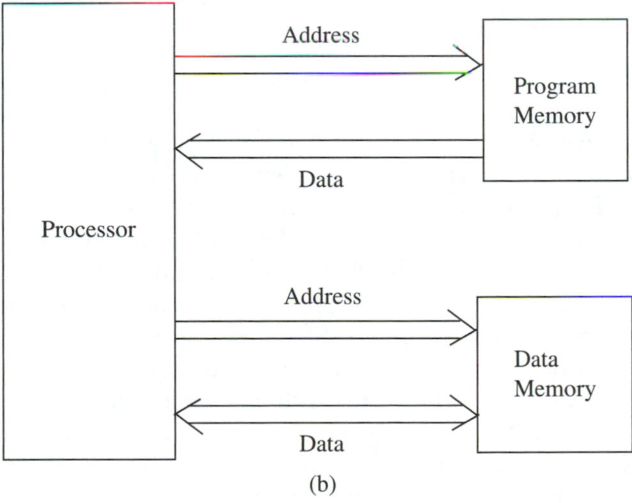


**Figure 4.8(b)** The bus structure of Harvard architecture

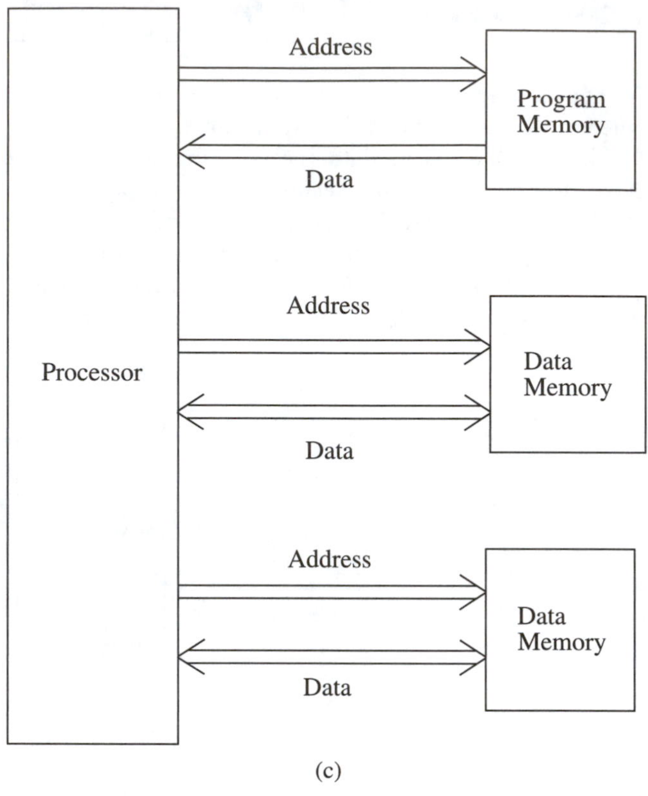


**Figure 4.8(c)** The bus structure for the architecture with one program memory and two data memories

the *Harvard architecture*, is shown in Figure 4.8(b). In fact, even this may not solve the problem completely. For example, the multiplication operation requires two operands to be fetched from the memory; one may be a data sample and the other, a coefficient. Even with separate memories for the program and data, it is not possible to fetch the two operands required for the multiplication along with the program instruction, and the processor has to wait for the second operand. It would therefore be required to provide dual data memories (for data and filter coefficients, for example) in addition to program memory and provide each with a separate bus for the processor to access them simultaneously. Figure 4.8(c) shows a possible bus structure of this type. As we can see, this will require a lot of hardware and interconnections to implement, thereby increasing the cost. Therefore, a compromise solution needs to be found to strike a balance between the hardware complexity and speed requirement of the multiplication operation, which is the most critical DSP operation in terms of the overall speed of algorithm implementation.

### 4.4.1 On-Chip Memory

A compromise between having multiple memories with individual buses for each and having fewer memories and buses is to provide some of the memories along with their buses on-chip. Even though the processor has to make simultaneous accesses to all the memories, only some of these are to the memories external to the DSP, thereby reducing the interconnection requirements to external devices.

On-chip memories help in running DSP algorithms faster than when the memories are located off-chip. This is because on-chip memories can have dedicated address and data buses unlike off-chip memories, whose buses are often multiplexed to reduce the pin count on the DSP. There are several issues related to the design of on-chip memories; two of these are considered next.

#### Speed

The on-chip memories should match the speeds of the ALU operations in order to maintain the single-cycle instruction execution requirement of the DSP. However, this is not a serious constraint because execution times of complex arithmetic operations such as multiplication are generally longer than memory access times. In fact, very often, more memory accesses than one are possible within a single instruction cycle, as will be explained later.

#### Size

Size is a major constraint for on-chip memories. In a given area of a DSP chip as many DSP functions as possible must be packed in order to get the best possible performance. On the other hand, the more area occupied by the on-chip memory, the less will be the area available for the other functions. The sizes of the on-chip memories are optimized taking into account the speed advantage, but without compromising any essential features required on the DSP.

### 4.4.2 Organization of the On-Chip Memory

Ideally, the entire memory required to implement a DSP algorithm should reside on-chip. This means, that the on-chip memory should be partitioned into program and data spaces. If necessary, the data memory should be further divided into separate areas for storing data samples, coefficients, and results. This way, an instruction with two operands can be fetched and executed and the result saved all in a single cycle. Writing the program and data into the on-chip memories is done before the program execution. Likewise, the results are read off the on-chip memory after the program execution is completed. However, this scheme is not practical because the different memory blocks and their buses take an enormous amount of chip area, thereby limiting the scope of other functions that are to be provided on the chip. There are several other ways in which the on-chip memory can be organized efficiently and in a cost-effective manner.

1. Many DSP algorithms require repeated executions of a single instruction such as the multiply and accumulate or a loop consisting of a few instructions. The result is normally saved only after the repetitions are completed. It is, therefore, sufficient to provide only two blocks of on-chip memories to hold the operands required for the execution of the instructions. The instruction or instructions required to carry out the repetitive calculations can reside in the external memory and, once fetched, can be repetitively used by keeping them in an instruction cache. Since the result is to be saved less frequently, there is no need to provide a separate memory for this purpose.
2. On-chip memories can be designed such that they can be accessed more than once in an instruction cycle. This way, fewer memory blocks can serve to hold the program, the operands, and results. This means that their access times should be sufficiently small to match the timing requirements of single-cycle instruction execution. Considering the advances made in memory design technology, it is possible to integrate dual-access on-chip memories on today's commercial DSPs. For example, let us assume that there are two on-chip memories and two buses in a DSP device. If each of these memories is fast enough to be accessed twice in each instruction cycle, execution of a multiply instruction that includes an instruction fetch, two operand fetches, and a memory access to save the result can be completed in one clock cycle.
3. On-chip memories can be configured for different uses at different times depending on the requirements. For example, if a DSP has two blocks of on-chip memory, ordinarily one of them will be configured as program memory and the other as the data memory. However, for execution of instructions, which requires two operands to be fetched simultaneously, they can both be configured as data memories. The instruction itself can be fetched from an external memory or it can reside in an on-chip cache.

In addition to program memory and data memories, DSP architecture should provide for a separate stack that can be directly accessed by the program counter. This provision can considerably reduce the overheads during the subroutine and interrupt calls and returns. If the cost becomes an issue in the choice of access times required for memories in a multiple memory system, it is preferable to provide faster memories for those segments that are more frequently accessed than the others.

## 4.5 Data Addressing Capabilities

The data processed by a digital signal-processing scheme typically consist of signal samples and filter coefficients. An efficient way of accessing data while performing computations can go a long way in the overall performance of an implementation. The provision of flexibility in accessing data helps in writing efficient programs for various applications. The data addressing capability of a programmable DSP device is provided by means of its addressing modes. The addressing modes that can enhance DSP implementations consist of immediate, register, direct, and indirect addressing modes. We now discuss each of these modes. These modes are summarized in Table 4.1.

**Table 4.1** Summary of DSP Addressing Modes

| Addressing Mode | Operand | Sample Format | Operation |
|---|---|---|---|
| Immediate | Immediate value | *ADD #imm* | *#imm* + A → A |
| Register | Register contents | *ADD reg* | *reg* + A → A |
| Direct | Memory address contents | *ADD mem* | *mem* + A → A |
| Indirect | Memory contents with address in the register | *ADD *addrreg* | **addrreg* + A → A |

Notations used in describing the operation in the table:
*#imm* = value represented by *imm*,
*reg* = contents of register *reg*,
*mem* = contents of memory location with address *mem*, and
**addrreg* = contents of memory location whose address is the contents of address register *addrreg*,
→ represents the transfer from left to right.

### 4.5.1 Immediate Addressing Mode

The capability to include data as part of the instruction is provided by the immediate addressing mode. For example, a DSP processor may allow the programmer to write the instruction

*ADD #imm*

to add the value represented by *imm* to the accumulator register, A. In other words, the operation

$$\#imm + \mathrm{A} \rightarrow \mathrm{A}$$

is implemented. In such an addressing mode data has to be a fixed number known at the time of writing instructions. Filter coefficients are examples of this kind of data.

### 4.5.2 Register Addressing Mode

In the register addressing mode a processor register provides the operand. Using this addressing mode the DSP processor may provide an instruction

$$ADD\ reg$$

to implement

$$reg + \mathrm{A} \rightarrow \mathrm{A}$$

### 4.5.3 Direct Addressing Mode

In the direct addressing mode a memory operand is specified by providing its memory address. For instance a DSP processor may allow an instruction

$$ADD\ mem$$

to implement

$$mem + \mathrm{A} \rightarrow \mathrm{A}$$

A signal sample stored in a memory location can be accessed using direct addressing mode. This mode, however, requires an explicit knowledge of the memory address, *mem*.

### 4.5.4 Indirect Addressing Mode

In the indirect addressing mode an operand is accessed using a pointer. A pointer is typically a register that holds the address of the location where the operand resides. For example, to add to the accumulator, A, the content of the memory location whose address is held in *addrreg*, the following instruction is implemented:

$$ADD\ {*addrreg}$$

which means

$$*addrreg + \mathrm{A} \rightarrow \mathrm{A}$$

In order to use this addressing mode, *addrreg* needs to be loaded before the use. Any memory location can be accessed by simply changing the register contents.

The indirect addressing mode can be enhanced by providing an automatic capability to manipulate the pointer register just before (pre) or just after (post) the use. The pointer register may be incremented or decremented. It may also be possible to add or subtract the contents of another register (offset register) provided in the architecture. This leads to the following enhanced indirect addressing modes:

*Post_increment* addressing mode,
*Post_decrement* addressing mode,
*Pre_increment* addressing mode,
*Pre_decrement* addressing mode,
*Post_offset_add* addressing mode,
*Post_offset_subtract* addressing mode,
*Pre_offset_add* addressing mode, and
*Pre_offset_subtract* addressing mode.

These enhanced indirect addressing modes are summarized in Table 4.2.

**Table 4.2** Enhancements to Indirect Addressing Mode

| Addressing Mode | Sample Format | Operation |
|---|---|---|
| *Post_increment* | *ADD *addrreg+* | A ← A + **addrreg*, *addrreg* ← *addrreg* + 1 |
| *Post_decrement* | *ADD *addrreg−* | A ← A + **addrreg*, *addrreg* ← *addrreg* − 1 |
| *Pre_increment* | *ADD + *addrreg* | *addrreg* ← *addrreg* + 1, A ← A + **addrreg* |
| *Pre_decrement* | *ADD − *addrreg* | *addrreg* ← *addrreg* − 1, A ← A + **addrreg* |
| *Post_add_offset* | *ADD *addrreg, offsetreg+* | A ← A + **addrreg*, *addrreg* ← *addrreg* + *offsetreg* |

*(continued)*

**Table 4.2** Continued

| Addressing Mode | Sample Format | Operation |
|---|---|---|
| *Post_subtract_offset* | ADD **addrreg, offsetreg–* | A ← A + **addrreg*, *addrreg* ← *addrreg* – *offsetreg* |
| *Pre_add_offset* | ADD *offsetreg+*, **addrreg* | *addrreg* ← *addrreg* + *offsetreg*, A ← A + **addrreg* |
| *Pre_subtract_offset* | ADD *offsetreg–*, **addrreg* | *addrreg* ← *addrreg* – *offsetreg*, A ← A + **addrreg* |

In order to realize the indirect addressing mode and its enhanced versions in a DSP architecture, additional hardware operating in conjunction with its addressing unit is required. For example to provide *pre_offset_add* addressing mode, an adder and another register to hold the offset are needed. It also means extra time for operand accessing or, alternatively, the need for computing the operand address using a dedicated address arithmetic unit working in parallel with the main arithmetic unit.

▷ **Example 4.9** What are the memory addresses of the operands in each of the following cases of indirect addressing modes? In each case, what will be the content of the *addrreg* after the memory access? Assume that the initial contents of the *addrreg* and the *offsetreg* are 0200h and 0010h, respectively.

a. ADD **addrreg–*

b. ADD + **addrreg*

**Table 4.3** Solution for Example 4.9

| Instruction | Addressing Mode | Operand Address | Contents of *addrreg* after the Memory Access |
|---|---|---|---|
| a | *Post_decrement* | 0200h | 0200h – 1h = 01FFh |
| b | *Pre_increment* | 0200h + 1h = 0201h | 0201h |
| d | *Pre_add_offset* | 0200h + 10h = 0210h | 0210h |
| d | *Post_subtract_offset* | 0200h | 0200h – 10h = 01F0h |

c. ADD *offsetreg+*, **addrreg*

d. ADD **addrreg*, *offsetreg–*

Solution The solution is given in Table 4.3.

---

### 4.5.5 Special Addressing Modes

In addition to the addressing modes mentioned earlier, special addressing modes are provided in the architecture of a DSP to implement real-time signal processing and to compute DFT using FFT algorithms. Real-time signal processing is enhanced by the provision of a circular buffer and the addressing mode that goes with it. The FFT implementation requires data to be accessed in a nonsequential, yet regular, manner. The data for FFT is accessed by what is called as *bit-reversed index*. A bit-reversed addressing mode is generally provided in the architecture to support FFT implementations. Similarly, to process two-dimensional data, it will be advantageous to provide a special addressing mode that can help access data organized in a matrix form. Now we consider two of these special addressing modes.

#### Circular Addressing Mode

The provision of a circular buffer allows one to handle a continuous stream of incoming data samples. In a circular buffer, successive data samples are stored in sequential buffer locations until the end of the buffer is reached. After reaching the end we start all over from the beginning of the buffer. This process can go on forever as long as the data samples get processed in a timely manner at a rate faster than the incoming data. To access a data sample from a circular buffer, a *circular addressing mode* is of great help. The implementation of such an addressing mode in hardware requires three registers: a pointer register (PNTR) to keep track of current address, a start address register (SAR) to hold the start address of the buffer, and an end address register (EAR) to hold the end address of the buffer. The pointer register should have the capability of getting incremented/decremented. Different forms of the indirect addressing mode for the pointer register are required in order to update the pointer for different applications. The pointer-updating algorithm is given in Figure 4.9.

The different cases that are encountered during the updating process of the pointer are shown in Figure 4.10. These cases are:

1. SAR $<$ EAR, and updated PNTR $>$ EAR
2. SAR $<$ EAR, and updated PNTR $<$ SAR
3. SAR $>$ EAR, and updated PNTR $>$ SAR
4. SAR $>$ EAR, and updated PNTR $<$ EAR

The buffer size in the first two cases $=$ (EAR $-$ SAR $+$ 1) and in the last two it is $=$ (SAR $-$ EAR $+$ 1).

```
; Pointer Updating Algorithm for the Circular Addressing Mode

Updated PNTR ← PNTR ± increment
If SAR < EAR
   and if Updated PNTR > EAR, then
          New PNTR ← Updated PNTR - Buffer size
   and if Updated PNTR < SAR, then
          New PNTR ← Updated PNTR + Buffer size
If SAR > EAR
   and if Updated PNTR > SAR, then
          New PNTR ← Updated PNTR - Buffer size
   and if Updated PNTR < EAR, then
          New PNTR ← Updated PNTR + Buffer size
Else
          New PNTR ← Updated PNTR
```

**Figure 4.9** Register pointer updating algorithm for circular buffer addressing mode. SAR = start address register contents, EAR = end address register contents, PNTR = pointer

▷ **Example 4.10** A DSP has a circular buffer with the start and the end addresses as 0200h and 020Fh, respectively. What would be the new values of the address pointer of the buffer if, in the course of address computation, it gets updated to (a) 0212h, (b) 01FCh?

Solution

$$\text{The buffer length} = 020\text{Fh} - 0200\text{h} + 1 = 10\text{h}$$

a. The new value of the pointer is updated value − buffer length, i.e., 0212h − 0010h = 0202h.

b. The new value of the pointer is updated value + buffer length, i.e., 01FCh + 0010h = 020Ch.

▷ **Example 4.11** Repeat the problem of Example 4.10 if the start and end addresses of the circular buffer are 0210h and 0201h, respectively.

Solution

a. The new value of the pointer is the updated value − buffer length, i.e., 0212h − 0010h = 0202h.

b. The new value of the pointer is the updated value + buffer length, i.e., 01FCh + 0010h = 020Ch.

Note that these values are the same as those in the previous example. This shows that in a circular buffer, the address pointer wraps around to point to an address inside the buffer, irrespective of whether the buffer start address is higher or the end address is higher.

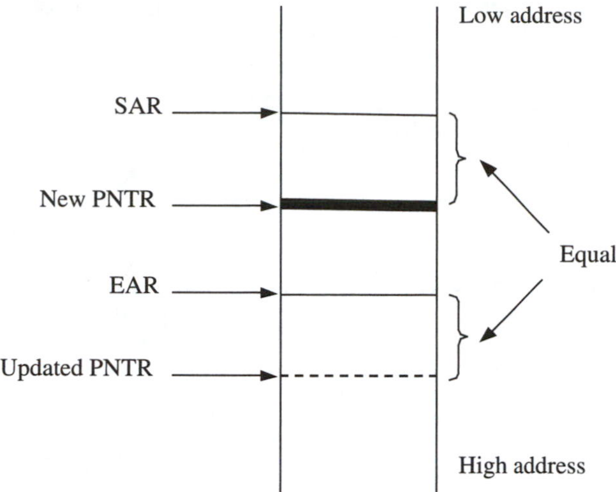


Case 1: SAR < EAR, and Updated PNTR > EAR

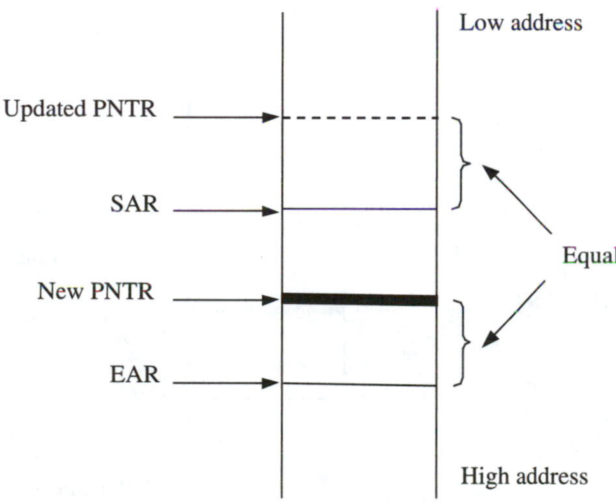


Case 2: SAR < EAR, and Updated PNTR < SAR

**Figure 4.10** Different cases that arise in updating the pointer in circular buffer addressing mode *(continued)*

## Bit-Reversed Addressing Mode

Special data access capability is needed in the FFT algorithm implementation. In the algorithm called *decimation in time* (DIT) FFT, the naturally ordered data needs to be accessed according to the indices, as shown in Table 4.4 for

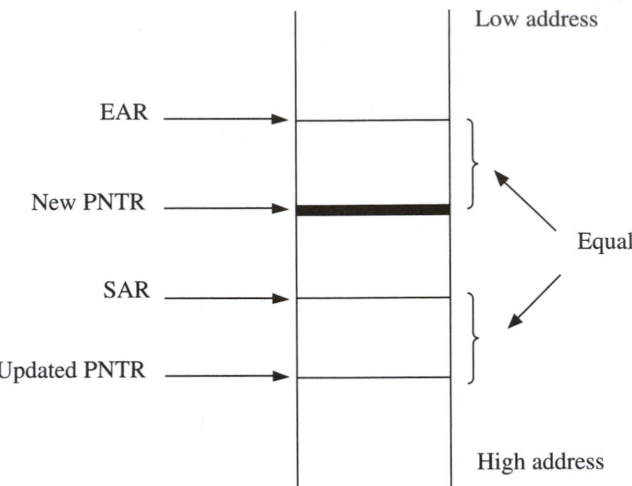


Case 3: SAR > EAR, and Updated PNTR > SAR

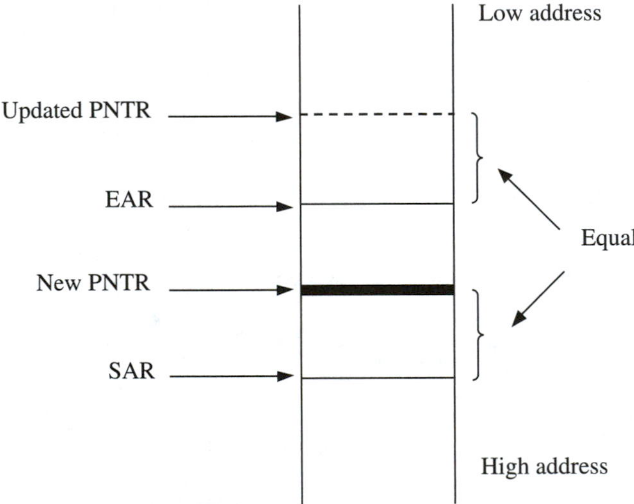


Case 4: SAR > EAR, and Updated PNTR < EAR

**Figure 4.10** Continued

an 8-point FFT. That is, in the case of an 8-point FFT, the input data $x(0)$, $x(1)$, $x(2)$, $x(3)$, $x(4)$, $x(5)$, $x(6)$, and $x(7)$ need to be accessed in the order $x(0)$, $x(4)$, $x(2)$, $x(6)$, $x(1)$, $x(5)$, $x(3)$, and $x(7)$. The interesting point is that the indices describing the order of data access can be obtained as follows: start

**Table 4.4** Index Computation Using Bit-Reversed Addressing Mode for an 8-point FFT

| Input Index (natural order) | Output Index (bit-reversed order) |
|---|---|
| 000 = 0 | 000 = 0 |
| 001 = 1 | 100 = 4 |
| 010 = 2 | 010 = 2 |
| 011 = 3 | 110 = 6 |
| 100 = 4 | 001 = 1 |
| 101 = 5 | 101 = 5 |
| 110 = 6 | 011 = 3 |
| 111 = 7 | 111 = 7 |

with index 0, obtain each current index by adding (in a special way) half the size of the FFT to the corresponding previous index, i.e.,

$$\text{Current index} = \text{previous index} + B(1/2(\text{FFT size})) \tag{4.8}$$

The addition however, is different in the sense that during addition the carry must propagate from the most significant to the least significant bit.

The *reverse-carry-add* operation can be provided in the architecture to implement this special addressing mode. The architecture will require a register to keep track of the index at any time in addition to the capability to propagate the carry in the reverse direction during the add operation in order to generate the next index to be used to access data. To provide this capability in parallel with the instruction execution, a special address generation unit is employed.

▷ **Example 4.12** Compute the sequence in which the input data should be ordered for a 16-point DIT FFT.

Solution Assuming that the first sample is located at address 0, the next sample should be located at address 0 + B(length of FFT/2) = 0 + 8 = 8. This address can be arrived at by carrying out binary addition with reverse carry propagation as follows:

Initial address in binary = 0000

Half the length of the FFT in binary = 1000

Next address (add with reverse carry propagation) = 1000

To compute the address of the third sample, repeat the operation.

Initial address in binary = 1000

Half the length of the FFT in binary = 1000

Next address (add with reverse carry propagation) = 0100

The process is repeated until the addresses of all the 16 samples are computed. Table 4.5 gives the results.

**Table 4.5** Solution for Example 4.12

| Sample Number | Binary Address | Hexa-decimal Address |
|---|---|---|
| 1 | 0000 | 0 |
| 2 | 1000 | 8 |
| 3 | 0100 | 4 |
| 4 | 1100 | C |
| 5 | 0010 | 2 |
| 6 | 1010 | A |
| 7 | 0110 | 6 |
| 8 | 1110 | E |
| 9 | 0001 | 1 |
| 10 | 1001 | 9 |
| 11 | 0101 | 5 |
| 12 | 1101 | D |
| 13 | 0011 | 3 |
| 14 | 1011 | B |
| 15 | 0111 | 7 |
| 16 | 1111 | F |

## 4.6 Address Generation Unit

The function of the address generation unit is to provide the addresses of the operands required to carry out the DSP operations. Since many instructions, such as the multiply instruction, require more than one operand for their execution, the address generation unit should work fast enough to provide the addresses within the time constraints imposed by the instruction execution requirements.

Further, in a DSP implementation, the address generation unit may be required to perform some computation of its own in order to arrive at the operand addresses. This is because of the need for the various enhancements to the indirect addressing mode as well as some special addressing modes, such as the circular addressing mode and the bit-reversed addressing mode. These special features were discussed in Section 4.5. In order to carry out the computations required for the specialized addressing modes the address generation unit in a DSP implementation is provided with a separate arithmetic unit of its own. This way, address computation overhead is removed from the main ALU, thereby allowing it to perform more efficiently.

Address generation typically involves one of the following operations:

1. Getting a new value from an immediate operand, a register, or a memory location.
2. Incrementing or decrementing the current address.
3. Adding or subtracting an offset to the current address.
4. Adding or subtracting an offset to the current address, comparing the new address with the limits defined for a circular addressing mode, and generating a new address as per the circular addressing mode algorithm.
5. Generating a new address from the current address by applying the bit-reversed addressing mode algorithm.

The hardware necessary to carry out the various operations listed above may consist of the following: an ALU; registers to store the current value, the offset, and the new value; registers to store the limits of the circular buffer; logic to implement the circular addressing mode; and the logic to implement the bit-reversed addressing mode. The block diagram of a typical addressing unit is shown in Figure 4.11.

## 4.7 Programmability and Program Execution

A programmable DSP device needs to provide programming capability similar to that of a microprocessor. It should be possible to write programs involving branching, loops, and subroutines. The branching capability is needed in order to alter conditionally or unconditionally the normal execution sequence. The looping operation is desirable in order to repeat a section of the program the desired number of times. The subroutine handling instructions provide the capability to develop structured software.

The implementation of repeat capability should be hardware based so that it can be programmed with minimal or zero overhead. For instance, a counter is needed to keep track of the number of times the execution of a block of instructions remains to be repeated. A dedicated register for this purpose can enhance the performance. Repeat is an operation that is needed in the implementation of many DSP algorithms, and hence its hardware implementation has a direct bearing on the overall performance of a DSP scheme.

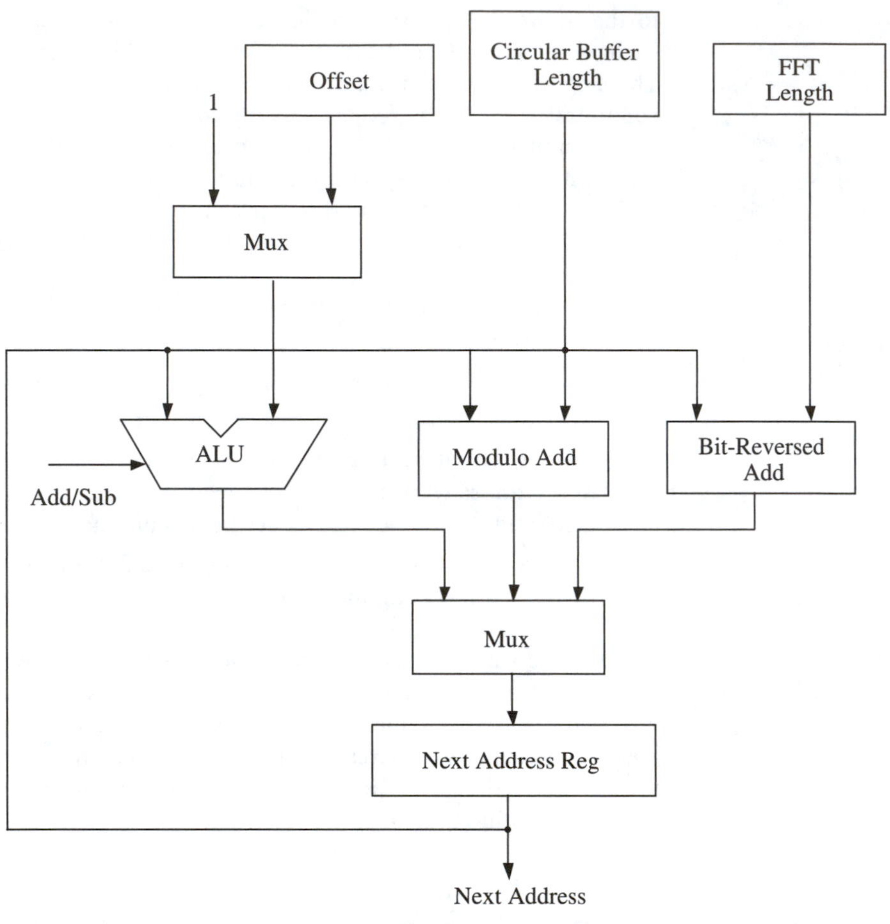


**Figure 4.11** Block diagram of an address generation unit

The subroutine implementation requires saving the return address in the stack. In a general-purpose microprocessor, a part of the memory is used to implement the stack. This means that to save the return address as well as to restore it on return, the processor requires to carry out memory read and write operations using the system data bus. These operations add to the overhead and make the overall program execution slow, thereby lowering the performance. For a DSP device, it is desirable that a last-in–first-out (LIFO) buffer directly interfaces to the program counter (instruction pointer) to save the return address. This approach avoids the use of the system bus for accessing the stack and thus speeds up the subroutine branching as well as its return.

### 4.7.1 Program Control

Like microprocessors, a DSP requires a control unit, which provides the necessary control and timing signals for proper execution of instructions. In microprocessors, the control unit is generally implemented by means of a microcoded sequencer. Each instruction of the microprocessor is broken down into several microinstructions and stored in a microstore as a microcode. Whenever one of the instructions is to be executed, the corresponding microcode is called from the microstore and executed, in a manner very similar to the execution of subroutines in a program. This type of control unit is easy to design and implement and uses less hardware. However, it is not very fast since execution of each instruction requires several accesses to the microstore. For a DSP, on the other hand, the speed of execution of instructions is a critical issue. For this reason the design of various building blocks is optimized for speed. In a DSP, the microcoded control unit is replaced by a hardwired design. In a hardwired design, the control unit is designed as a single, comprehensive, hardware unit taking into account the complete instruction set of the DSP. Although the hardware complexity is high and the design is not easy to change to incorporate additional features, this works much faster compared to the microcoded design and reduces the overhead for the instruction execution time.

### 4.7.2 Program Sequencer

The program sequencer, which is a part of the control unit, generates instruction addresses in the sequence needed to access instructions. Normally, instructions are executed in the order in which they are stored in the memory. However, there are several exceptions to this normal flow. Examples are subroutines, loops, and branching. The program sequencer hardware computes the instruction address under various conditions.

After fetching each instruction from the program memory, the sequencer generates the address from which the next instruction is to be fetched. The next address is from one of the following sources:

1. The program counter, which is incremented after each instruction fetch.
2. The instruction register, which holds the address of the instruction in branching, looping, and subroutine calls.
3. The interrupt vector table, in the case of interrupt service routines.
4. The stack, which holds the return addresses in the case of return from subroutines, return from interrupt service routines, and end of loops.

Figure 4.12 shows the block diagram of a program sequencer. The program sequencer, in effect, acts as a multiplexer, which selects the address of the next

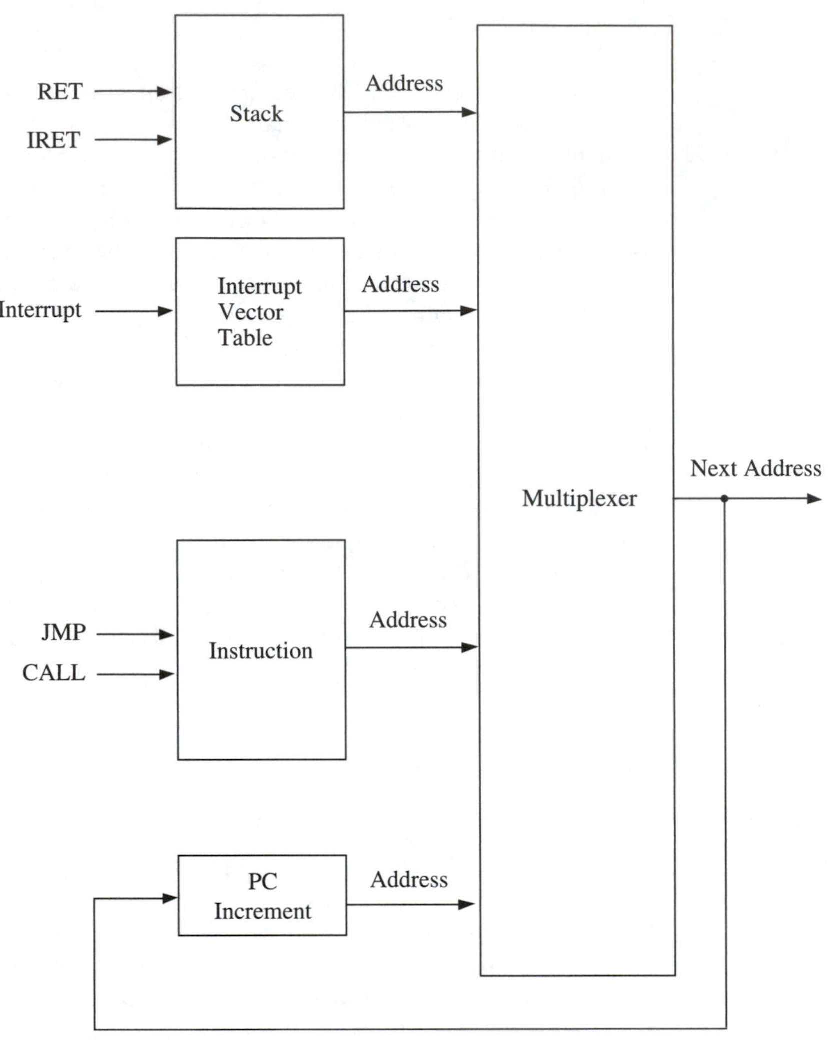


**Figure 4.12** A conceptual diagram of a program sequencer

instruction to be obtained from one of the sources listed above. In order to carry out this task, several hardware features are incorporated in the program sequencer. The program counter has to be updated after every fetch. Circuitry is provided for this purpose. Counters are provided to hold the counts in the case of loop and repeat instructions. Stacks push the return addresses for subroutines and interrupt service routines and while executing loops and repeat instructions. The program sequencer also requires a logic block to test conditions under which jump and loop instructions are executed as well as to determine when to terminate loop and repeat instructions. This logic, called the *condition logic*, tests various arithmetic conditions by means of status flags to decide if conditional jump and loop instructions are to be executed. This logic also monitors repeat and loop counters to determine when these have to be terminated to return to the normal program flow.

## 4.8 Speed Issues

Fast execution of algorithms is an essential requirement of a digital signal-processing architecture. In order to meet this requirement, DSP architecture must include features that facilitate high speed of operation and large throughputs. Many of these features are possible due to advances in VLSI technology and design innovations. In this section, we will discuss some of these features and see how they can increase the execution speed of the DSP architecture. We shall also discuss certain trade-offs between speed and performance in relation to some of these features.

### 4.8.1 Hardware Architecture

Functions such as multiplication, scaling, loops and repeats, and special addressing modes are essential for signal-processing algorithms. The architectures designed for the signal-processing applications should implement these functions in the quickest possible time. This is achieved by hardware units, which are specially designed to implement these functions. For example, conventional microprocessors implement the multiplication by means of a microprogram (microcode) using the well-known shift and add algorithm. This approach takes a large number of clock cycles to implement. In order to increase the speed of the operations considerably, parallel multipliers have been used to carry out the entire multiplication in a single clock cycle. Thanks to breakthroughs in VLSI technology, this is possible today. Similar hardware solutions have also been found to implement the other functions mentioned earlier to reduce overheads and to increase the speed. Such methods typically replace the slow microprogrammed solutions used in conventional microprocessors.

Harvard architecture, which separates the program and data memories with separate buses for each, increases the speed of execution of programs considerably. Dual data memories with individual buses for each help in accessing dual operands simultaneously.

Multiple external memories require multiple buses external to the DSP. In addition to being expensive, external buses are slow for program access and execution. By providing on-chip memories and an instruction cache, program execution is speeded up considerably. Further, these on-chip memories can also be accessed twice in a clock cycle, thereby reducing the number of separate memories and buses required in a device.

In addition to the hardware issues mentioned earlier, there are many techniques used in DSP architectures to increase their speed of operation. We shall consider two of these techniques: parallelism and pipelining.

### 4.8.2 **Parallelism**

A very major requirement to achieve high speed of operation in DSP architecture is the provision of parallelism. Parallelism may mean several things. One is the provision of functional units, which may operate in parallel and increase the throughput. For example, instead of the same arithmetic unit being used to do computations on data and address, a separate address arithmetic unit can be provided to take care of address computations. This frees up the main arithmetic unit to concentrate on data computations alone and thereby increases the throughput. Another example, which was discussed earlier, is the provision of multiple memories and multiple buses to fetch an instruction and operands simultaneously. In short, there are many functional blocks operating simultaneously for each of the most commonly used DSP operations, such as add, multiply, shift, etc. This way, algorithms can perform more than one operation at the same time, such as adding while carrying out a multiply, shifting while reading data from memory, etc.

Availability of multiple functional units can increase the speed of the DSP architectures. They should be exploited to their full potential by structuring the instructions to carry out the required operations in parallel. This requires complex hardware to control these units, and the controller is hardwired rather than microprogrammed in order to ensure high speed. The architecture should be such that instructions and data required for a computation are fetched from the memory simultaneously.

An ideal parallelism in the DSP architecture with regard to the multiply and accumulate operation, which is the most used operation in DSP implementations, should be able to accomplish the following operations in a single clock cycle:

- Fetch instructions and multiple data required for the computation
- Shift data as they are fetched in order to accomplish scaling
- Carry out a multiplication operation on the fetched data
- Add the product to the previously computed result in the accumulator
- Save the accumulator contents in the memory storage, if required, and
- Compute new addresses for the instruction and data required for the next operation

### 4.8.3 Pipelining

An architectural feature to increase the speed of the DSP algorithm is pipelining. In a pipelined architecture, an instruction to be executed is broken into a number of steps. A separate unit of the architecture performs each of these steps. When the first of these units performs the first step on the current instruction, the second unit will be performing the second step on the previous instruction, the third unit will be performing the third step on the instruction prior to that, etc. If $p$ steps were required to complete the execution of each instruction, it would take $p$ units of time for the complete execution of each instruction. However, since all the units will work all the time, one output will flow out of the architecture at the end of each time unit, and the throughput can be maintained as one instruction per unit time. A problem with this approach is dividing each instruction into steps taking equal amounts of time to perform and designing the architectural units accordingly. In practice, however, this may not be entirely possible and the slowest unit decides the throughput. A second problem is the extra time required at the start of algorithm execution, as the pipeline has to be filled before the result of the first instruction can start to flow out. This initial delay in units of time, called the *pipeline latency*, is related to the number of units in the pipeline. Likewise, when there is a change in the instruction sequence, as in the case of a branch or a loop, the pipeline needs to be cleared before the steps of the new instruction can be loaded into the pipeline, thereby causing a delay. This condition can, however, be avoided, at the cost of additional hardware to anticipate the branch instruction ahead of time and not filling the pipeline beyond the branch instruction. As an example, let us assume that the execution of an instruction can be broken into five steps: instruction fetch, instruction decode, operand fetch, execute, and save the result. Figure 4.13 shows how a pipelined

| Time Slot | Step 1 | Step 2 | Step 3 | Step 4 | Step 5 | Result |
|---|---|---|---|---|---|---|
| $t_0$ | Inst 1 | | | | | |
| $t_1$ | Inst 2 | Inst 1 | | | | |
| $t_2$ | Inst 3 | Inst 2 | Inst 1 | | | |
| $t_3$ | Inst 4 | Inst 3 | Inst 2 | Inst 1 | | |
| $t_4$ | Inst 5 | Inst 4 | Inst 3 | Inst 2 | Inst 1 | Inst 1 complete |
| $t_5$ | Inst 6 | Inst 5 | Inst 4 | Inst 3 | Inst 2 | Inst 2 complete |
| • | • | • | • | • | • | • |

**Figure 4.13** Pipelining for speeding up the execution of an instruction

processor will handle this. For the sake of simplicity we will assume that all the steps take equal amounts of time.

As we can see from the figure, the output corresponding to the first instruction is available after 5 units of time. However, once the result starts to come out, we get an output after each unit of time. In other words, the steady-state throughput of the system is one instruction per unit time.

### 4.8.4 System Level Parallelism and Pipelining

The parallelism and pipelining concepts explained in the last two subsections can be extended to the implementation of DSP algorithms. Consider the example of an 8-tap (8 coefficients) FIR filter given by

$$y(n) = \sum_{i=0}^{7} h(i)x(n-i) \tag{4.9}$$

The filter can be implemented in many ways depending on the number of multipliers and accumulators available. Let us look at some of these implementations.

#### Implementation Using a Single MAC Unit

If only one multiplier and accumulator is available, it must be used 8 times to compute the eight product terms in Eq. 4.9 and find their sum. Figure 4.14(a) shows such an implementation. Each input sample is delayed from the previous sample by $8T$, where $T$ is the time taken by the multiplier and accumulator to compute one product term and add it to the previously accumulated sum in the accumulator. Input samples and the filter coefficients are fed to the multiplier through multiplexers, which are controlled such that the correct combination of a sample and the corresponding filter coefficient are fed to the multiplier at a given time. As each product term is generated, it is added to the previously accumulated sum in the MAC unit. After all the eight product terms are accumulated, the MAC contents are available as the output. Output $y(n)$ is available $8T$ units of time after $x(n)$ is made available to the filter. At this time, a new sample $x(n+1)$ is applied to the filter. The filter then uses eight samples, namely, $x(n+1)$, $x(n)$, $x(n-1)$, ..., $x(n-6)$ to compute $y(n+1)$ after another $8T$ units of time. Thus, this implementation can take in a fresh input sample once every $8T$ units of time and generate an output sample at the same rate. In other words, the maximum sampling rate that this filter implementation can handle is $1/8T$.

$x(n)$ $x(n-1)$ $x(n-2)$ $x(n-3)$ $x(n-4)$ $x(n-5)$ $x(n-6)$ $x(n-7)$

8T 8T 8T 8T 8T 8T 8T

Multiplexer

MAC Unit

$y(n)$

Multiplexer

$h(0)$ $h(1)$ $h(2)$ $h(3)$ $h(4)$ $h(5)$ $h(6)$ $h(7)$

(a)

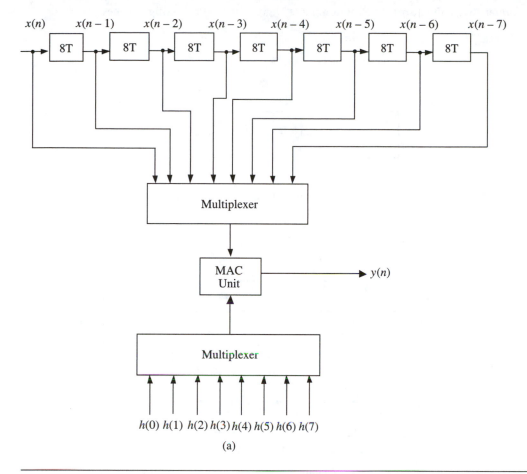

**Figure 4.14(a)** Single MAC implementation of an 8-tap FIR filter

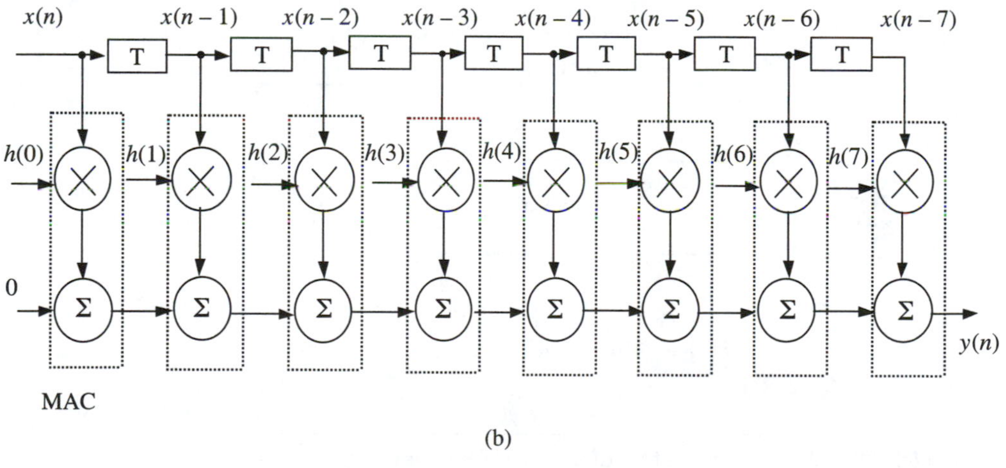


**Figure 4.14(b)** Pipelined implementation of an 8-tap FIR filter using eight MACs

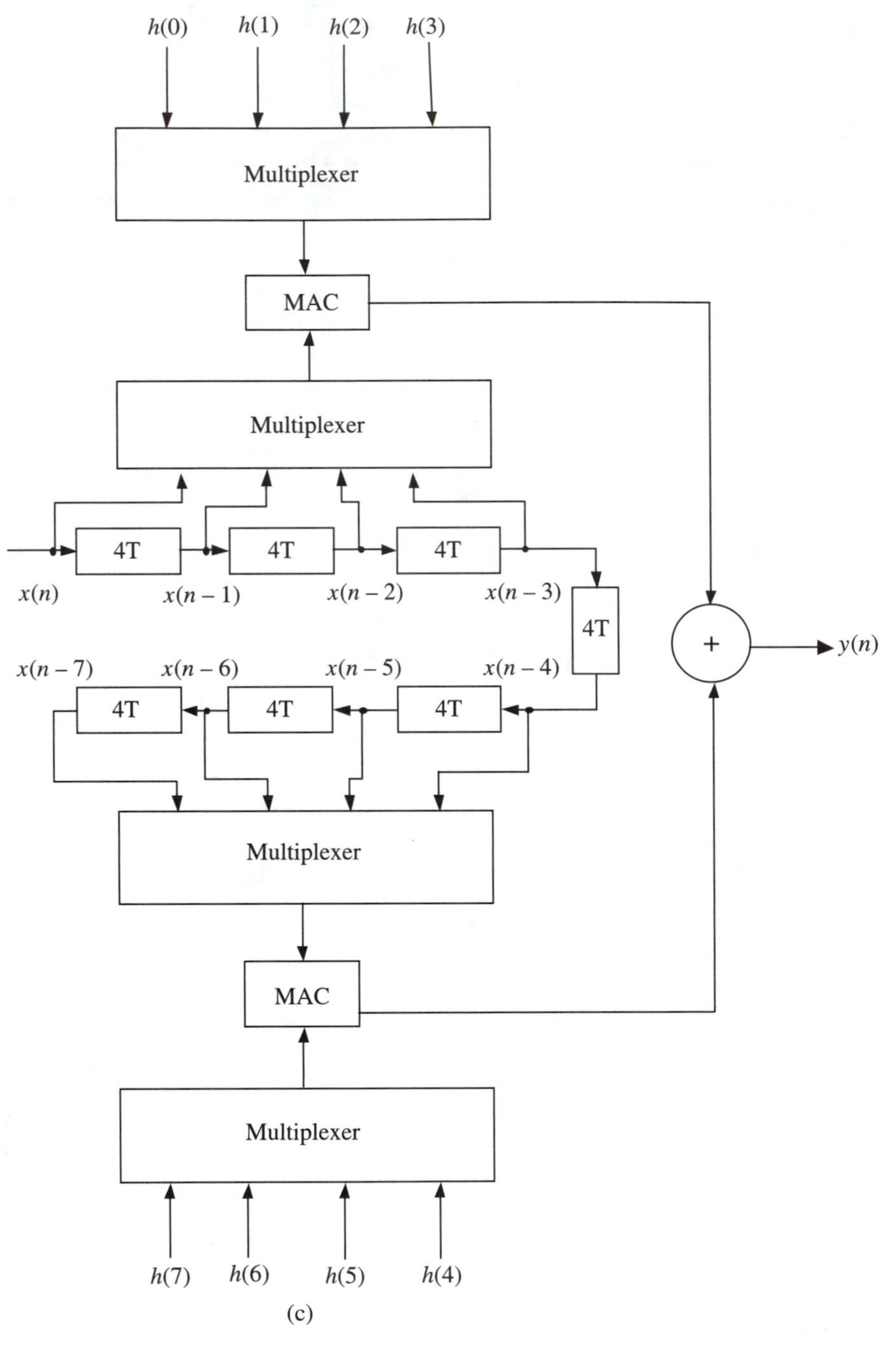


**Figure 4.14(c)** Parallel implementation of an 8-tap FIR filter using two MAC units

### Pipelined Implementation Using Eight Multipliers and Eight Accumulators

The implementation of the FIR filter of Eq. 4.9 can be speeded up if more multipliers and accumulators are available. Let us assume that there are eight multipliers and eight accumulators connected in a pipelined structure, as shown in Figure 4.14(b). Each multiplier computes one product term and passes it on to the corresponding accumulator, which in turn adds it to the summation passed on from the previous accumulator. Since all the multipliers and accumulators work all the time, a new output sample is generated once every $T$ units of time. This is the time required by the multiplier and accumulator to compute one product term and add it to the sum passed on from the previous stage of the pipeline. This implementation can take in a new input sample once every $T$ units of time and generate an output sample at the same rate. In other words, this filter implementation works 8 times faster than the simple one MAC implementation.

### Parallel Implementation Using Two MAC Units

A third implementation of the FIR filter of Eq. 4.9 is shown in Figure 4.14(c). This implementation uses two MAC units and an adder at the output. Each MAC computes four of the eight product terms in Eq. 4.9. Input samples and the filter coefficients are fed to the MACs using multiplexers that are controlled such that correct combinations of samples and the corresponding filter coefficients are fed to the two MACs at any given time. If $T$ time units are required to compute one pair of products and add them to the previously accumulated sum in the MAC units, it will require $4T$ units of time to generate the final output by adding the outputs of the two MACs. At this time, a new input sample can be applied to the filter for computation of the next output sample. The speed of this implementation is 2 times that of one MAC implementation of Figure 4.14(a) and one fourth of that of the pipelined eight-multiplier, eight-accumulator implementation of Figure 4.14(b). The maximum rate at which input samples can be applied to this filter implementation is 2 times that of the first implementation and one fourth that of the second.

**Table 4.6** Performance Summary of Different Implementations of an 8 -tap FIR Filter

| Type of Implementation | Maximum Sample Rate | Maximum Throughput |
|---|---|---|
| One MAC | $1/8T$ | One sample in $8T$ units of time |
| Pipelined (8 Multipliers and 8 Adders) | $1/T$ | One sample in $T$ units of time |
| Two MAC | $1/4T$ | One sample in $4T$ units of time |

$T$ = MAC time

Table 4.6 summarizes the performance of the three implementations described above. The example shows that it is possible to achieve higher-speed implementation by the use of parallelism and/or pipelining. This, however, increases the hardware complexity.

## 4.9 Features for External Interfacing

It is important for a DSP device to be able to communicate with the outside world. The outside world provides the signal to be processed and receives the processed signal. Therefore, most of the peripherals used with conventional microprocessors are also needed in a DSP system. These peripherals include interfaces for interrupts, direct memory access, serial I/O, and parallel I/O. In addition, since DSP is a digital device that is expected to process analog signals, conversions from analog-to-digital and digital-to-analog representations need to be carried out outside the device. From signal interfacing viewpoint, a DSP device should be capable of handling commonly available serial and parallel signal converters. All these features require the availability of appropriate address, data, and control signals to set up interfaces with the peripherals. The inclusion of a timer in the architecture is also very desirable to implement events at regular intervals, such as periodically initiating an A/D converter to start the conversion. A timer should be able to interrupt the processor to get its attention when needed so that the data acquisition can go on in the background simultaneously with the execution of the signal-processing program.

## 4.10 Summary

In this chapter, architectural features of programmable DSP devices have been examined based on the most frequently used DSP operations. Computational building blocks and other functional units have been described along with examples of implementations. Bus architecture and memory organization are explained to show how they help in realizing fast implementations of DSP algorithms. Trade-off between complexity and speed has also been discussed to show how the architectural features of programmable DSP devices can be optimized for efficient implementations.

In summary, the following is a list of architectural features of a programmable DSP device that should be evaluated before implementing an algorithm:

- Data representation format: fixed-point, floating-point formats and data word length for accuracy and dynamic range.
- Computational capability: an ALU with a hardware multiplier and shifters for scaling.

- Harvard architecture: provision of separate memories for program and data to fetch instructions and data simultaneously.
- On-chip memories: provision of on-chip program and data memories to avoid bus contention and to speed up program execution.
- Addressing modes: data addressing capabilities including indirect, indexed, circular buffer, and bit-reversed addressing modes.
- Programmability: programming capabilities including subroutines, branching, loops and repeats.
- Hardwired control: fast implementation of sequencing and control for single-cycle instruction execution.
- Parallelism: multiple functional units for parallel implementation of different functions such as simultaneous execution of an arithmetic operation and an address computation.
- Pipelining: simultaneous operation of different stages of an instruction execution by splitting it into steps handled by individually designed units.
- Interfacing: provision to interface serial devices such as A/D and D/A converters, parallel I/O, interrupt, and direct memory access.

## References


1. Allen, J. "Computer Architecture for Digital Signal Processing," *IEEE Proceedings*, Vol. 73, pp. 852–873, May 1985.
2. Lee, E. A. "Programmable DSP Architectures: Part I," *IEEE ASSP Magazine*, pp. 4–19, October 1988.
3. Lee, E. A. "Programmable DSP Architectures: Part II," *IEEE ASSP Magazine*, pp. 4–14, October 1989.
4. Kung, S. Y. *VLSI Array Processors*, Englewood Cliffs, NJ, Prentice Hall, 1988.
5. Higgin, R. J. *Digital Signal Processing in VLSI*, Englewood Cliffs, NJ, Prentice Hall, 1990.
6. Kung, S. Y., Whitehouse, H. T., and Kailath, T. *VLSI and Modern Signal Processing*, Englewood Cliffs, NJ, Prentice Hall, 1985.
7. Braun, E. L. *Digital Computer Design*, New York, Academic Press, 1963.
8. Baugh, C. R., and Wooley, B. A. "A 2's Complement Parallel, Array Multiplication Algorithm," *IEEE Trans. Computers*, Vol. C-22, pp. 1045–1047, December 1973.
9. Lapsley, P., Bier, J., Shoham, A., and Lee, E. A. *DSP Processor Fundamentals: Architectures and Features*, Piscataway, NJ, IEEE Press, 1997.

**10.** Eyre, J., and Bier, J. "DSP Processors Hit the Mainstream," *Computer*, pp. 51–59, August 1998.

**11.** Bates, A., and Paterson-Stephens, I. *The DSP Handbook: Algorithms, Applications and Design Techniques*, Englewood Cliffs, NJ, Prentice Hall, 2002.

## Assignments

**4.1** What distinguishes a digital signal processor from a general-purpose microprocessor with regard to basic capabilities?

**4.2** Specify the basic architecture required to implement the following operations so that they can be executed in the least possible time:

a. $(x_1 + jy_1)(x_2 + jy_2)$

b. $(0.5x_1 + 4x_2)/256$

**4.3** Draw a structure similar to that of Figure 4.1(b) for an $8 \times 8$ unsigned binary multiplier.

**4.4** How will you implement an $8 \times 8$ multiplier using $4 \times 4$ multipliers as the building blocks?

**4.5** Suggest a scheme to implement a multiplier to multiply two complex numbers using the multiplier shown in Figure 4.1(b) as the building block.

**4.6** Draw a structure based on Eq. 4.7 to multiply two 4-bit signed numbers, A and B.

**4.7**

a. Assuming the availability of a single 16-bit data bus, how many memory accesses will be required to access two 16-bit operands from the memory, multiply them, and save the 32-bit product back in the memory?

b. Suggest a suitable hardware scheme to implement the multiplication specified in part (a).

**4.8** Figure 4.3(b) shows the structure of a 4-bit barrel shifter. The switches shown connect each input bit to one of the output lines, depending on the number of bits to be shifted. Suggest a suitable hardware scheme for the switches and redraw Figure 4.3(b) by replacing the switches with its hardware. Also show how the control inputs control the switches to achieve the desired shift.

**4.9** What should be the minimum width of the accumulator in a DSP device that receives 10-bit A/D samples and is required to add 64 of them without causing an overflow?

**4.10**

a. What is meant by overflow in an arithmetic computation? How is an overflow condition detected in an ALU?

b. By means of numerical examples using 8-bit, 2's complement numbers, illustrate the conditions of (i) no overflow, (ii) overflow, (iii) no underflow, and (iv) underflow resulting from arithmetic operations in an ALU. In each case, verify if the circuit of Figure 4.6 can detect the condition.

**4.11** Suggest the memory architecture required for a DSP device to implement each of the following algorithms:

a. $N$-tap FIR filter

b. $2^M$-point FFT

c. autocorrelation of a segment of $N$ samples

d. crosscorrelation of two sequences of $N$ samples each

**4.12** Figure 4.8(c) allows for an instruction and two operands to be fetched simultaneously from the memory to the DSP to execute a multiply instruction in a single cycle. However, to save the result in memory, one more memory access is required. Can you specify an architecture that allows the result to be written back to the memory in the same cycle?

**4.13** Identify the addressing modes of the operands in each of the following instructions (AR stands for address register):

*ADD* #1234h

*ADD* 1234h

*ADD *AR+*

*ADD offsetaddr−, *AR*

**4.14** What is the bit-reversed sequence of 32 samples $x_0, x_1, x_2, \ldots, x_{31}$ as obtained by sampling a signal?

**4.15** Table 4.4 shows how bit reversing is done for 8 points. A similar algorithm can be used for any $2^n$ points. Specify using a block diagram how it can be implemented in hardware.

**4.16** How will you organize samples and filter coefficients using a circular buffer addressing scheme to implement a 32-tap FIR filter given by

$$y(n) = \sum_{k=0}^{31} b_k x(n-k)$$

**4.17** When a two-dimensional array of data such as a matrix is organized in a memory with linear (or one-dimensional) addressing, it is usually arranged in a row-ordered format. That is, all the elements of the first row are placed first in successive memory locations, starting with the very first location. This is followed by the elements of the second row, and so on, until all the elements of all the rows are arranged. Write a pseudocode to compute the address of any given element of this matrix, say, the element $(i, j)$, assuming that there are $N$ rows and $M$ columns in the matrix.

**4.18** Suggest a hardware architecture for the addressing unit that computes the two-dimensional address described in Problem 4.17 without the overhead required for computing it in software.

**4.19** Given below is the pseudocode of a software loop normally used in a general-purpose microprocessor for repetitive execution of an arithmetic operation.

Modify the code for a DSP with zero-overhead looping hardware:

Load count register

Back: Get operands; Compute; Update pointers

Decrement Count

If Count is not zero then jump Back

Proceed

**4.20** Explain the difference between a single-instruction, zero-overhead hardware looping and multiple-instruction, zero-overhead hardware looping in terms of architectural requirements and the performance.

**4.21** What is the difference between a microcoded program control and a hard-wired program control? Why is the latter preferred for DSP implementations?

**4.22** List the major architectural features used in a digital signal processor to achieve high speed of program execution.

**4.23** What architectural features are required in a DSP device to implement an FIR filter with $N$ taps so that a steady-state throughput of one output sample per cycle is achieved?

**4.24** List the essential peripherals required to implement the following DSP systems:

A speech processing system

A biomedical instrumentation system

An image processing system

# Chapter 5

# Programmable Digital Signal Processors

## 5.1 Introduction

In Chapter 4, we learned about the architectural requirements of digital signal processors. In this chapter, we first examine the basic architectures of three commonly used commercial DSP families and see how they incorporate the various features discussed in Chapter 4. We then study in detail, the Texas Instruments' TMS320C54xx processors, which, while retaining all the features of the basic architecture, provide a number of additional features for improved speed and performance. These devices will be used in the later chapters of this book to illustrate programming and interfacing concepts. The topics covered in this chapter are as follows:

Commercial digital signal-processing devices
The architecture of TMS320C54xx digital signal processors
Data addressing modes of TMS320C54xx processors
Memory space of TMS320C54xx processors
Program control in TMS320C54xx processors
TMS320C54xx instructions and programming
On-chip peripherals of TMS320C54xx processors
Interrupts of TMS320C54xx processors
Pipeline operation of TMS320C54xx processors

## 5.2 Commercial Digital Signal-Processing Devices

There are several families of commercial DSP devices. Right from the early eighties, when these devices began to appear in the market, they have been used in numerous applications, such as communication, control, computers, instrumentation, and consumer electronics. The architectural features and the processing power of these devices have been constantly upgraded based on the advances in technology and the application needs. However, in their basic versions, most of them have Harvard architecture, a single-cycle hardware multiplier, an address generation unit with dedicated address registers, special addressing modes, on-chip memories with off-chip expansion capability, hardware support for loops, and on-chip peripheral interfaces.

Of the various families of programmable DSP devices that are commercially available, the three most popular ones are those from Texas Instruments, Motorola, and Analog Devices. Texas Instruments was one of the first to come out with a commercial programmable DSP with the introduction of its TMS32010 in 1982. This was followed in 1984 by TMS32020, which had many additional features compared to TMS32010, and in 1985 by TMS320C25 [1] with a speed improvement by a factor of 2 when compared to the TMS32020. Since then, TMS320C25 has been used widely in many communication, control, and instrumentation applications. Likewise, around the same time, Motorola introduced DSP 56000 [2], and Analog Devices, ADSP 2100 [3]. Both of these devices have features, speed, and performance comparable to those of TMS320C25 and have also been used in many similar applications as the Texas Instruments' device.

Over the years, each of these families has evolved into several devices to fit different application needs and constant demands for improved performance and speed. Although these improvements have been brought about by an increase in the number of features with better performance, there have been no major changes in the basic architectures of these DSP devices. Therefore, we consider the architectures of TMS320C25, DSP 56000, and ADSP 2100 in order to get an insight into how the various features discussed in Chapter 4 are incorporated in typical commercial DSP devices. Figures 5.1–5.3 show the basic architectures of the three processors respectively. Table 5.1 summarizes these features for the three processors. Architectures and features of these devices will form the basis for exploring the more advanced architecture of the TMS320C54xx processors in the subsequent sections of this chapter.

## 5.3 Data Addressing Modes of TMS320C54xx Digital Signal Processors

TMS320C54xx processors retain the basic Harvard architecture of their predecessor, TMS320C25, but have several additional features, which improve their performance over it. Figure 5.4 shows a functional block diagram of TMS320C54xx processors. They have one program and three data memory spaces with separate buses, which provide simultaneous accesses to a program instruction and two data operands and enables writing of a result at the same time. Part of the memory is implemented on-chip and consists of a combination of ROM, dual-access RAM, and single-access RAM. Transfers between the memory spaces are also possible.

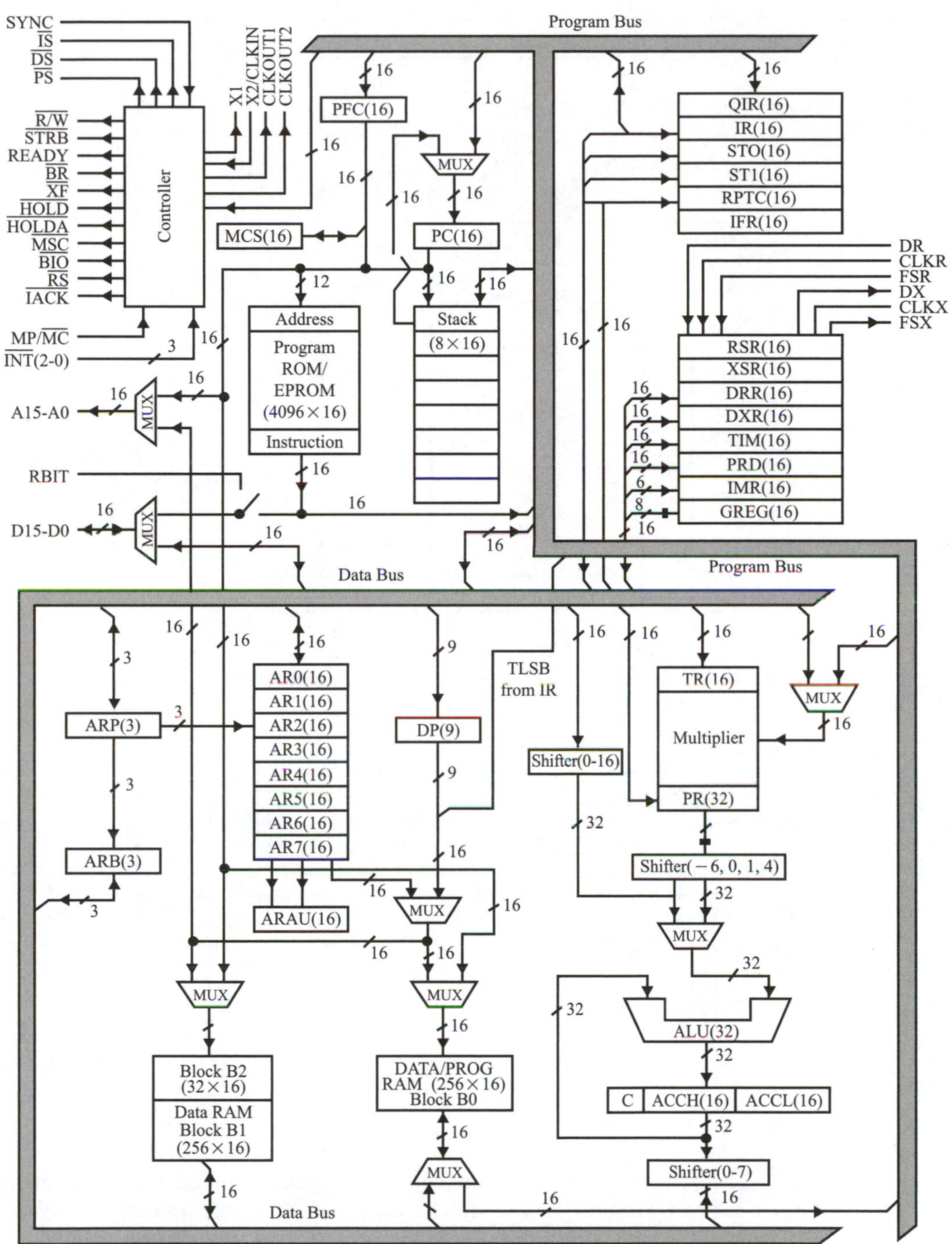


**Figure 5.1** Architecture of the Texas Instruments' TMS320C25 signal processor

(Courtesy of Texas Instruments Inc.)

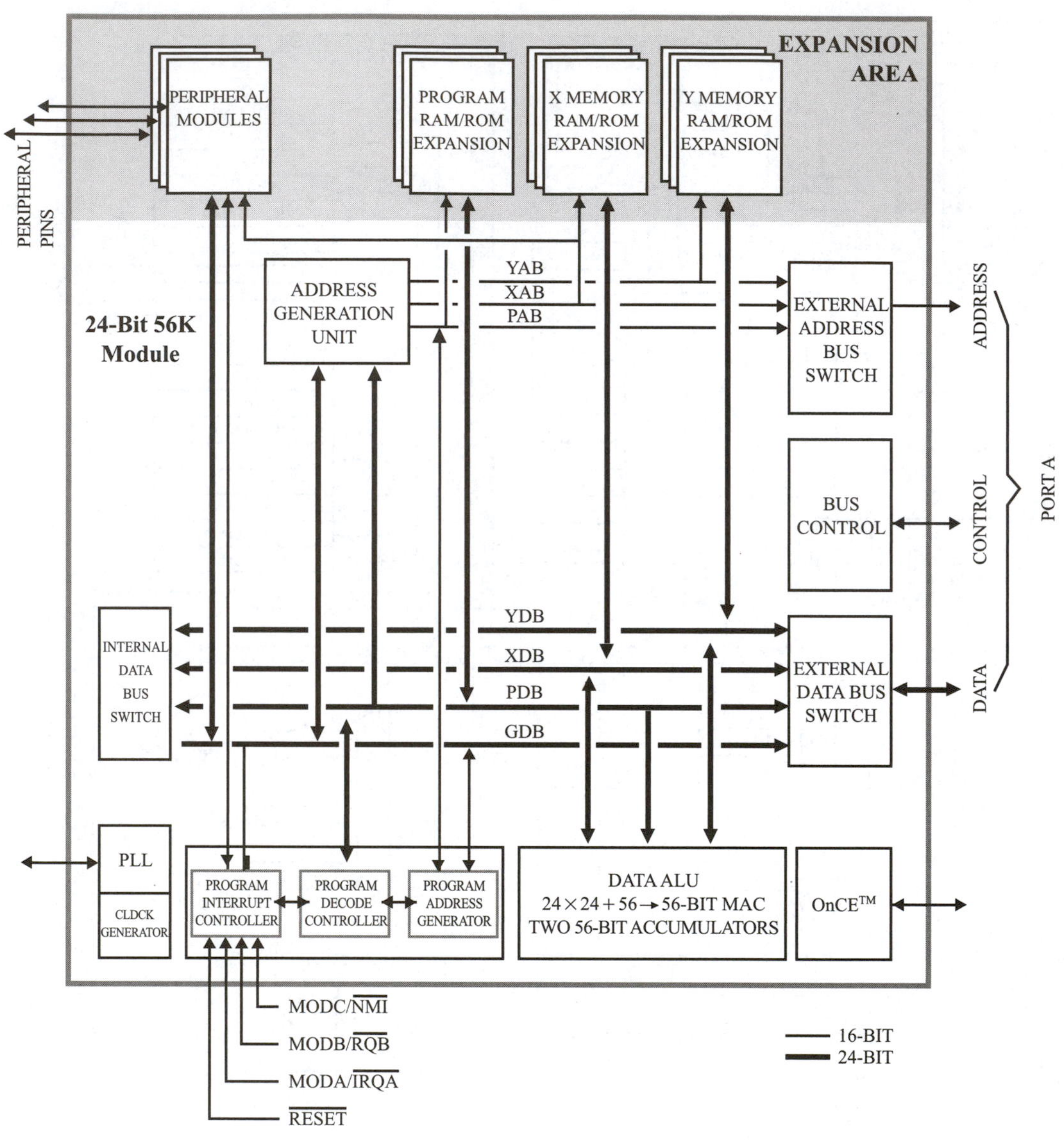


**Figure 5.2** Architecture of Motorola's DSP 56000 signal processor

(Courtesy of Motorola Inc.)

The central processing unit (CPU) of TMS320C54xx processors consists of a 40-bit arithmetic logic unit (ALU), two 40-bit accumulators, a barrel shifter, a $17 \times 17$ multiplier, a 40-bit adder, data address generation logic (DAGEN) with its own arithmetic unit, and a program address generation logic (PAGEN). These major functional units are supported by a number of registers and logic in the architecture.

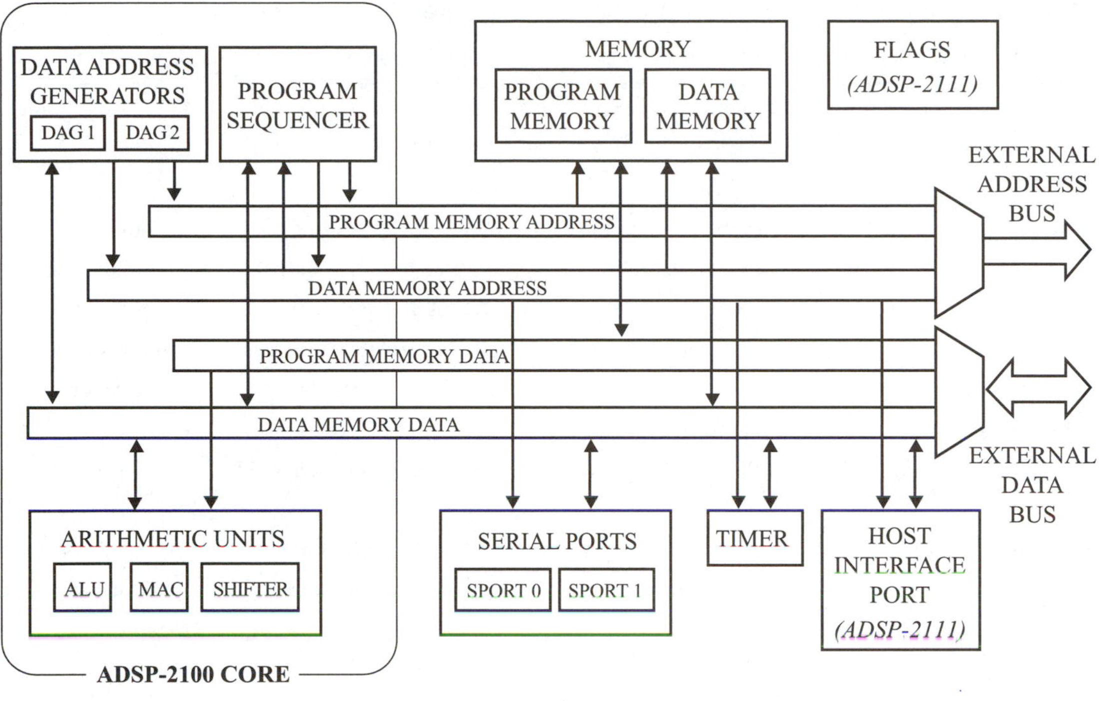


**Figure 5.3** Architecture of the Analog Devices' ADSP 2100 signal processor

(Courtesy of Analog Devices Inc.)

A powerful instruction set with a hardware-supported, single-instruction repeat and block repeat operations, block memory move instructions, instructions that pack two or three simultaneous reads, and arithmetic instructions with parallel store and load make these devices very efficient for running high-speed DSP algorithms.

Several peripherals, such as a clock generator, a hardware timer, a wait state generator, parallel I/O ports, and serial I/O ports, are also provided on-chip. These peripherals make it convenient to interface the signal processors to the outside world.

In the following sections, we examine in detail the various architectural features of the TMS320C54xx family of processors [4, 5].

## 5.3.1 Bus Structure

The performance of a processor gets enhanced with the provision of multiple buses to provide simultaneous access to various parts of memory or peripherals. The '54xx architecture is built around four pairs of 16-bit buses with each pair consisting of an address bus and a data bus. As shown in Figure 5.4,

**Table 5.1** Summary of the Architectural Features of Three Fixed-Point DSPs

| Architectural Feature | TMS320C25 | DSP 56000 | ADSP 2100 |
|---|---|---|---|
| Data representation format | 16-bit fixed point | 24-bit fixed point | 16-bit fixed point |
| Hardware multiplier | 16 × 16 | 24 × 24 | 16 × 16 |
| ALU | 32 bits | 56 bits | 40 bits |
| Internal buses | 16-bit program bus | 24-bit program bus | 24-bit program bus |
| | 16-bit data bus | 2 × 24-bit data buses | 16-bit data bus |
| | | 24-bit global data bus | 16-bit result bus |
| External buses | 16-bit program/data bus | 24-bit program/data bus | 24-bit program bus |
| | | | 16-bit data bus |
| On-chip memory | 544 words RAM | 512 words PROM | — |
| | 4K words ROM | 2 × 256 words data RAM | |
| | | 2 × 256 words data ROM | |
| Off-chip memory | 64K words program | 64K words program | 16K words program |
| | 64K words data | 2 × 64K words data | 16K words data |
| Cache memory | — | — | 16 words program |
| Instruction cycle time | 100 nsec. | 97.5 nsec. | 125 nsec. |
| Special addressing modes | Bit reversed | Modulo | Modulo |
| | | Bit reversed | Bit reversed |
| Data address generators | 1 | 2 | 2 |
| Interfacing features | Synchronous serial I/O DMA | Synchronous and asynchronous serial I/O DMA | DMA |

these are the program bus pair (PAB, PB), which carries the instruction code from the program memory, and three data bus pairs (CAB, CB; DAB, DB; and EAB, EB), which interconnect the various units within the CPU. In addition, the pairs CAB, CB and DAB, DB are used to read from the data memory, while the pair EAB, EB carries the data to be written to the memory. The '54xx can generate up to two data-memory addresses per cycle using the two auxiliary register arithmetic units (ARAU0 and ARAU1) in the DAGEN block. This enables accessing two operands simultaneously.

### 5.3.2 Central Processing Unit (CPU)

The '54xx CPU is common to all the '54xx devices. The '54xx CPU contains a 40-bit arithmetic logic unit (ALU); two 40-bit accumulators (A and B); a

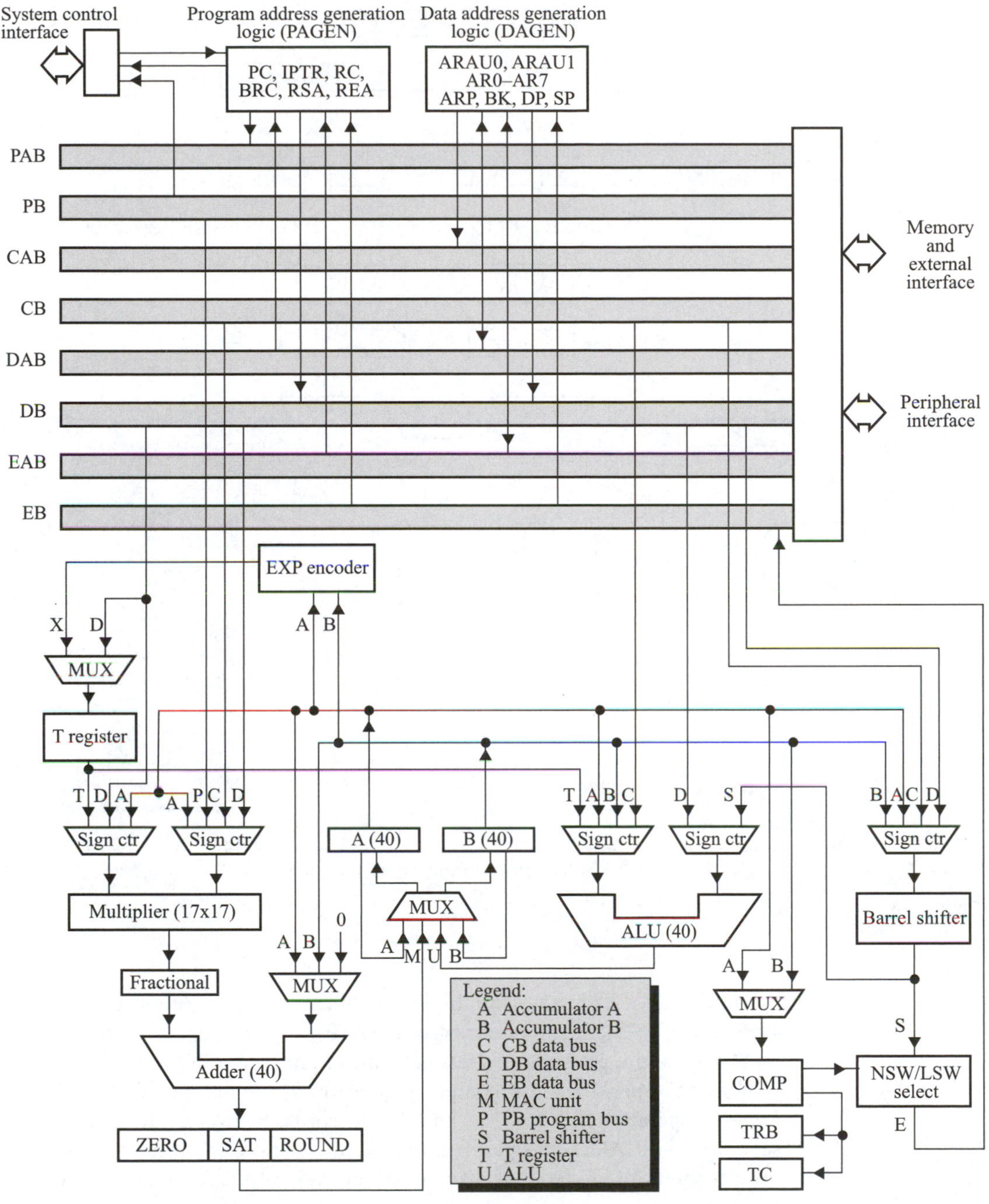


**Figure 5.4** Functional architecture for TMS320C54xx processors

(Courtesy of Texas Instruments)

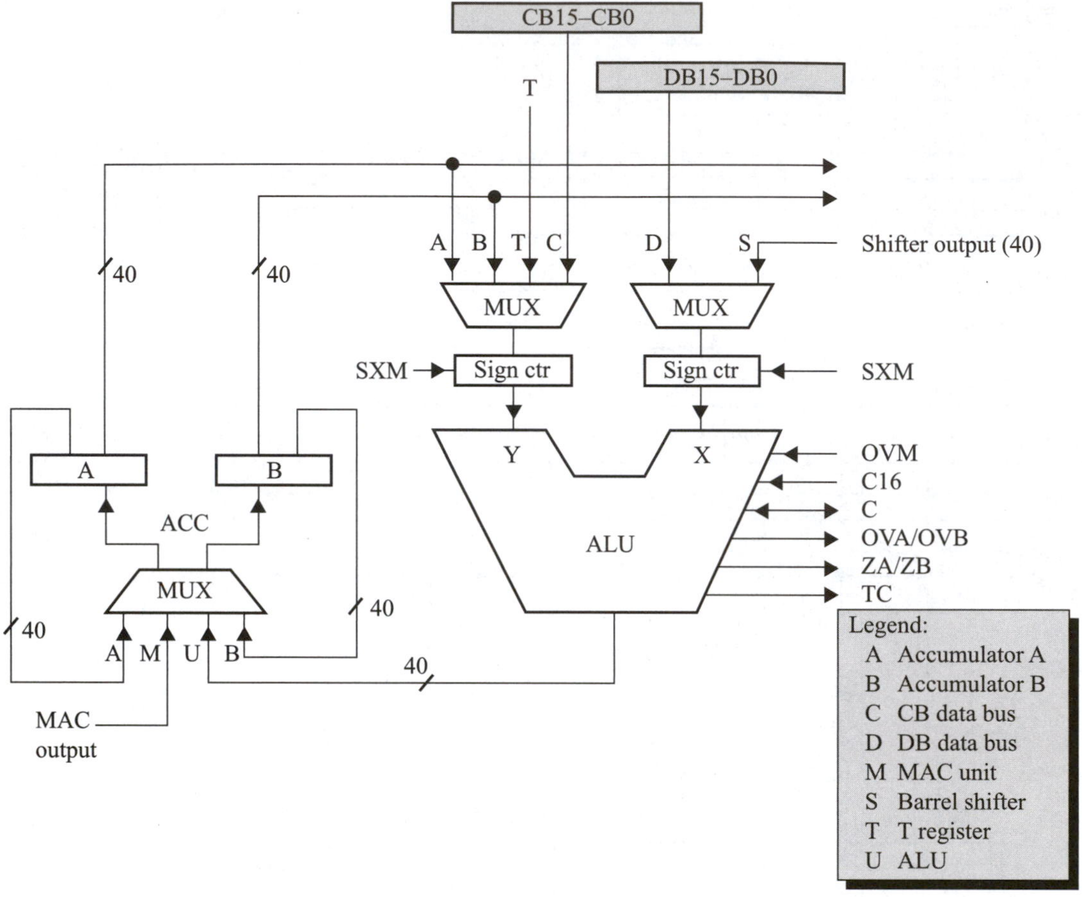


**Figure 5.5** Functional diagram of the central processing unit of the TMS320C54xx processors (Courtesy of Texas Instruments Inc.)

barrel shifter; a 17 × 17-bit multiplier; a 40-bit adder; a compare, select and store unit (CSSU); an exponent encoder (EXP); a data address generation unit (DAGEN); and a program address generation unit (PAGEN).

The ALU performs 2's complement arithmetic operations and bit-level Boolean operations on 16-, 32-, and 40-bit words. It can also function as two separate 16-bit ALUs and perform two 16-bit operations simultaneously. Figure 5.5 shows the functional diagram of the ALU of the TMS320C54xx family of devices.

Accumulators A and B store the output from the ALU or the multiplier/adder block and provide a second input to the ALU. Each accumulator is divided into three parts: guard bits (bits 39–32), high-order word (bits 31–

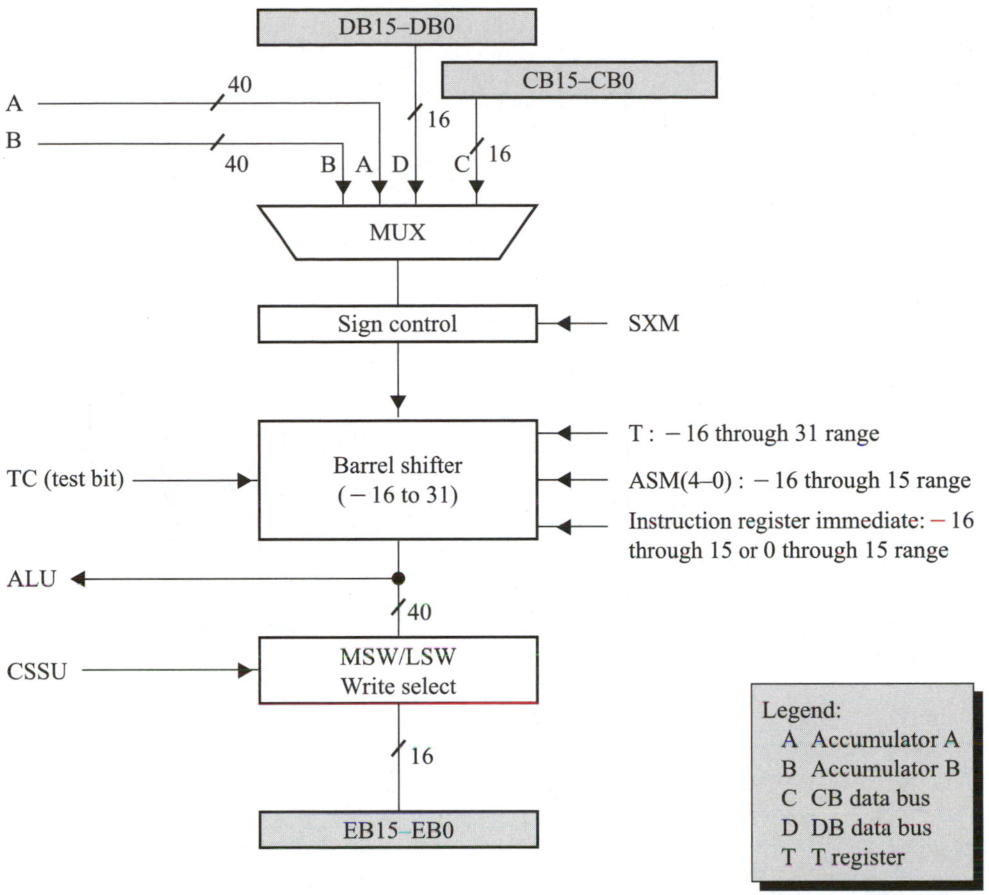


**Figure 5.6** Functional diagram of the barrel shifter of the TMS320C54xx processors

(Courtesy of Texas Instruments Inc.)

16), and low-order word (bits 15–0), which can be stored and retrieved individually.

The barrel shifter provides the capability to scale the data during an operand read or write. No overhead is required to implement the shift needed for the scaling operations. The '54xx barrel shifter can produce a left shift of 0 to 31 bits or a right shift of 0 to 16 bits on the input data. The shift requirements are defined in the shift count field of the instruction, the shift count field of status register ST1, or in the temporary register T. Figure 5.6 shows the functional diagram of the barrel shifter of TMS320C54xx processors.

The barrel shifter and the exponent encoder normalize the values in an accumulator in a single cycle. The LSBs of the output are filled with 0s, and

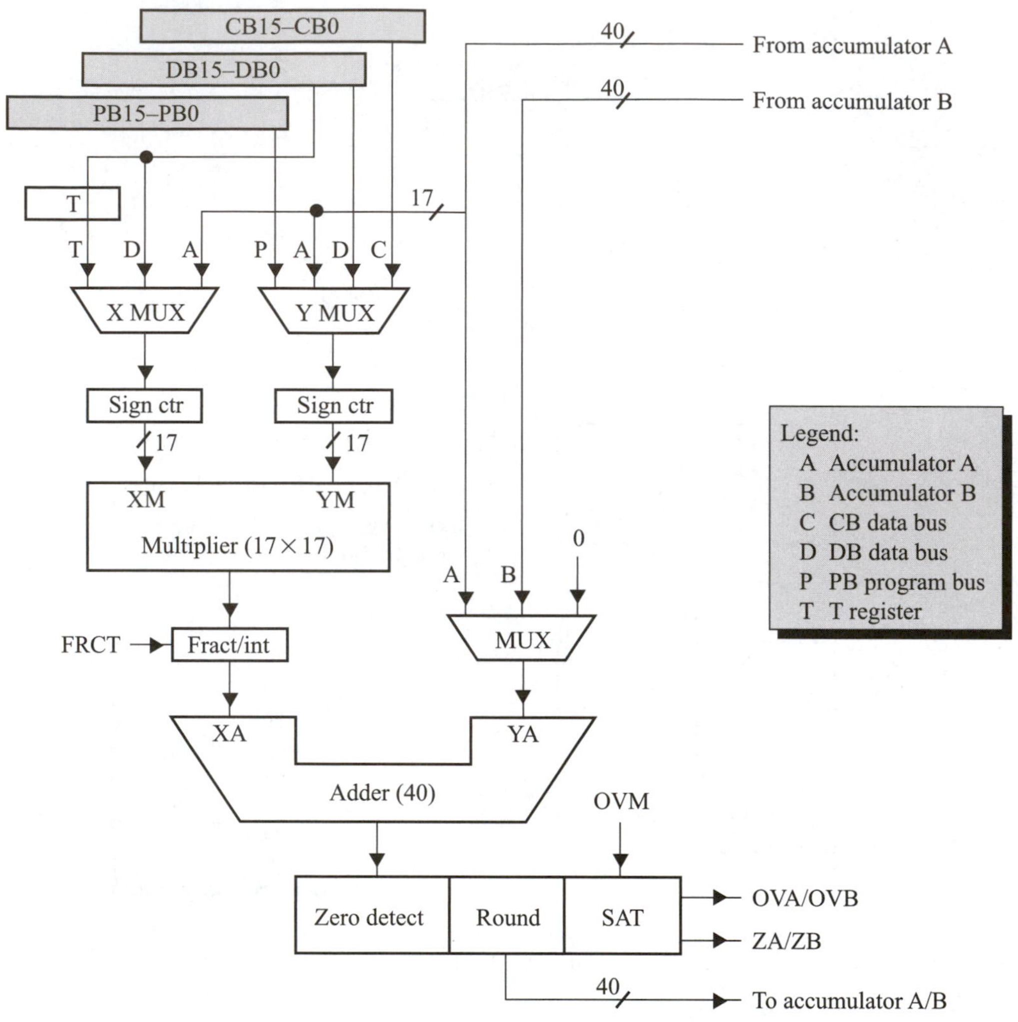


**Figure 5.7** Functional diagram of the multiplier/adder unit of TMS320C54xx processors

(Courtesy of Texas Instruments Inc.)

the MSBs can be either zero filled or sign extended, depending on the state of the sign-extension mode bit in the status register ST1. Additional shift capabilities enable the processor to perform numerical scaling, bit extraction, extended arithmetic, and overflow prevention operations.

The kernel of the DSP device architecture is its multiplier/adder unit. The multiplier/adder unit of TMS320C54xx devices performs $17 \times 17$ 2's-complement multiplication with a 40-bit addition effectively in a single instruction cycle. In addition to the multiplier and adder, the unit consists of control logic for integer and fractional computations and a 16-bit temporary storage register, T. Figure 5.7 shows the functional diagram of the multiplier/adder unit of TMS320C54xx processors.

The compare, select, and store unit (CSSU) is a hardware unit specifically incorporated to accelerate the add/compare/select operation. This operation is essential to implement the *Viterbi* algorithm used in many signal-processing applications.

The exponent encoder unit supports the EXP instruction, which stores in the T register the number of leading redundant bits of the accumulator content. This information is useful while shifting the accumulator content for the purpose of scaling.

### 5.3.3 Internal Memory and Memory-Mapped Registers

The amount and the types of memory of a processor have direct relevance to the efficiency and the performance obtainable in implementations with the processor. The ’54xx memory is organized into three individually selectable spaces: program, data, and I/O spaces. All ’54xx devices contain both RAM and ROM. RAM can be either dual-access type (DARAM) or single-access type (SARAM). The on-chip RAM for these processors is organized in pages having 128 word locations on each page.

The ’54xx processors have a number of CPU registers to support operand addressing and computations. The CPU registers and peripheral registers are all located on page 0 of the data memory. Figures 5.8(a) and (b) show the internal CPU registers and peripheral registers with their addresses. Figure 5.8(c) shows the processor mode status (PMST) register that is used to configure the processor. It is a memory-mapped register located at address 1Dh on page 0 of the RAM. The peripheral registers are covered in subsequent chapters.

A part of on-chip ROM may contain a bootloader and look-up tables for functions such as sine, cosine, $\mu$-law, and A-law. Details of the memory space of TMS320C54xx processors are discussed in Section 5.5.

## 5.4 Data Addressing Modes of TMS320C54xx Processors

Data addressing modes provide various ways to access operands to execute instructions and place results in the memory or the registers. The ’54xx devices offer seven basic addressing modes: immediate addressing, absolute addressing, accumulator addressing, direct addressing, indirect addressing, memory-mapped register addressing, and stack addressing.

| NAME | ADDRESS | | DESCRIPTION |
|---|---|---|---|
| | DEC | HEX | |
| IMR | 0 | 0 | Interrupt mask register |
| IFR | 1 | 1 | Interrupt flag register |
| — | 2–5 | 2–5 | Reserved for testing |
| ST0 | 6 | 6 | Status register 0 |
| ST1 | 7 | 7 | Status register 1 |
| AL | 8 | 8 | Accumulator A low word (15–0) |
| AH | 9 | 9 | Accumulator A high word (31–16) |
| AG | 10 | A | Accumulator A guard bits (39–32) |
| BL | 11 | B | Accumulator B low word (15–0) |
| BH | 12 | C | Accumulator B high word (31–16) |
| BG | 13 | D | Accumulator B guard bits (39–32) |
| TREG | 14 | E | Temporary register |
| TRN | 15 | F | Transition register |
| AR0 | 16 | 10 | Auxiliary register 0 |
| AR1 | 17 | 11 | Auxiliary register 1 |
| AR2 | 18 | 12 | Auxiliary register 2 |
| AR3 | 19 | 13 | Auxiliary register 3 |
| AR4 | 20 | 14 | Auxiliary register 4 |
| AR5 | 21 | 15 | Auxiliary register 5 |
| AR6 | 22 | 16 | Auxiliary register 6 |
| AR7 | 23 | 17 | Auxiliary register 7 |
| SP | 24 | 18 | Stack pointer register |
| BK | 25 | 19 | Circular buffer size register |
| BRC | 26 | 1A | Block repeat counter |
| RSA | 27 | 1B | Block repeat start address |
| REA | 28 | 1C | Block repeat end address |
| PMST | 29 | 1D | Processor mode status (PMST) register |
| XPC | 30 | 1E | Extended program page register |
| — | 31 | 1F | Reserved |

(a)

**Figure 5.8(a)** Internal memory-mapped registers of TMS320C54xx signal processors

(Courtesy of Texas Instruments Inc.)

## 5.4.1 Immediate Addressing

In this mode, the instruction contains the specific value of the operand. The operand can be short (3, 5, 8, or 9 bits in length) or long (16 bits in length). The instruction syntax for short operands occupies one memory location,

| NAME | ADDRESS DEC | HEX | DESCRIPTION |
|---|---|---|---|
| DRR20 | 32 | 20 | McBSP 0 Data Receive Register 2 |
| DRR10 | 33 | 21 | McBSP 0 Data Receive Register 1 |
| DXR20 | 34 | 22 | McBSP 0 Data Transmit Register 2 |
| DXR10 | 35 | 23 | McBSP 0 Data Transmit Register 1 |
| TIM | 36 | 24 | Timer Register |
| PRD | 37 | 25 | Timer Period Register |
| TCR | 38 | 26 | Timer Control Register |
| — | 39 | 27 | Reserved |
| SWWSR | 40 | 28 | Software Watt-State Register |
| BSCR | 41 | 29 | Bank-Switching Control Register |
| — | 42 | 2A | Reserved |
| SWCR | 43 | 2B | Software Watt-State Control Register |
| HPIC | 44 | 2C | HPI Control Register (HMODE = 0 only) |
| — | 45–47 | 2D–2F | Reserved |
| DRR22 | 48 | 30 | McBSP 2 Data Receive Register 2 |
| DRR12 | 49 | 31 | McBSP 2 Data Receive Register 1 |
| DXR22 | 50 | 32 | McBSP 2 Data Transmit Register 2 |
| DXR12 | 51 | 33 | McBSP 2 Data Transmit Register 1 |
| SPSA2 | 52 | 34 | McBSP 2 Subbank Address Register |
| SPSD2 | 53 | 35 | McBSP 2 Subbank Data Register |
| — | 54–55 | 36–37 | Reserved |
| SPSA0 | 56 | 38 | McBSP 0 Subbank Address Register |
| SPSD0 | 57 | 39 | McBSP 0 Subbank Data Register |
| — | 58–59 | 3A–3B | Reserved |
| GPIOCR | 60 | 3C | General-Purpose I/O Control Register |
| GPIOSR | 61 | 3D | General-Purpose I/O Status Register |
| CSIDR | 62 | 3E | Device ID Register |
| — | 63 | 3F | Reserved |
| DRR21 | 64 | 40 | McBSP 1 Data Receive Register 2 |
| DRR11 | 65 | 41 | McBSP 1 Data Receive Register 1 |
| DXR21 | 66 | 42 | McBSP 1 Data Transmit Register 2 |
| DXR11 | 67 | 43 | McBSP 1 Data Transmit Register 1 |
| — | 68–71 | 44–47 | Reserved |
| SPSA1 | 72 | 48 | McBSP 1 Subbank Address Register |
| SPSD1 | 73 | 49 | McBSP 1 Subbank Data Register |
| — | 74–83 | 4A–53 | Reserved |
| DMPREC | 84 | 54 | DMA Priority and Enable Control Register |
| DMSA | 85 | 55 | DMA Subbank Address Register |

**Figure 5.8(b)** Peripheral registers for the TMS320C5416 processor

(Courtesy of Texas Instruments Inc.)

(*continued*)

| | | | |
|---|---|---|---|
| DMSDI | 86 | 56 | DMA Subbank Data Register with Autoincrement‡ |
| DMSDN | 87 | 57 | DMA Subbank Data Register |
| CLKMD | 88 | 58 | Clock Mode Register (CLKMD) |
| — | 89–95 | 59–5F | Reserved |

(b)

**Figure 5.8(b)** Continued

| 15–7 | 6 | 5 | 4 | 3 | 2 | 1 | 0 |
|---|---|---|---|---|---|---|---|
| IPTR | MP/$\overline{\text{MC}}$ | OVLY | AVIS | DROM | CLKOFF† | SMUL† | SST† |

†These bits are only supported on C54x devices with revision A or later, or on C54x devices numbered C548 or greater.

(c)

**Figure 5.8(c)** Processor mode status (PMST) register of TMS320C54xx processors

(Courtesy of Texas Instruments Inc.)

whereas that for long operands occupies two memory locations. This addressing mode can be used to initialize registers and memory locations. Examples of instructions using this addressing mode are

```
LD #20, DP     ; This accomplishes #20 → DP
RPT #0FFFFh    ; This accomplishes #FFFFh → RC
```

### 5.4.2 Absolute Addressing

In this mode, the instruction contains a specific address. The specified address may be for a data memory location (*dmad* addressing), a program memory location (*pmad* addressing), a port address (*PA* addressing), or a location in the data space specified directly (*(*lk*) addressing). Examples of instructions using this mode of addressing are

```
MVKD 1000h, *AR5   ; 1000h → AR5 (dmad addressing)
MVPD 1000h, *AR7   ; 1000h → *AR7 (pmad addressing)
PORTR 05h, *AR3    ; 05h → *AR3 (PA addressing)
LD *(1000h), A     ; *(1000h) → A (*(lk) addressing)
```

### 5.4.3 Accumulator Addressing

This mode uses the accumulator contents as the address and is used to move data between a program memory location and a data memory location. Examples of instructions in this mode are READA and WRITA. READA transfers a word from a program-memory location specified by accumulator A to a data-memory location. WRITA transfers a word from a data-memory location to a program-memory location specified by accumulator A.

Here is an example:

```
READA *AR2  ; This accomplishes *A → *AR2
```

### 5.4.4 Direct Addressing

In the direct addressing mode, the 16-bit address of the data-memory location is formed by combining the lower 7 bits of the data-memory address contained in the instruction with a base address given by the data-page pointer (DP) or the stack pointer (SP). Figure 5.9 shows the operation of the direct addressing mode of TMS320C54xx processors.

Using this form of addressing, one can access a page of 128 contiguous locations without changing the DP or the SP. The compiler mode bit (CPL), located in the status register ST1, is used to select between the two pointers

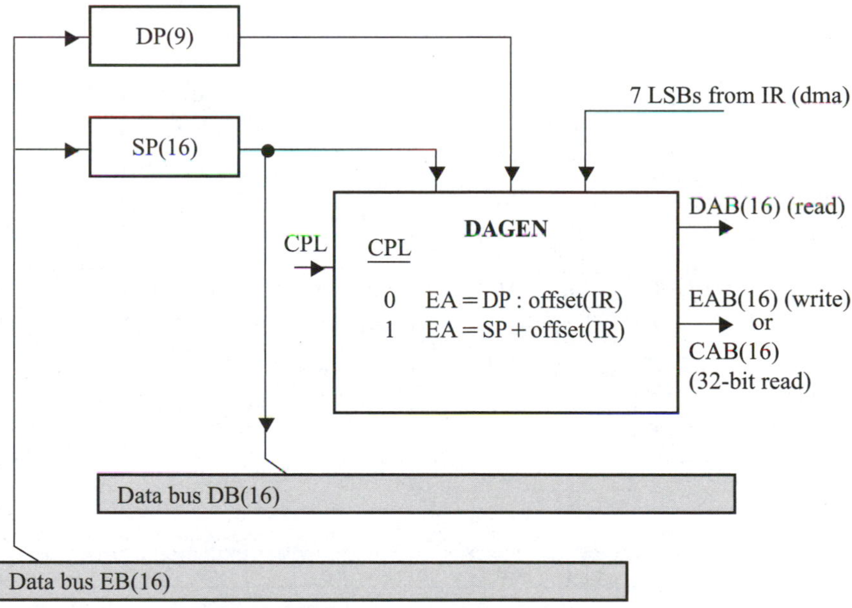


**Figure 5.9** Block diagram of the direct addressing mode for TMS320C54xx processors (Courtesy of Texas Instruments Inc.)

used to generate the address. CPL = 0 selects DP and CPL = 1 selects SP. For example, when CPL = 0, to add the contents of the memory location 0 on page 4 in the data memory to accumulator B, we can use the instruction sequence:

```
LD #4, DP   ; DP = 4 = upper 9 bits of address
ADD=0, B    ; Lower 7 bits of the address
```

With this example the contents of the first locations on data page 4 (memory address 0200h) are added to accumulator B.

It should be remembered that when SP is used instead of DP, the effective address is computed by adding the 7-bit offset to SP.

### 5.4.5 Indirect Addressing

In indirect addressing, any location in the data space can be accessed by means of an address contained in an auxiliary register. The ’54xx devices have eight 16-bit auxiliary registers (AR0–AR7). Indirect addressing is used when

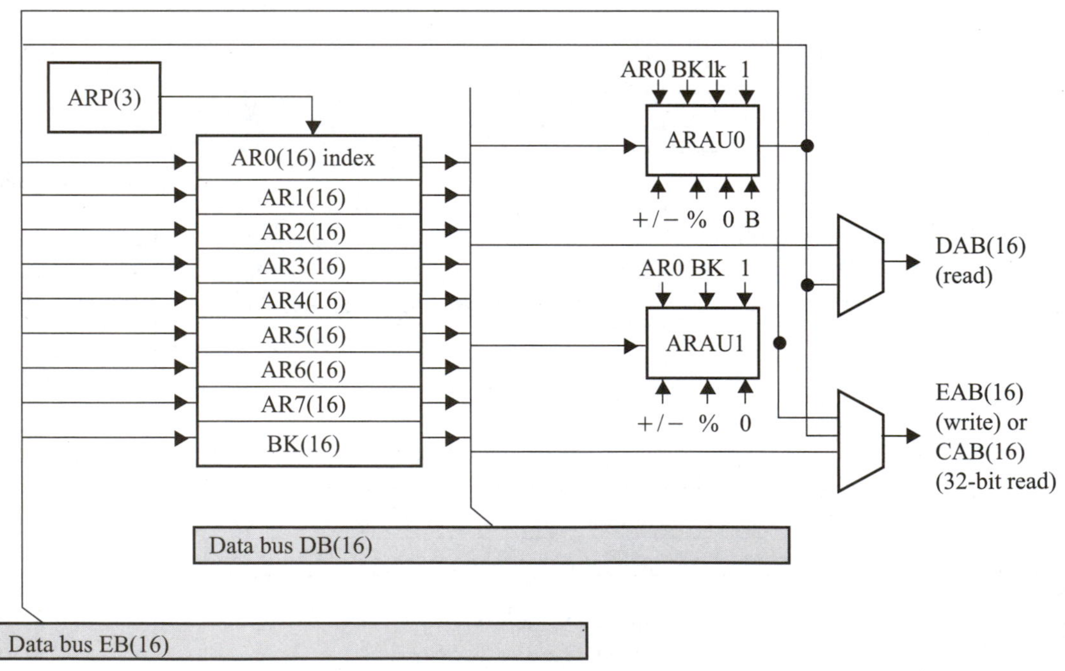


**Figure 5.10** Block diagram for the indirect addressing mode of TMS320C54xx processors

(Courtesy of Texas Instruments Inc.)

there is a need to step through a sequence of locations in the memory in fixed-sized steps.

Two auxiliary register arithmetic units (ARAU0 and ARAU1) are used to modify the contents of the auxiliary registers for the indirect addressing mode. They perform unsigned, 16-bit arithmetic operations. The auxiliary registers can be loaded with an immediate value, loaded via the data bus, and modified by the indirect addressing field of any instruction that supports indirect addressing or by the modify auxiliary register (MAR) instruction and used as loop counters.

Figure 5.10 shows how ARAUs are used to generate an address in the indirect addressing mode using a single data-memory operand. An address can be modified before or after accessing the location or can be left unchanged. Modification can be by incrementing or decrementing the address by 1, adding a 16-bit offset, or indexing with the value in AR0. Each of these modifications may be carried out either before or after accessing the memory location. Table 5.2 gives the operand syntax and the corresponding ARAU operations for the single operand indirect addressing mode.

▷ **Example 5.1** Assuming the current contents of AR3 to be 200h, what will be its contents after each of the following TMS320C54xx addressing modes is used? Assume that the contents of AR0 are 20h.

a. *AR3 + 0

b. *AR3 − 0

c. *AR3+

d. *AR3−

e. *AR3

f. *+AR3(40h)

g. *+AR3(−40h)

**Solution**

a. AR3 ← AR3 + AR0;
AR3 = 200h + 20h = 220h.

b. AR3 ← AR3 − AR0;
AR3 = 200h − 20h = 1E0h.

c. AR3 ← AR3 + 1;
AR3 = 200h + 1 = 201h.

d. AR3 ← AR3 − 1;
AR3 = 200h − 1 = 1FFh.

e. AR3 is not modified.
AR3 = 200.

f. AR3 ← AR3 + 40h;
AR3 = 200h + 40h = 240h.

g. AR3 ← AR3 − 40h;
AR3 = 200h − 40h = 1C0h.

---

**Table 5.2** Indirect Addressing Options with a Single Data-Memory Operand

| Operand Syntax | Operation |
|---|---|
| *ARx | addr ← ARx |
| *ARx+ | addr ← ARx |
| | ARx ← ARx + 1 |
| *ARx− | addr ← ARx |
| | ARx ← ARx − 1 |
| *+ARx | ARx ← ARx + 1 |
| | addr ← ARx |
| *ARx + 0 | addr ← ARx |
| | ARx ← ARx + AR0 |
| *ARx − 0 | addr ← ARx |
| | ARx ← ARx − AR0 |
| *ARx + 0B | addr ← ARx |
| | ARx ← B(ARx + AR0) |
| *ARx − 0B | addr ← ARx |
| | ARx ← B(ARx − AR0) |
| *ARx + % | addr ← ARx |
| | ARx ← circ(ARx + 1) |
| *ARx − % | addr ← ARx |
| | ARx ← circ(ARx − 1) |
| *ARx + 0% | addr ← ARx |
| | ARx ← circ(ARx + AR0) |
| *AR0 − 0% | addr ← ARx |
| | ARx ← circ(ARx − AR0) |
| *(lk) | addr ← lk |
| *ARx(lk) | addr ← ARx + lk |
| *+ARx(lk) | ARx ← ARx + lk |
| | addr ← ARx |
| *+ARx(lk)% | ARx ← circ(ARx + lk) |
| | addr ← ARx |

## Circular Addressing

Many fast real-time algorithms, such as convolution, correlation, and FIR filters, require the implementation of a circular buffer in memory. A circular buffer is a sliding window containing the most recent data. As new data come in, the buffer overwrites the oldest data. An indirect addressing mode with circular address modification allows implementation of circular buffers.

The circular-buffer size register (BK) specifies the size of the circular buffer. A circular buffer must start on an *N*-bit boundary; that is, the *N* LSBs of the base address of the circular buffer must be 0. For example, a 31-word circular buffer must start at an address whose five LSBs are 0 and the value 30 must be loaded into BK. Similarly, a 48-word circular buffer must start at an address whose six LSBs are 0 and the value 47 must be loaded into BK.

The algorithm for circular addressing works as follows:

```
If 0 ≤ index + step < BK:  index = index + step;
else if index + step ≥ BK:  index = index + step - BK;
else if index + step < 0:  index = index + step + BK.
```

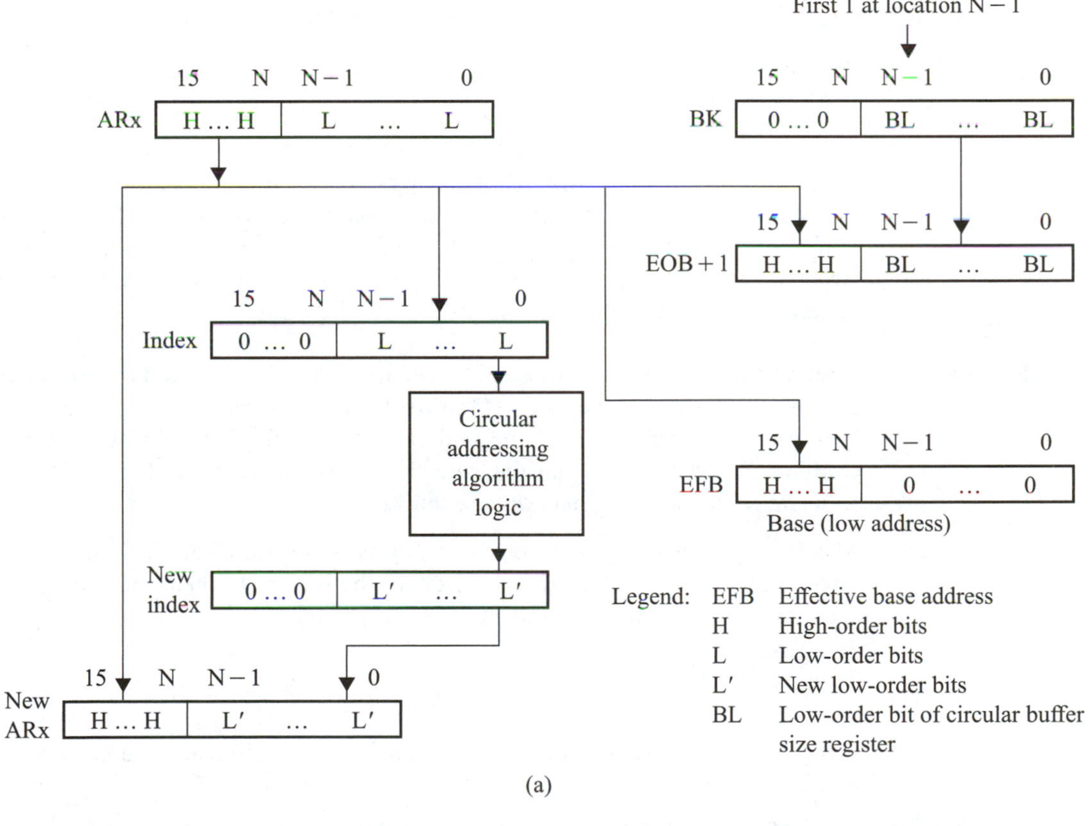


**Figure 5.11(a)** Block diagram of the circular addressing mode for TMS320C54xx processors

(Courtesy of Texas Instruments Inc.)

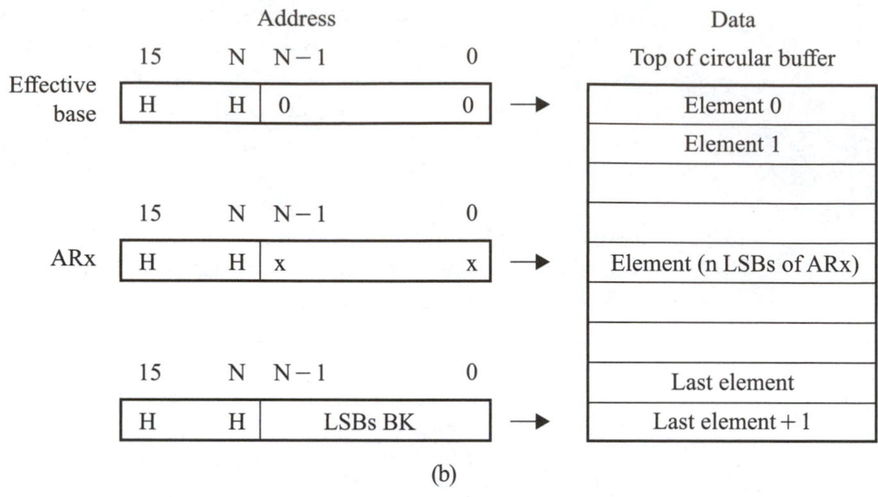


**Figure 5.11(b)** Circular addressing mode implementation in TMS320C54xx processors (Courtesy of Texas Instruments Inc.)

Figure 5.11(a) illustrates the relationships between BK, the auxiliary register ARx (the pointer), the bottom of the circular buffer, the top of the circular buffer, and the index into the circular buffer. Figure 5.11(b) shows how the circular buffer is implemented and illustrates the relationship between the generated values and the elements in the circular buffer.

▷ **Example 5.2** Assume that the register AR3 with contents 1020h is selected as the pointer for the circular buffer. Let BK = 40h to specify the circular buffer size as 41h. Determine the start and the end addresses for the buffer. What will be the contents of register AR3 after the execution of the instruction LD *AR3 + 0%, A, if the contents of register AR0 are 0025h?

Solution AR3 = 1020h means that currently it points to location 1020h. Making the lower 6 bits zeros gives the start address of the buffer as 1000h. Replacing the same bits with the BK gives the end address as 1040h.

The instruction

LD *AR3 + 0%, A

modifies AR3 by adding AR0 to it and applying the circular modification. It yields

$$AR3 = \text{circ}(1020h + 0025h) = \text{circ}(1045h) = 1045h - 40h = 1005h.$$

Thus the location 1005h is the one pointed to by AR3.

### Bit-Reversed Addressing

Bit-reversed addressing is used in FFT algorithms. In this addressing mode, AR0 specifies one half of the size of the FFT. An auxiliary register points to the physical location of a data value. The address of the next location is generated by adding, in a bit-reversed manner, AR0 and the other specified auxiliary register. In the bit-reversed addition, the carry bit propagates from left to right, instead of right to left as in the regular add.

▷ **Example 5.3** Assuming the current contents of AR3 to be 200h, what will be its contents after each of the following TMS320C54xx addressing modes is used? Assume that the contents of AR0 are 20h.

a. *AR3 + 0B

b. *AR3 − 0B

Solution

a. AR3 ← AR3 + AR0 with reverse carry propagation;
AR3 = 200h + 20h (with reverse carry propagation) = 220h.

b. AR3 ← AR3 − AR0 with reverse carry propagation;
AR3 = 200h − 20h (with reverse carry propagation) = 23Fh.

---

### Dual-Operand Addressing

Dual data-memory operand addressing is used for instructions that simultaneously perform two reads (32-bit read) or a single read (16-bit read) and a parallel store (16-bit store) indicated by two vertical bars, ||. These instructions access operands using indirect addressing mode.

If in an instruction with a parallel store the source operand and the destination operand point to the same location, the source is read before writing to the destination. Only 2 bits are available in the instruction code for selecting each auxiliary register in this mode. Thus, just four of the auxiliary registers, AR2–AR5, can be used, The ARAUs, together with these registers, provide the capability to access two operands in a single cycle. Figure 5.12 shows how an address is generated using dual data-memory operand addressing.

## 5.4.6 Memory-Mapped Register Addressing

Memory-mapped register addressing is used to access the memory-mapped registers without affecting either the current data-page pointer (DP) value or the current stack-pointer (SP) value. This mode works for both direct and indirect addressing. Taking only the seven least significant bits of the 16-bit direct address or the value of the auxiliary register used for indirect addressing, the required address is generated.

For example, if AR1 is used indirectly to point to a memory-mapped register using the memory-mapped register addressing mode and its contents are

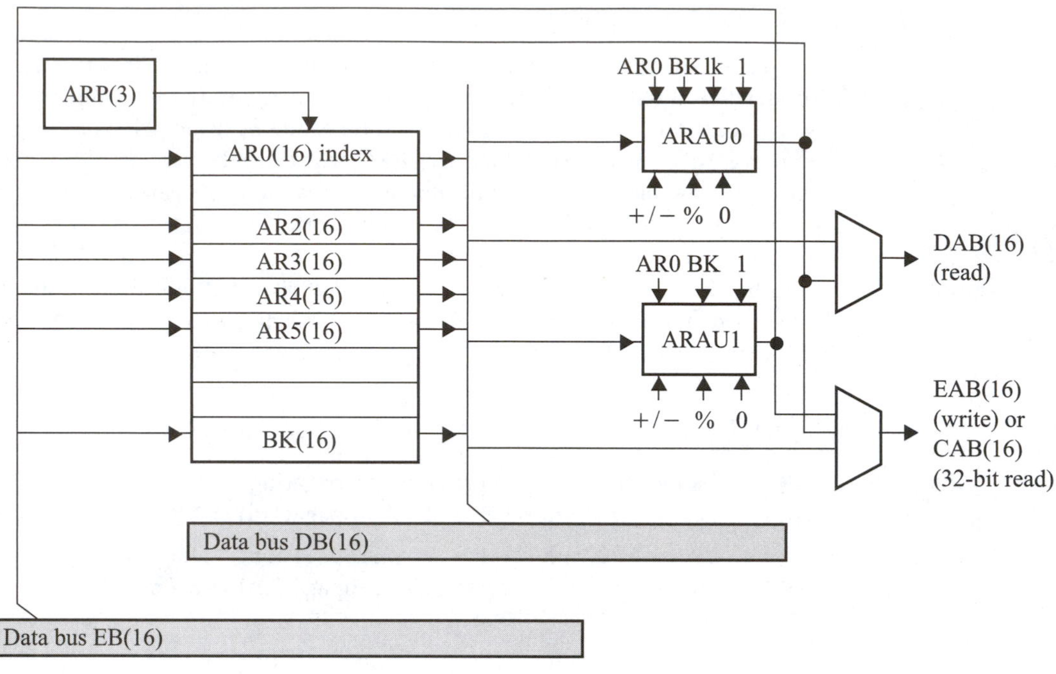


**Figure 5.12** Block diagram of the indirect addressing mode of TMS320C54xx processors using dual memory operands

(Courtesy of Texas Instruments Inc.)

3825h, then AR1 points to the timer period register (PRD), since the seven LSBs of AR1 are 25h, which is the address of the PRD register. After execution, AR1 contains 0025h.

Consider the following instruction as another example:

```
LDM AR4, A
```

In this case the data stored at 0014h, which is the memory address of AR4, is loaded onto A.

### 5.4.7 Stack Addressing

The stack is used to store the return address during the servicing of interrupts and invoking of subroutines. It can also be used to pass parameters to subroutines during program execution. The stack is filled from the highest to the lowest memory address and emptied from the lowest to the highest address.

A 16-bit stack pointer (SP) is used to address the stack location at a given instance. SP points to the last element stored onto the stack. Instructions that access the stack for saving and recovering data on the stack consist of PUSHD, PUSHM, POPD, and POPM.

## 5.5 Memory Space of TMS320C54xx Processors

TMS320C54xx processors provide for a total of 128K words of memory extendable up to 8192K words. This includes both program memory and data memory. Within this space, RAM (both single access and dual access), ROM, EPROM, EEPROM, or memory-mapped peripherals may reside either on- or off-chip. The program memory space is used to store program instructions and the tables used in the execution of programs. The data-memory space is used to store data required to run programs and for external memory-mapped peripherals. Figures 5.13(a) and (b) show memory maps for the basic and extended memories of the TMS320C5416 processor.

The size of the data memory is 64K words, part of which is on-chip DARAM. The device automatically accesses the on-chip RAM when the address is within its range. Memory-mapped registers are also part of the data-memory space.

The program memory is organized into 128 pages, each of 64K word size. Page 0 is part of the basic 128K space, and pages 1 to 127 are extended pages. Out of the 64K words on page 0, 4K words are on-chip ROM. The remaining space on page 0 as well as the extended space consist of DARAM and SARAM, both on-chip and off-chip, as shown in Figures 5.13(a) and (b). The 4K on-chip ROM space contains a GSM EFR speech codec table, a bootloader, $\mu$-law and A-law expansion tables, a sine look-up table, and an interrupt vector table.

The MP/$\overline{\text{MC}}$, OVLY, and DROM bits located in the processor mode status register (PMST) are used to enable and disable on-chip memories in the program and data spaces. The functions of these bits are described in Table 5.3.

▷ **Example 5.4** What is the configuration of on-chip DARAM, on-chip SARAM, and ROM if MP/$\overline{\text{MC}}$ = 0, OVLY = 1, and DROM = 0 for TMS320C5416?

Solution

a. Since MP/$\overline{\text{MC}}$ = 0, 16K on-chip ROM is enabled as program memory at address c000h-feffh.

b. Since OVLY = 1, DARAM is mapped on to the program memory space at address 0080h-7fffh. Memory at addresses 000h-007fh is reserved for memory-mapped registers and the scratch pad purpose.

c. Since DROM = 0, ROM is not mapped on to the data memory.

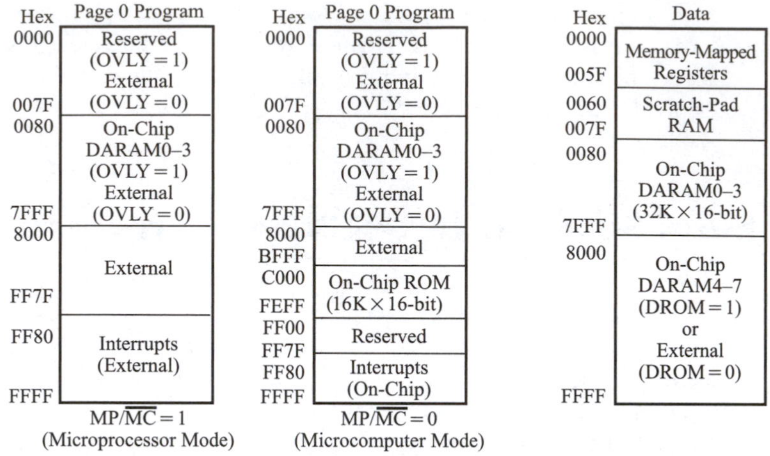


Address ranges for on-chip DARAM in data memory are: DARAM0: 0080h–1FFFh; DARAM1: 2000h–3FFFh DARAM2: 4000h–5FFFh; DARAM3: 6000h–7FFFh DARAM4: 8000h–9FFFh; DARAM5: A000h–BFFFh DARAM6: C000h–DFFFh; DARAM7: E000h–FFFFh

(a)

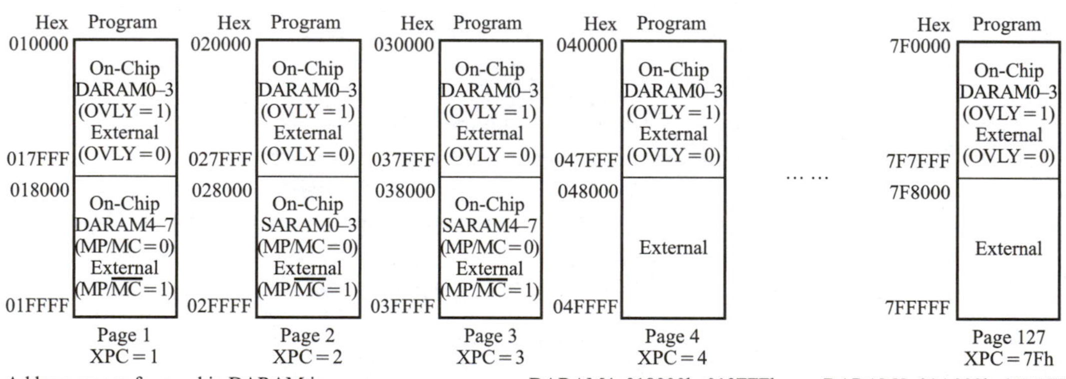


Address ranges for on-chip DARAM in program memory are: DARAM4: 018000h–019FFFh; DARAM5: 01A000h–01BFFFh DARAM6: 01D0000h–01DFFFh; DARAM7: 01E000h–01FFFFh

Address ranges for on chip SARAM in program memory are: SARAM0: 028000h–029FFFh; SARAM1: 02A000h–02BFFFh SARAM2: 02C000h–02DFFFh; SARAM3: 02E000h–02FFFFh SARAM4: 038000h–039FFFh; SARAM5: 03A000h–03BFFFh SARAM6: 03C000h–03DFFFh; SARAM7: 03E000h–03FFFFh

(b)

**Figure 5.13** Memory map for the TMS320C5416 processor

(Courtesy of Texas Instruments Inc.)

**Table 5.3** Processor Bits for Configuring the On-Chip Memories

| PMST Bit | Logic | On-chip Memory Configuration |
|---|---|---|
| MP/$\overline{\text{MC}}$ | 0 | ROM enabled |
| | 1 | ROM not available |
| OVLY | 0 | RAM in data space |
| | 1 | RAM in program space (except page 0) |
| DROM | 0 | ROM not in data space |
| | 1 | ROM in data space |

▷ **Example 5.5** Repeat Example 5.4 if MP/$\overline{\text{MC}}$ = 1, OVLY = 1, and DROM = 1.

Solution

a. Since MP/$\overline{\text{MC}}$ = 1, TMS320C5416 is in microprocessor mode, the 16K ROM is off-chip in the program memory space.

b. Since OVLY = 1, DARAM is mapped on to the program memory space at address 0080h-7fffh. Memory at addresses 0000h-007fh is reserved for memory-mapped registers and the scratch pad purpose.

c. Since DROM = 1, 16K ROM is mapped on to the on-chip data memory at address c000h-feffh and memory from ff00h-ffffh is left for reserved purpose.

## 5.6 Program Control

The program control unit of TMS320C54xx processors contains the program counter (PC), the program counter-related hardware, hardware stack, repeat counters, and status registers. The PC addresses the program memory, either on-chip or off-chip, and is loaded in one of several ways, depending on the sequence of instructions being executed. These are

- Sequential: PC ← PC + 1.
- Branch: The PC is loaded with the immediate value following the branch instruction.
- Subroutine call: The PC is loaded with the immediate value following the call instruction.
- Interrupt: The PC is loaded with the address of the appropriate interrupt vector.
- Instructions such as BACC, CALA, etc.: The PC is loaded with the contents of the accumulator low word.

- End of a block repeat loop: The PC is loaded with the contents of the block repeat program address start register.
- Return: The PC is loaded from the top of the stack.

The program counter-related hardware PAGEN provides for the above options. The stack is used to save and restore the PC value during subroutine calls and interrupts. It can also be used to save and restore the accumulator low word or a data-memory value when required.

The TMS320C54xx processors provide hardware support for repetitive execution of either a single instruction or a block of instructions. Repeat counters are used for this purpose.

A single instruction can be repeated $N + 1$ times by loading the value $N$ in the repeat counter register (RC). Likewise, a block of instructions can be repeated $N + 1$ times by loading the value $N$ in the block repeat counter register (BRC).

## 5.7 TMS320C54xx Instructions and Programming

TMS320C54xx architecture supports an instruction set consisting of a large number of instructions [6]. Many of these are similar to the instructions for general-purpose microprocessors. However, the TMS320C54xx instruction set consists of a number of instructions that are specifically designed to carry out the numerically intensive signal-processing operations efficiently. In this section, we shall summarize the instruction set of the TMS320C54xx processors. In particular, we shall discuss those instructions that are frequently used to implement DSP algorithms and illustrate their use by means of sample programs.

### 5.7.1 Summary of the Instruction Set of TMS320C54xx Processors

TMS320C54xx assembly language instructions can be classified into the following categories based on their functions:

#### Load and Store Operations

- Load instructions; Examples: LD, LDM
- Store instructions; Examples: ST, STM
- Conditional store instructions; Examples: CMPS, STRCD
- Parallel load and store instructions; Example: ST||LD

- Parallel load and multiply instructions; Example: LD||MAC
- Parallel store and add/subtract instructions; Examples: ST||ADD, ST||SUB
- Parallel store and multiply instructions; Examples: ST||MPY, ST||MAC
- Miscellaneous load-type and store-type instructions; Examples: MVDD, MVPD

### Arithmetic Operations

- Add instructions; Examples: ADD, ADDC
- Subtract instructions; Examples: SUB, SUBB
- Multiply instructions; Examples: MPY, MPYA
- Multiply–accumulate instructions; Examples: MAC, MACD
- Multiply–subtract instructions; Examples: MAS, MASA
- Double (32-bit operand) instructions; Examples: DADD, DSUB
- Application-specific instructions; Examples: EXP, LMS

### Logical Operations

- AND instructions; Examples: AND, ANDM
- OR instructions; Examples: OR, ORM
- XOR instructions; Examples: XOR, XORM
- Shift instructions; Examples: ROL, SFTL
- Test instructions; Examples: BIT, CMPM

### Program-Control Operations

- Branch instructions; Examples: B, BACC
- Call instructions; Examples: CALL, CALA
- Interrupt instructions; Examples: INTR, TRAP
- Return instructions; Examples: RET, FRET
- Repeat instructions; Examples: RPT, RPTB
- Stack-manipulating instructions; Examples: PUSHD, POPD
- Miscellaneous program-control instructions; Examples: IDLE, RESET

For detailed descriptions of these and other instructions, the reader is referred to the Texas Instruments' TMS320C54xx DSP Reference Set, Volume 2: *Mnemonic Instruction Set* [6]. We shall now discuss a few of these instructions in detail.

### Multiply Instruction (MPY)

This instruction can take several forms. One such form is

> MPY *Xmem*, *Ymem*, *dst*; where *Xmem* and *Ymem* are dual data-memory operands and *dst* is accumulator A or B.

The instruction multiplies a data-memory value by another data-memory value and stores the result in accumulator A or B. The register T is loaded with the Xmem value in the read-memory phase.

$$\text{dst} \leftarrow (\text{Xmem}) \times (\text{Ymem}); \quad \text{T} \leftarrow (\text{Xmem})$$

In the indirect addressing mode, the instruction can also modify the contents of the auxiliary registers used for indirect addressing.

▷ **Example 5.6** Describe the operation of the following MPY instructions:

a. MPY 13, B

b. MPY #01234, A

c. MPY *AR2−, *AR4 + 0, B

**Solution** Instruction (a) multiplies the current contents of the T register by the contents of the data-memory location 13 in the current data page. The result is placed in the accumulator B.

Instruction (b) multiplies the current contents of the T register by the constant 1234 and places the result in the accumulator A.

Instruction (c) multiplies the contents of memory pointed by AR2 by the contents of memory pointed by AR4. The result is placed in the accumulator B. During this instruction execution, register T is loaded with the contents of the same data-memory location pointed by AR2. AR2 is then decremented by 1 and AR4 is updated by adding to it the contents of AR0.

---

### Multiply and Accumulate Instruction (MAC)

This instruction is an improvement over the MPY instruction. One of the several forms that this instruction can take is

> MAC *Xmem*, *Ymem*, *src*, *dst*; where *Xmem* and *Ymem* are dual data-memory operands and *src* and *dst* are accumulators A and B.

The instruction multiplies a data-memory value by another data-memory value and adds the product to the contents of the source, which may be either of the two accumulators A and B. The result is stored in the other accumulator. The register T is loaded with the Xmem value.

$$dst \leftarrow (Xmem) \times (Ymem) + (src); \quad T \leftarrow (Xmem)$$

Similar to the MPY instruction, this instruction can modify the contents of auxiliary registers used in indirect addressing.

▷ **Example 5.7** Describe the operation of the following MAC instructions:

a. MAC *AR5+, #1234h, A

b. MAC *AR3−, *AR4+, B, A

Solution Instruction (a) multiplies the contents of the data-memory location pointed by AR5 by the constant 1234h and adds the product to the contents of the accumulator A. During the execution, register T is loaded with the content of the data-memory location pointed by AR5. AR5 is then incremented by 1.

Instruction (b) multiplies the contents of the data memory pointed by AR3 by the contents of the data memory pointed by AR4. The contents of the accumulator B are added to the product and the result is placed in the accumulator A. The register T is loaded with the contents of the same data-memory location pointed by AR3. AR3 is then decremented by 1 and AR5 is incremented by 1.

The MAC instruction is used for computing the sum of a series of product terms.

---

### Multiply and Subtract Instruction (MAS)

This instruction is similar to the MAC instruction. One form of this instruction is

> MAS *Xmem*, *Ymem*, *src*, *dst*; where *Xmem* and *Ymem* are dual data-memory operands and *src* and *dst* are accumulators A and B.

The instruction multiplies a data-memory value by another data-memory value and subtracts the product from the contents of the source, which may be either of the two accumulators A and B. The result is stored in the other accumulator. The register T is loaded with the Xmem value in the read-memory phase.

$$dst \leftarrow (src) - (Xmem) \times (Ymem); \quad T \leftarrow (Xmem)$$

In the indirect mode, in addition to the multiply operation, the instruction can modify the contents of the auxiliary registers used for indirect addressing.

▷ **Example 5.8** Describe the operation of the following MAS instruction:

```
MAS *AR3-, *AR4+, B, A
```

Solution This instruction multiplies the contents of the data memory pointed by AR3 by the contents of the data memory pointed by AR4. The product is subtracted from the contents of the accumulator B and the result is placed in the accumulator A. During this instruction, register T is loaded with the contents of the same data-memory location pointed by AR3. AR3 is then decremented by 1 and AR5 incremented by 1.

The MAS instruction is used for computing butterflies in FFT implementation.

### Multiply, Accumulate, and Delay Instruction (MACD)

This instruction carries out all the functions of the MAC instruction and, in addition, copies the contents of the current data-memory address to the next higher data-memory address. However, the two operands of the multiplier are required to be a single data-memory value and a program-memory value. This feature is equivalent to implementing the $z^{-1}$ delay encountered in digital signal-processing algorithms. For this reason, the MACD instruction is often used for implementing FIR filters. The format and all other features of the MACD instruction are same as those of the MAC instruction.

### Repeat Instruction (RPT)

The format of this instruction is

```
      RPT Smem  ; Smem is a single data-memory operand
or    RPT #k    ; k is a short or a long constant.
```

The instruction loads the operand in the repeat counter, RC. The instruction following the RPT instruction is repeated $k + 1$ times, where $k$ is the initial value of the RC.

Due to the dedicated hardware support, the repeat instruction is used to repeat an instruction a given number of times without any penalty for looping. It may be used to compute the sum of products as required in the implementation of FIR filters.

▷ **Example 5.9** Explain what is accomplished by the following instruction sequence:

```
RPT  #2
MAC  *AR1+, *AR2-, A
```

Solution The first instruction loads the register RC with 2. This number is the repeat count for the next MAC instruction. The MAC instruction executes three times. It multiplies and accumulates in A the data locations contents pointed to by the registers AR1 and AR2. After each multiply and add the pointer AR1 is incremented and pointer AR2 is decremented.

### Block Repeat Instruction (RPTB)

RPTB instruction has the format

> RPTB *pmad*, where *pmad* is the program memory address denoting the end of the block of instructions to be repeated.

This instruction is similar to the RPT instruction, except that it repeats a block of code a given number of times without any penalty for looping. One more than the number of times the block of instructions is to be repeated is initially loaded into the memory-mapped block repeat counter register, BRC.

## 5.7.2 Programming Examples

We now look at a few sample programs written for the TMS320C54xx signal processors. These programs particularly illustrate the use of some of the signal-processing instructions and the addressing modes to access data operands.

▷ **Example 5.10** Write a program to find the sum of a series of signed numbers stored at successive locations in the data memory and place the result in the accumulator A, i.e.,

$$A = \sum_{i=410h}^{41fh} dmad(i) \tag{5.1}$$

Solution The TMS320C54xx program for this example is shown in Figure 5.14. AR1 is used as the pointer to the numbers and AR2 as the counter for the numbers. The program initializes the accumulator to 0, sets AR1 to 410h to point to the first number and AR2 to the initial count. This will be used to track the number of processed locations at each step of execution. Sign-extension mode is selected to handle signed numbers. The program adds each number in turn to the accumulator, increments the pointer and decrements the counter. The process is repeated until the count in AR2 reaches 0. At the end of the program, the accumulator A has the sum of the numbers in location s 410h to 41fh.

---

▷ **Example 5.11** Write a program to compute the sum of three product terms given by the equation

$$y(n) = h_0x(n) + h_1x(n-1) + h_2x(n-2) \tag{5.2}$$

where $x(n)$, $x(n-1)$ and $x(n-2)$ are data samples stored at three successive

```
*************************************************************
*
* This program computes the signed sum of data memory locations
* from address 410h to 41fh. The result is placed in A.
*
*    A = dmad(410h) + dmad(411h) + ··· dmad(41fh)
*
*************************************************************

    .mmregs
    .global _c_int00

    .text

_c_int00:
    STM   #10H, AR2      ; Initialize counter AR2 = 10h
    STM   #410H, AR1     ; Initialize pointer AR1 = 410h
    LD    #0H, A         ; Initialize sum A = 0
    SSBX  SXM            ; Select sign extension mode

START:
    ADD   *AR1+, A       ; Add the next data value
    BANZ  START, *AR2-   ; Repeat if not done
    NOP                  ; No operation

    .end
```

**Figure 5.14** TMS320C54xx program for Example 5.10

data-memory locations and $h_0$, $h_1$, and $h_2$ are constants stored at three other successive locations in the data memory. The result $y(n)$ is to be stored in the data memory. Use direct addressing mode to access the data memory.

Solution Let $h_0$, $h_1$, and $h_2$ be stored starting at address h, and $x(n)$, $x(n-1)$, and $x(n-2)$ starting at address 310h in the data memory. Product terms $h_0x(n)$, $h_1x(n-1)$, and $h_2x(n-2)$ are computed using the MPY instruction by moving one of the operands to register T and accessing the other operand directly from the data memory. Note that the data-page pointer, DP, needs to be initialized before using the direct addressing mode to access the operand. Product terms are computed in A or B and added. When all the three multiplications are done, the result accumulated in B is stored in the data memory $y(n)$. Since $y(n)$ is 32 bits long, it is saved at two successive locations labeled as y, with the lower 16 bits at memory location y and the higher 16 bits at the next memory location. The TMS320C54xx program for this example is shown in Figure 5.15.

```
**********************************************************************
*
* This program computes multiply and accumulate using direct addressing
* mode.
*
*   y(n) = h(0)x(n) + h(1)x(n-1) + h(2)x(n-2)
*
*   h(0), h(1), and h(2) are stored in data-memory locations starting at
*   location h and x(n), x(n-1), and x(n-2) are stored in data-memory
*   locations starting at location x. y(n) is saved in data-memory
*   location y (low 16 bits) and y + 1 (high 16 bits).
*
**********************************************************************

    .global _c_int00

x   .usect "Input Samples", 3
y   .usect "Output", 2
h   .usect "Coefficients", 3

    .text

_c_int00:
    SSBX  SXM      ; Select sign extension mode
    LD    #h, DP   ; Select the data page for coefficients
    LD    @h, T    ; Get the coefficient h(0)
    LD    #x, DP   ; Select the data page for input samples
    MPY   @x, A    ; A = x(n)*h(0)

    LD    #h, DP   ; Select the data page for coefficients
    LD    @h+1, T  ; Get the coefficient h(1)
    LD    #x, DP   ; Select the data page for input signals
    MPY   @x+1, B  ; B = x(n-1)*h(1)

    ADD   A, B     ; B = x(n)*h(0) + x(n-1)*h(1)

    LD    #h, DP   ; Select the data page for coefficients
    LD    @h+2, T  ; Get the coefficient h(2)
    LD    #x, DP   ; Select the data page for input signals
    MPY   @x+2, A  ; A = x(n-2)*h(3)

    ADD   A, B     ; B = x(n)*h(0) + x(n-1)*h(1) + x(n-2)*h(3)

    LD    #y, DP   ; Select the data page for output
    STL   B, @y    ; Save low part of output
    STH   B, @y+1  ; Save high part of output
    NOP            ; No operation

    .end
```

**Figure 5.15** TMS320C54xx program for Example 5.11

▷ **Example 5.12** Repeat the problem of Example 5.11 using the indirect addressing mode to access data.

Solution In this example, let us use the auxiliary register AR2 to address the data using the indirect addressing mode. AR2 is initialized to 310h, the location where $x(n)$ is stored, and is advanced to the next address after each multiply opera-

```
**********************************************************************
*
* This program computes multiply and accumulate using indirect
* addressing mode.
*
*   y(n) = h(0)x(n) + h(1)x(n-1) + h(2)x(n-2)
*
*   h(0), h(1), and h(2) are stored in data-memory locations starting at
*   location h, x(n), x(n-1), and x(n-2) are stored in data-memory
*   locations 310h, 311h, & 312h resp. y(n) is saved in data-memory
*   location 313h (low 16 bits) and 314h (high 16 bits)
*
**********************************************************************

    .global _c_int00

h   .int 10, 20, 30

    .text

_c_int00:
    SSBX  SXM              ; Select sign extension mode
    STM   #310H, AR2       ; Initialize pointer AR2 for x(n) stored at
                             310H
    STM   #h, AR3          ; Initialize pointer AR3 for coefficients

    MPY   *AR2+, *AR3+, A  ; A = x(n)*h(0)

    MPY   *AR2+, *AR3+, B  ; B = x(n-1)*h(1)

    ADD   A, B             ; B = x(n)*h(0) + x(n-1)*h(1)

    MPY   *AR2+, *AR3+, A  ; A = x(n-2)*h(2)

    ADD   A, B             ; B = x(n)*h(0) + x(n-1)*h(1) + x(n-2)*h(2)

    STL   B, *AR2+         ; Save low part of result
    STH   B, *AR2+         ; Save high part of result
    NOP                    ; No operation

    .end
```

**Figure 5.16** TMS320C54xx program for Example 5.12

tion. AR3 is used as the pointer to access coefficients starting at h. At the end of three multiply operations, AR2 points to 313h, the address at which the lower 16 bits of $y(n)$ are to be stored. The TMS320C54xx program for this example is shown in Figure 5.16.

---

▷ **Example 5.13** Repeat the problem of Example 5.11 by using the MAC instruction.

```
***********************************************************************
*
* This program computes multiply and accumulate using the MAC
* instruction
*
*   y(n) = h(0)x(n) + h(1)x(n-1) + h(2)x(n-2)
*
*   where, h(0), h(1), and h(2) are in the program-memory locations
*   starting at h, x(n), x(n-1), and x(n-2) are in data-memory locations
*   starting at x. y(n) is to be saved in location y (low 16 bits) and
*   y + 1 (high 16 bits).
*
***********************************************************************

    .global _c_int00

    .data
    .bss x, 3
    .bss y, 2

h   .int 10, 20, 30

    .text

_c_int00:
    SSBX  SXM               ; Select sign extension mode
    STM   #x, AR2           ; Initialize AR2 to point to x(n)
    STM   #h, AR3           ; Initialize AR3 to point to h(0)
    LD    #0H, A            ; Initialize result in A = 0

    RPT   #2                ; Repeat the next operation 3 times
    MAC   *AR2+, *AR3+, A   ; y(n) computed

    STM   #y, AR2           ; Select the page for y(n)
    STL   A, *AR2+          ; Save the low part of y(n)
    STH   A, *AR2+          ; Save the high part of y(n)
    NOP                     ; No operation

    .end
```

**Figure 5.17** The TMS320C54xx program for Example 5.13

Solution The MAC instruction multiplies the contents of two data-memory locations and adds the result to the previous contents of the accumulator being used. (Note that only auxiliary registers AR2–AR5 can be used.) This instruction is repeated twice using RPT instruction. After each MAC instruction the auxiliary registers, which are being used, should be incremented by 1. Finally, the result is stored in the memory location pointed by "*y*" using STL instruction first for the lower 16 bits and then using STH instruction for the higher 16 bits. The TMS32054Cxx program for this example is shown in Figure 5.17.

## 5.8 On-Chip Peripherals

On-chip peripherals facilitate interfacing with external devices such as modems and analog-to-digital converters. They also provide certain features that are required for implementing real time systems using the processors. All the '54xx devices have the same CPU, but different on-chip peripherals are available in different devices. These peripherals include general-purpose I/O pins, a software-programmable wait-state generator, hardware timer, host port interface (HPI), clock generator, and serial ports. Of these, the general-purpose I/O and the software-programmable wait-state generator are described in Chapter 9 on parallel peripheral devices. The timer, the host port interface, clock generator, and serial ports are briefly described below. The tables in Appendix A give details of the information required for programming these on-chip peripherals.

### 5.8.1 Hardware Timer

The timer is an on-chip down counter that can be used to generate a signal to initiate an interrupt or to initiate any other process. The timer consists of three memory-mapped registers—TIM, PRD, and TCR. A logical block diagram of the timer circuit is shown in Figure 5.18. The timer register (TIM) is a 16-bit memory-mapped register that decrements at every pulse from the prescaler block (PSC). The timer period register (PRD) is a 16-bit memory-mapped register whose contents are loaded onto the TIM whenever the TIM decrements to zero or the device is reset ($\overline{\text{SRESET}}$). The timer can also be independently reset using the TRB signal. The timer control register (TCR) is a 16-bit memory-mapped register that contains status and control bits. Table 5.4 shows the functions of the various bits in the TCR. The prescaler block is also an on-chip counter. Whenever the prescaler bits count down to 0, a clock

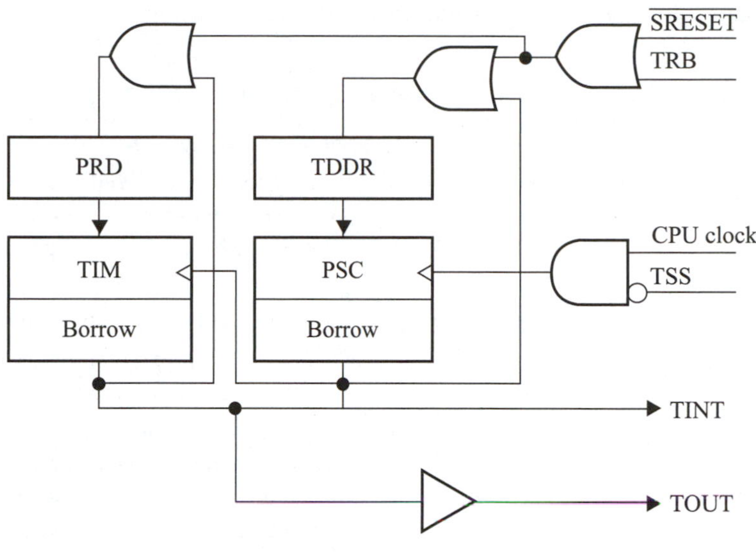


**Figure 5.18** Logical block diagram of timer circuit

(Courtesy of Texas Instruments Inc.)

pulse is given to the TIM register that decrements the TIM register by 1. The TDDR bits contain the divide-down ratio, which is loaded onto the prescaler block after each time the prescaler bits count down to 0. That is to say that the 4-bit value of TDDR determines the divide-by ratio of the timer clock with respect to the system clock. In other words, the TIM decrements either at the rate of the system clock or at a rate slower than that as decided by the value of the TDDR bits. TOUT and TINT are the output signals generated as the TIM register decrements to 0. TOUT can trigger the start of the conversion signal in an ADC interfaced to the DSP. The sampling frequency of the ADC determines how frequently it receives the TOUT signal. TINT is used to generate interrupts, which are required to service a peripheral such as a DRAM controller periodically. The timer can also be stopped, restarted, reset, or disabled by specific status bits.

### 5.8.2 Host Port Interface (HPI)

The host port interface (HPI) is a unit that allows the DSP to interface to an 8-bit or a 16-bit host device or a host processor. The HPI communicates with the host independently of the DSP. The HPI features allow the host to interrupt the DSP, or vice versa, when required. The interface contains minimal

**Table 5.4** Function of Various Bits in the TCR Registers

| Bit | Name | Reset Value | Function |
|---|---|---|---|
| 15–12 | Reserved | — | Reserved; always read as 0. |
| 11 | Soft | 0 | Used in conjunction with the Free bit to determine the state of the timer when a breakpoint is encountered in the HLL debugger. When the Free bit is cleared, the Soft bit selects the timer mode.<br>Soft = 0 The timer stops immediately.<br>Soft = 1 The timer stops when the counter decrements to 0. |
| 10 | Free | 0 | Used in conjunction with the Soft bit to determine the state of the timer when a breakpoint is encountered in the HLL debugger. When the Free bit is cleared, the Soft bit selects the timer mode.<br>Free = 0 The Soft bit selects the timer mode.<br>Free = 1 The timer runs free regardless of the Soft bit. |
| 9–6 | PSC | — | Timer prescaler counter. Specifies the count for the on-chip timer. When PSC is decremented past 0 or the timer is reset, PSC is loaded with the contents of TDDR and the TIM is decremented. |
| 5 | TRB | — | Timer reload. Resets the on-chip timer. When TRB is set, the TIM is loaded with the value in the PRD and the PSC is loaded with the value in TDDR. TRB is always read as a 0. |
| 4 | TSS | 0 | Timer stop status. Stops or starts the on-chip timer. At reset, TSS is cleared and the timer immediately starts timing.<br>TSS = 0 The timer is started.<br>TSS = 1 The timer is stopped. |
| 3–0 | TDDR | 0000 | Timer divide-down ratio. Specifies the timer divide-down ratio (period) for the on-chip timer. When PSC is decremented past 0, PSC is loaded with the contents of TDDR. |

*(Courtesy of Texas Instruments Inc.)*

external logic, so that a system with a host and a DSP can be designed without increasing the hardware on the board. The HPI interfaces to the PC parallel ports directly. A generic block diagram of the HPI is shown in Figure 5.19.

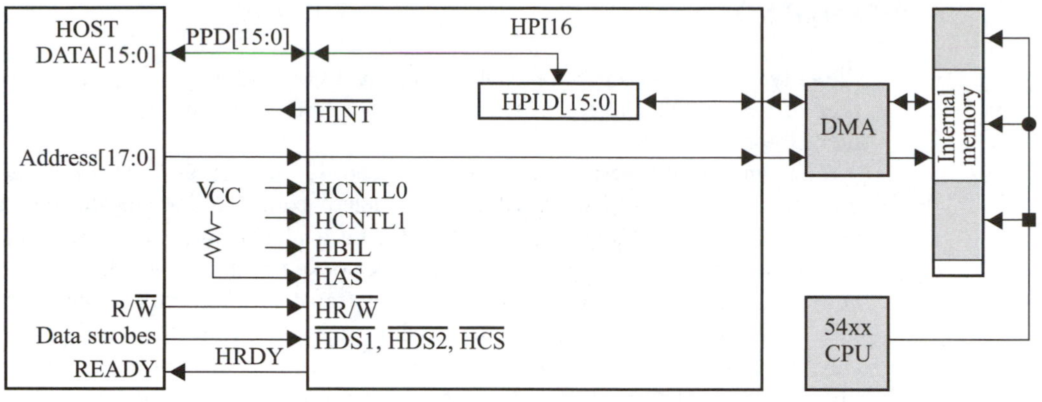


**Figure 5.19** A generic diagram of the host port interface (HPI)

(Courtesy of Texas Instruments Inc.)

Important signals in the HPI are as follows:

- The 16-bit data bus and the 18-bit address bus.
- The host interrupt, HINT, for the DSP to signal the host when its attention is required.
- HRDY, a DSP output indicating that the DSP is ready for transfer.
- HCNTL0 and HCNTL1, control signals that indicate the type of transfer to carry out. The transfer types are data, address, etc.
- HBIL. If this is low it indicates that the current byte is the first byte; if it is high, it indicates that it is the second byte.
- HR/$\overline{\text{W}}$, indicates if the host is carrying out a read operation or a write operation.

By appropriately using these signals, the DSP device can be interfaced on a host such as a PC.

## 5.8.3 Clock Generator

The clock generator on TMS320C54xx devices has two options—an external clock and the internal clock. In the case of the external clock option, a clock source is directly connected to the device. The internal clock source option, on the other hand, uses an internal clock generator and a phase locked loop (PLL) circuit. The PLL, in turn, can be hardware configured or software programmed. Not all devices of the TMS320C54xx family have all these clock options; they vary from device to device.

### 5.8.4 Serial I/O Ports

Three types of serial ports are available on the '54xx devices, depending on the type of the device. These are synchronous, buffered, and time-division multiplexed ports.

The synchronous serial ports are high-speed, full-duplex ports that provide direct communication with serial devices, such as codec, and analog-to-digital (A/D) converters. A buffered serial port (BSP) is a synchronous serial port that is provided with an autobuffering unit and is clocked at the full clock rate. The autobuffering unit supports high-speed data transfers and reduces the overhead of servicing interrupts. A time-division multiplexed (TDM) serial port is a synchronous serial port that is provided to allow time-division multiplexing of the data. We will cover serial I/O in Chapter 10.

The functioning of each of these on-chip peripherals is controlled by memory-mapped registers assigned to the respective peripheral. Figure 5.8(b) gives the list of peripheral memory-mapped registers along with their addresses for the TMS320C54xx devices.

## 5.9 Interrupts of TMS320C54xx Processors

Many times, when the CPU is in the midst of executing a program, a peripheral device may require a service from the CPU. In such a situation, the main program may be interrupted by a signal generated by the peripheral device. This results in the processor suspending the main program in order to execute another program, called interrupt service routine, to service the peripheral device. On completion of the interrupt service routine, the processor returns to the main program to continue from where it left.

Interrupt may be generated either by an internal or an external device. It may also be generated by software. Not all interrupts are serviced when they occur. Only those interrupts that are called *nonmaskable* are serviced whenever they occur. Other interrupts, which are called *maskable* interrupts, are serviced only if they are enabled. There is also a priority to determine which interrupt gets serviced first if more than one interrupts occur simultaneously.

Almost all the devices of the TMS320C54xx family have 32 interrupts. However, the types and the number under each type vary from device to device. Some of these interrupts are reserved for use by the CPU. Figure 5.20 gives the types of interrupts, their locations, and priorities for TMS320C54xx processors.

A more detailed description of interrupts and how an interrupt is handled when it occurs is given in Chapter 9.

| | LOCATION | | | |
|---|---|---|---|---|
| NAME | DECIMAL | HEX | PRIORITY | FUNCTION |
| $\overline{\text{RS}}$, SINTR | 0 | 00 | 1 | Reset (hardware and software reset) |
| $\overline{\text{NMI}}$, SINT16 | 4 | 04 | 2 | Nonmaskable interrupt |
| SINT17 | 8 | 08 | — | Software interrupt #17 |
| SINT18 | 12 | 0C | — | Software interrupt #18 |
| SINT19 | 16 | 10 | — | Software interrupt #19 |
| SINT20 | 20 | 14 | — | Software interrupt #20 |
| SINT21 | 24 | 18 | — | Software interrupt #21 |
| SINT22 | 28 | 1C | — | Software interrupt #22 |
| SINT23 | 32 | 20 | — | Software interrupt #23 |
| SINT24 | 36 | 24 | — | Software interrupt #24 |
| SINT25 | 40 | 28 | — | Software interrupt #25 |
| SINT26 | 44 | 2C | — | Software interrupt #26 |
| SINT27 | 48 | 30 | — | Software interrupt #27 |
| SINT28 | 52 | 34 | — | Software interrupt #28 |
| SINT29 | 56 | 38 | — | Software interrupt #29 |
| SINT30 | 60 | 3C | — | Software interrupt #30 |
| $\overline{\text{INT0}}$, SINT0 | 64 | 40 | 3 | External user interrupt #0 |
| $\overline{\text{INT1}}$, SINT1 | 68 | 44 | 4 | External user interrupt #1 |
| $\overline{\text{INT2}}$, SINT2 | 72 | 48 | 5 | External user interrupt #2 |
| TINT, SINT3 | 76 | 4C | 6 | Timer interrupt |
| RINT0, SINT4 | 80 | 50 | 7 | McBSP #0 receive interrupt (default) |
| XINT0, SINT5 | 84 | 54 | 8 | McBSP #0 transmit interrupt (default) |
| RINT2, SINT6 | 88 | 58 | 9 | McBSP #2 receive interrupt (default) |
| XINT2, SINT7 | 92 | 5C | 10 | McBSP #2 transmit interrupt (default) |
| $\overline{\text{INT3}}$, SINT8 | 96 | 60 | 11 | External user interrupt #3 |
| $\overline{\text{HINT}}$, SINT9 | 100 | 64 | 12 | HPI interrupt |
| RINT1, SINT10 | 104 | 68 | 13 | McBSP #1 receive interrupt (default) |
| XINT1, SINT11 | 106 | 6C | 14 | McBSP #1 transmit interrupt (default) |
| DMAC4, SINT12 | 112 | 70 | 15 | DMA channel 4 (default) |
| DMAC5, SINT13 | 116 | 74 | 16 | DMA channel 5 (default) |
| Reserved | 120–127 | 78–7F | — | Reserved |

**Figure 5.20** Table for interrupt locations and priorities for TMS320C54xx processors

(Courtesy of Texas Instruments Inc.)

## 5.10 Pipeline Operation of TMS320C54xx Processors

The CPU of '54xx devices has a six-level-deep instruction pipeline. The six stages of the pipeline are independent of each other. This allows overlapping execution of instructions. During any given cycle, up to six different instructions can be active, each at a different stage of processing. The six levels of the pipeline structure are program prefetch, program fetch, decode, access, read, and execute.

1. During program prefetch, the program address bus, PAB, is loaded with the address of the next instruction to be fetched.
2. In the fetch phase, an instruction word is fetched from the program bus, PB, and loaded into the instruction register, IR. These two phases form the instruction fetch sequence.
3. During the decode stage, the contents of the instruction register, IR, are decoded to determine the type of memory access operation and the control signals required for the data-address generation unit and the CPU.

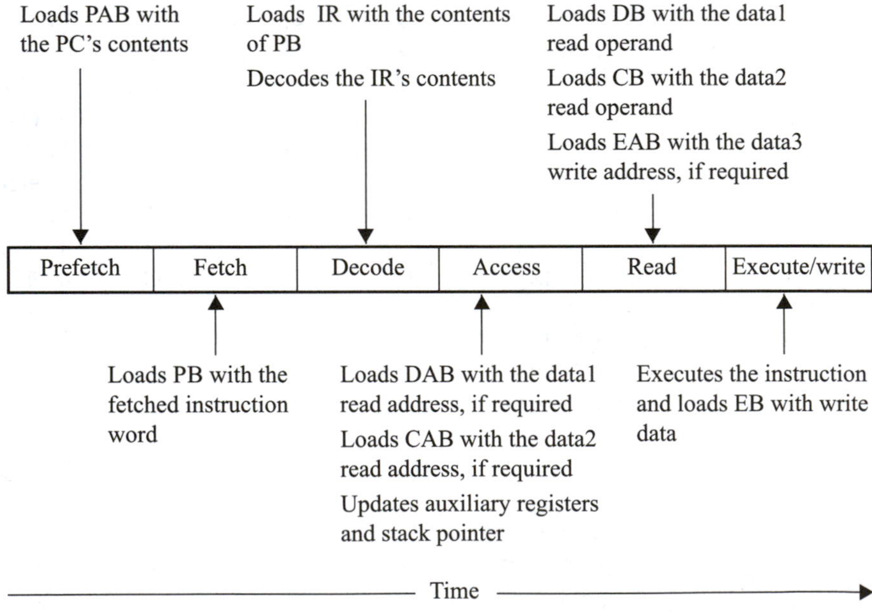


**Figure 5.21** Six-stage pipeline of TMS320C54xx execution

(Courtesy of Texas Instruments Inc.)

4. The access phase outputs the read operand's address on the data address bus, DAB. If a second operand is required, the other data address bus, CAB, is also loaded with an appropriate address. Auxiliary registers in indirect addressing mode and the stack pointer (SP) are also updated.
5. In the read phase the data operand(s), if any, are read from the data buses, DB and CB. This phase completes the two-phase read process and starts the two-phase write process. The data address of the write operand, if any, is loaded into the data write address bus, EAB.
6. The execute phase writes the data using the data write bus, EB, and completes the operand write sequence. The instruction is also executed in this phase.

Figure 5.21 shows the six stages of the pipeline and the events that occur in each stage. The following examples demonstrate how the TMS320C54xx pipeline works while executing instructions.

▷ **Example 5.14** Show the pipeline operation of the following sequence of instructions if the initial value of AR3 is 80 and the values stored in memory location 80, 81, 82 are 1, 2, and 3.

```
LD *AR3+, A
ADD #1000h, A
STL A, *AR3+
:
:
```

Solution Figure 5.22 is the solution to this example problem.

| Cycle | Prefetch | Fetch | Decode | Access | Read | Exec & Write | AR3 | A |
|---|---|---|---|---|---|---|---|---|
| 1 | LD | | | | | | 80 | X |
| 2 | ADD | LD | | | | | 80 | X |
| 3 | STL | ADD | LD | | | | 80 | X |
| 4 | | STL | ADD | LD | | | 81 | X |
| 5 | | | STL | ADD | LD | | 81 | 1 |
| 6 | | | | STL | | LD | 82 | 0001h |
| 7 | | | | | STL | ADD | 82 | 1001h |
| 8 | | | | | | STL | 82 | 1001h |

**Figure 5.22** Pipeline operation of the instruction sequence of Example 5.14

| Cycle | Prefetch | Fetch | Decode | Access | Read | Exec & Write | AR3 | AR1 | A | T |
|---|---|---|---|---|---|---|---|---|---|---|
| 1 | ADD | | | | | | 81 | 84 | 1 | X |
| 2 | LD | ADD | | | | | 81 | 84 | 1 | X |
| 3 | MPY | LD | ADD | | | | 81 | 84 | 1 | X |
| 4 | ADD | MPY | LD | ADD | | | 82 | 84 | 1 | X |
| 5 | | ADD | MPY | LD | ADD | | 82 | 85 | 1 | X |
| 6 | | | ADD | MPY | LD | ADD | 83 | 85 | 03 | 06 |
| 7 | | | | ADD | MPY | LD | 83 | 85 | 03 | 06 |
| 8 | | | | | ADD | MPY | 83 | 85 | 03 | 06 |
| 9 | | | | | | ADD | 83 | 85 | 15h | 06 |

**Figure 5.23** Pipeline operation of the instruction sequence of Example 5.15

▷ **Example 5.15** Show the pipeline operation of the following sequence of instructions if the initial values of AR1, AR3, A are 84, 81, 1 and the values stored in memory location 81, 82, 83, 84 are 2, 3, 4, 6. Also provide the values of registers AR3, AR1, T and accumulator, A, after completion of each cycle.

```
ADD *AR3+, A
LD *AR1+, T
MPY *AR3+, B
ADD B, A
⋮
```

Solution Figure 5.23 is the solution to this example problem.

## 5.11 **Summary**

In this chapter, we have looked at the architectural features of the commercially available programmable digital signal processors. In particular, we have studied in detail the following features of the Texas Instruments TMS320C54xx DSPs:

- Architecture of the processors, consisting of the bus structure, central processing unit (CPU), and internal memory organization

- Addressing modes, consisting of immediate addressing, absolute addressing, accumulator addressing, direct addressing, indirect addressing, memory-mapped addressing, and stack addressing
- Address-generation unit, including single-operand address modifications, circular address modifications, bit-reversed address modifications, and dual-operand address modifications
- Assembly language instructions, including signal processing-specific instructions and programming examples
- Memory organization
- On-chip peripherals
- Interrupts
- Pipeline operation

## References


1. TMS320C2x *User's Guide*, Texas Instruments, 1993.
2. DSP 56000/56001 Digital Signal Processor *User's Manual*, Motorola, 1993.
3. ADSP2101/2102 *User's Manual*, Analog Devices, 1993.
4. TMS320C54xx *DSP Reference Set*, Vols. 1 and 2, Texas Instruments, 1997.
5. TMS320VC5416 *DSP Data Manual*, Texas Instruments, 2002.
6. TMS320C54x *DSP Reference Set*, Vol. 2, Texas Instruments, 1997.


## Assignments

**5.1** How will you configure a TMS320C5416 processor to have the following on-chip memories? Specify the address range in each case.

On-chip DARAM: for program

On-chip ROM: for program

How much RAM for data will be available in the specified configuration?

**5.2** Explain the difference between the internal and external modes of clocking TMS320C54xx processors. How do you vary the clock frequency in each case?

**5.3** Identify the addressing mode of the source operand in each of the following instructions:

a. ADD *AR2, A

b. ADD *AR2+, A

c. ADD *AR2 + %, A

d. ADD #0ffh, A

e. ADD 1234h, A

f. ADD *AR2 + 0B, A

g. ADD *+AR2, A

**5.4** What will be the contents of accumulator A after the execution of the instruction

LD *AR4, 4, A

if the current AR4 points to a memory location whose contents are 8b0eh and the SXM bit of the status register ST1 is set?

**5.5** Write a sequence of TMS320C54xx instructions to configure a circular buffer with a start address at 0200h and an end address at 021fh with current buffer pointer (AR6) pointing to address 0205h.

**5.6** Write a TMS320C54xx program to compute the equation

$$y = mx + c$$

Assume that $x$ and $c$ are stored in the data memory and $m$ in the program memory. The result should be stored in the data memory.

**5.7** Write a TMS320C54xx program to implement second-order IIR filter equations

$$d(n) = x(n) + d(n-1)a_1 + d(n-2)a_2$$
$$y(n) = d(n)b_0 + d(n-1)b_1 + d(n-2)b_2$$

where $a_1$, $a_2$, $b_0$, $b_1$, $b_2$ are filter coefficients (integers), $x(n)$ is the latest input sample, $y(n)$ is the filtered output sample, and $d(n)$ is an intermediate result. You may assume that, during calculations, all signals remain within values represented by 16 bits.

**5.8** Write a TMS320C54xx program to read the cosine value of a variable from a table stored in the program memory and store it in the data memory. The variable is located at address VALUE in the data memory, and the cosine value should be stored at the same location. The cosine table is stored at address TABLE in the program memory.

**5.9** Write a TMS320C54xx program to read 100h words from the input port at address INPORT and store them in the data memory starting at address BUFFER.

**5.10** Write a TMS320C54xx program to mask the lower 6 bits of a word stored in the data memory and write the modified word back at the same location.

**5.11** What is the role of the interrupt pins in a DSP device? Are these the only means of interrupting a DSP program? How do you prevent a signal on an interrupt pin from interrupting a time-critical program being executed by the DSP?

**5.12** By means of a figure, explain the pipeline operation of the following sequence of TMS320C54xx instructions if the initial value of AR3 is 80 and the values stored in memory location 80, 81, 82 are 1, 2, and 3.

```
LD *AR3+, A
ADD *AR3+, A
STL A, *AR3+
    .
    .
```

# Chapter 6

## Development Tools for Digital Signal-Processing Implementations

### 6.1 Introduction

In the last chapter, we studied TMS320C54xx DSP's architecture and instructions, and we wrote a few simple programs to illustrate the use of its instructions. In this chapter, we introduce a development tool that can be used to implement and test DSP algorithms. This tool is the DSP System Design Kit, or DSK, for TMS320C54xx processors. It comes with the development software called the *Code Composer Studio* (CCS). We will briefly describe this tool and show how it can be used to develop DSP applications. Specifically, we discuss the following topics:

The DSP development tools
The DSP System Design Kit (DSK)
Software for development
The assembler and the assembly source file
The linker and memory allocation
The C/C$^{++}$ compiler
The Code Composer Studio (CCS)
DSP software development example

### 6.2 The DSP Development Tools

A development tool provides a hardware/software platform to implement and test a design. For implementing TMS320C54xx DSP designs, a range of systems exist with varying developmental capability and price tags. The least expensive developmental system is the DSP System Design Kit, or DSK, and the most expensive and also the most capable system is the Emulator. The medium-capability system is the Evaluation Module, or the EVM. Here, we limit our discussions to the use of the DSK for implementing and testing DSP algorithms. The DSK provides all the capabilities that a beginner needs to start implementing DSP schemes using TMS320C54xx DSP devices.

## 6.3 The DSP System Design Kit (DSK)

TMS320VC5416 DSK, or simply DSK, is a low-cost development tool that allows a student to explore TMS320C54xx DSP architecture and implement signal-processing algorithms. The DSK is specifically suitable for a beginner learning DSP implementations. It comes with a TMS320VC5416-based board, and DSK-specific development software. The DSK board can be connected to a PC using the universal serial bus (USB) cable, as shown in Figure 6.1. An embedded JTAG emulation logic on the DSK allows for code development and debug without the use of an external emulator. Four jacks for analog inputs (such as a microphone) and outputs (such as a speaker) provide interface to the outside world.

The board is shown in the block diagram of Figure 6.2. The DSK board is designed around a 16–160 MHz VC5416 DSP processor. The DSP device provides a 64K-word dual-access program/data RAM, a 64K-word single-access program RAM, and a 16K-word program ROM. In addition to the memories, it also provides three multichannel buffered serial ports (McBSPs), a DMA controller, 8/16-bit host port interface, and a timer. Additional external memory is provided with a 64K-word SRAM and a 256K-word flash memory on the DSK board.

The DSK uses the PCM3002 stereo codec consisting of a 16-bit analog-to-digital converter (ADC) and a 16-bit digital-to-analog converter (DAC). The codec provides the capability to convert an analog signal to a serial digital signal for the DSP's multichannel buffered serial port McBSP2 and to convert

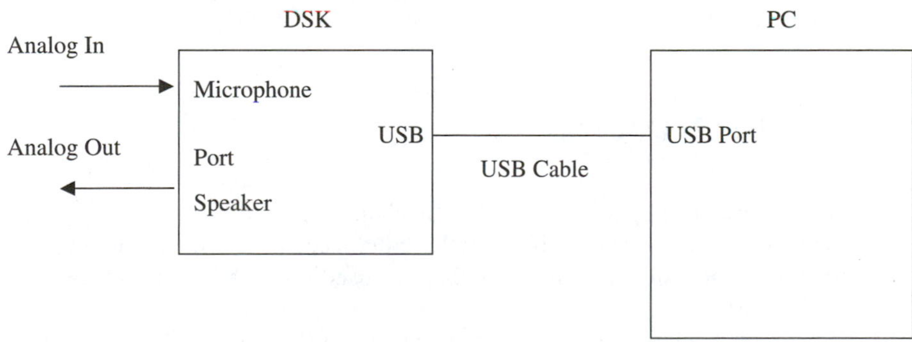


**Figure 6.1** Signal-processing configuration using the C5416 DSK

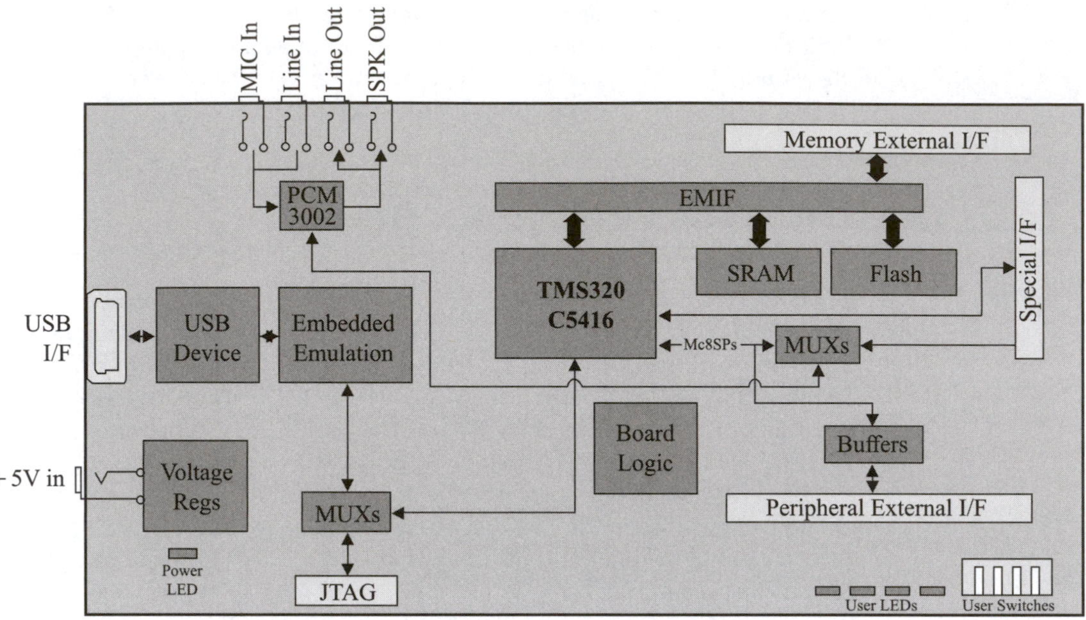


**Figure 6.2** Block diagram of the DSK board

(Courtesy of Texas Instruments Inc.)

the digital signal to analog for the analog output port. We consider the details of this interface in Chapter 10.

The other provisions on the board include three expansion connectors for memory, peripherals, and host interfaces. Four jumpers are provided to configure the board for various clock frequencies and running the DSP in microprocessor or microcomputer mode. A reset push button switch is provided to reset the board. The board uses 5V dc power supply. For more details on the DSK board hardware, the reader should consult reference [1] given at the end of this chapter.

## 6.4 Software for Development

The software development flow chart of Figure 6.3 describes the various languages, tools, and libraries that may be employed to develop an application. The flow chart also shows the files that are used and created in the development process.

The tools depicted in the flow chart consist of the compiler, the assembler, the linker, and the debugger. The utilities that may be needed consist of the archiver, the library builder, and the hex converter. The files encountered in

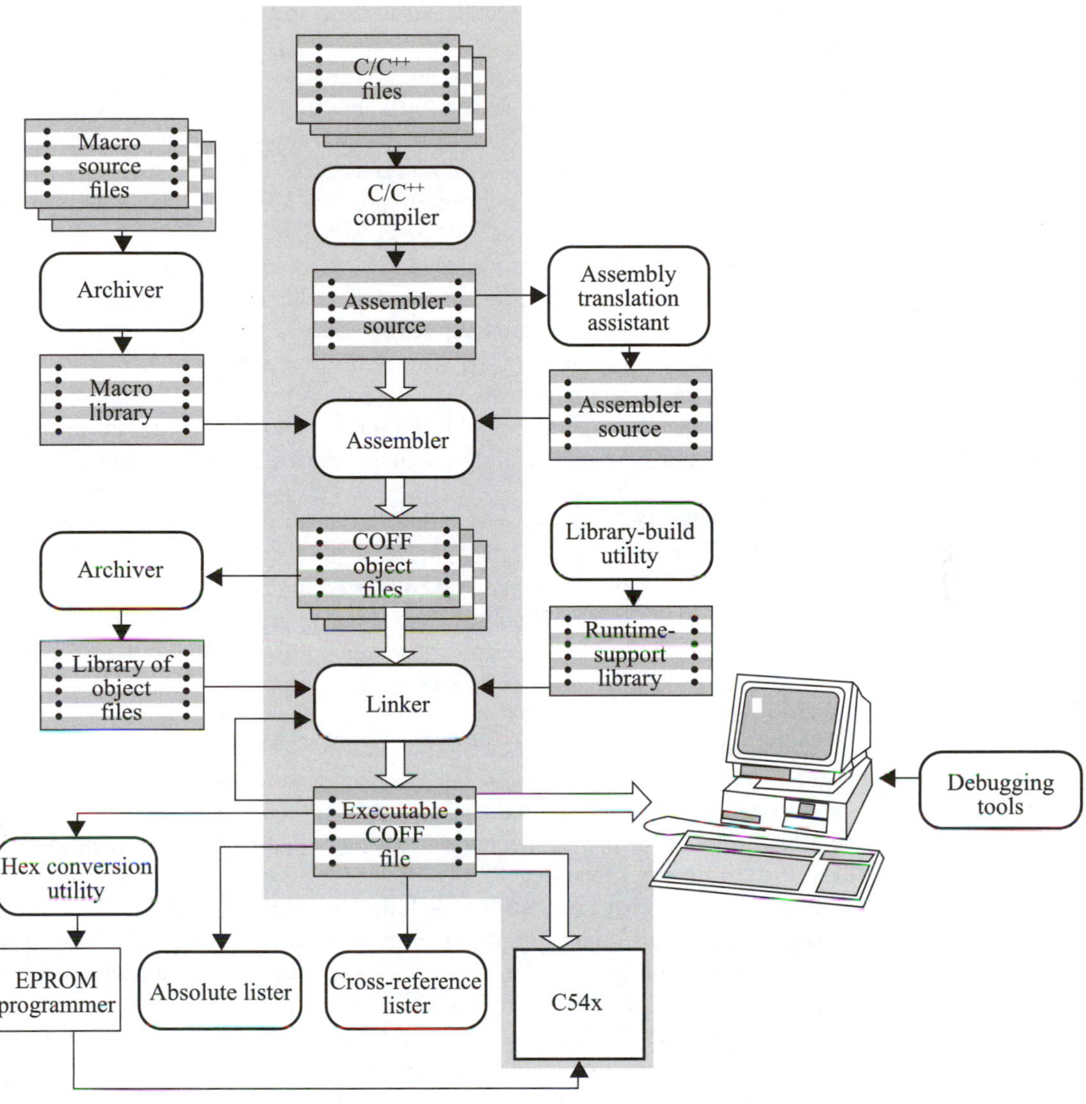


**Figure 6.3** Software development flow chart

(Courtesy of Texas Instruments Inc.)

the development process consist of source files, COFF object files, and a COFF executable file.

The **C compiler** translates a $C^{++}$ source file into a C54xx assembly language source file. To create a source file, a tool called the **editor** is needed. An editor may be any ASCII editor available on the PC, such as EDIT in DOS. The **assembler** translates assembler source files into COFF object files. Source files can contain instructions, assembler directives, and macros. The assembler directives are employed to control the assembly process, such as the source listing format, data alignment, and section contents. The **linker** combines the relocatable COFF object files and library modules into a single executable COFF object file. It creates the executable module by assigning symbols to memory locations and resolving symbol references.

The **archiver utility** collects a group of files into a single archive file. Macros can be combined to form a macro library. During assembly, the assembler searches the library and uses the needed macros. Archiver can also be used to combine a group of object files into an object library. The linker uses the object library to resolve external references during the linking process. The compiler package may include the **library-build utility**, which can be used to build runtime-support libraries. The **assembly translation assistant** utility can be used to convert an assembly language source file containing mnemonic instructions to an assembly language source file containing algebraic instructions.

TMS320C54xx DSP accepts executable COFF files as input. A **hex conversion utility** is used to convert a COFF object file into TI-tagged, Intel, Motorola, or Tektronix object formats. The converted file can be downloaded to an EPROM programmer. The **absolute lister** accepts linked object files as input and creates an absolute file as output. The created file has absolute rather than relative addresses. The **cross-reference lister** uses object files to produce a cross-reference.

The **debugging tool** provides a mechanism to download an executable program to the board and run it to verify its operation. More important, it is used to debug the program by using controlled execution and the monitoring support provided in the debugging environment. The DSK debugging tool is described in the next section.

In order to support application development using DSK, the DSK software provides host utilities and board drivers and libraries. The host utilities run on the host PC and provide functions to control the DSK board, whereas the target libraries are for the DSK board and provide functions to control the peripherals on the board. C54xx DSK host utilities provide the user with a way to use the board without having to write an application from scratch. These utilities support C54xx DSK board control, such as DSP reset, DSP application loading and execution, device configuration, status display, board confidence testing, and flash memory programming. The host utilities can be used to load and run any application or to configure and monitor the C54xx DSK device without writing the application to do it. Stand-alone embedded executable functions can be programmed into flash memory. For more information on these utilities, the reader should consult reference [2] given at the end of this chapter.

The board drivers for the C54xx DSK provide the low-level software interface. These drivers are not intended to be directly accessible for the user-mode applications. A Win32 DLL that provides a consistent API across all supported

Window platforms hides the details of accessing these drivers. The purpose of the board driver functions is to allow the user-mode DLL to access and control C54xx DSK. These functions provide a basic interface that gives access to the board in all supported Windows environments. The Win32 DLL provides intelligent processing and control functions that call kernel-mode board driver functions to access board resources and the PCI configuration data. The board libraries provide functions for board initialization as well as initialization and control of on-board peripherals.

## 6.5 The Assembler and the Assembly Source File

A program written in an assembly language is called an *assembly source* program. An example of such a program is shown in Figure 6.4. This program is essentially the same as the one in the last chapter, except that a few new

```
*******************************************************************
*
* This program computes the signed sum of 16 data memory
* locations starting at Number. The result is to be placed in A.
*
*******************************************************************

    .mmregs
    .global _c_int00

    .data

Number:
    .int 5, 14, -7, 22, -25, 4, 2, 0, 6, 33, 4, 11, 12, -12, 8, 16

    .text

_c_int00:
    stm   #10h, AR2       ; Init counter AR2 = 16
    stm   #Number, AR1    ; Init pointer AR1 to first number
    ld    #0h, A          ; Initialize sum A = 0
    ssbx  SXM             ; Select sign extension mode

START:
    add   *AR1+, A        ; Add the next data value
    banz  START, *AR2-    ; Repeat if not done
    nop                   ; No operation, just for debugging

    .end
```

**Figure 6.4** An assembly source program for TMS320C54xx

directives statements have been added. The directive statements are for the host program that will be used to convert the source program to the machine program for execution on the processor. The program that does this conversion is called an *assembler*. The statements added in the program in Figure 6.4 are for the assembler that comes with DSK. Here, we will briefly discuss these statements. However, the reader is advised to consult reference [3] for complete details.

The instructions in the program are the processor instructions that we discussed in the last chapter. The labels such as START in Figure 6.4 refer to the memory addresses for the instructions. The statements starting with a star (*) are the comments to facilitate program understanding and do not produce any converted code. The statements that start with a dot (.) are called *directives*. A directive is not a processor instruction; it is an instruction to the assembler program to control the assembly process. For instance, the .int directive in the program of Figure 6.4 specifies to the assembler to allocate word-size memory locations and initialize them with the data specified after the directive. The memory allocation starts at the address to which the label "Number" refers.

The .mmregs directive defines memory-mapped registers of the processor. For instance, AR0 register refers to a specific memory location after assembling and this reference or definition is provided by the .mmregs directive. The .global directive declares the specified label visible to other program modules. The .data and .text are called *section directives*. These are provided to define data and code sections of a program. For instance, starting at .data till .text, the allocation is to the data section. Starting at .text, the allocation is to the code section. Finally, the .end directive specifies the end of the source file.

There are many other directives that facilitate the assembly process of converting instructions and allocation of code and data. The reader is advised to look these up in reference [3].

## 6.6 The Linker and Memory Allocation

The linker is another program that is also a part of the development system. It is needed to allocate the user program and its sections to actual physical memory on the target, such as the DSK board. It provides a way by which we can use the resources of the hardware in view of the program that we intend to test. Another important use of the linker is to allow a programmer to write an application in modules. The linker combines these modules into a single machine program for the hardware execution on the DSP device.

Typically, a command file is used to define the connection between the hardware resources and the program sections. An example of a command file for the program in Figure 6.4 is shown in Figure 6.5. Memory is defined as consisting of two pages, PAGE 0 and PAGE 1. PAGE 0 refers to the program

```
/*
 *  ======== example6p1.cmd ========
 *
 */

MEMORY
{
    PAGE 0: IPROG:  origin = 0x1000,  len = 0x3000
    PAGE 1: IDATA:  origin = 0x400,   len = 0x100
}

SECTIONS
{
    .text:  { } > IPROG PAGE 0
    .data:  { } > IDATA PAGE 1
}
```

**Figure 6.5** A command file for the program of Figure 6.4

memory; it starts at 0x1000 and has a length of 0x3000. PAGE 1 refers to the data memory starting at 0x400; it has a length of 0x100. These are valid memory locations in the DSK board. The sections of the program are assigned to exist in these two types of pages. For instance, .text is the code section and it is assigned to PAGE 0 or the program memory. Similarly, .data section is defined to be in the data memory or PAGE 1. For more on linker and memory allocation, the reader is advised to consult reference [3].

## 6.7 The C/C$^{++}$ Compiler

The DSK comes with a C/C$^{++}$ compiler that can be used to develop DSP applications using the high-level languages C and C$^{++}$. The compiler generates an assembly file that can be further converted with the assembler program to generate an object file for the linker. For information on developing C or C$^{++}$ programs, the reader is advised to consult an appropriate reference [4].

## 6.8 The Code Composer Studio (CCS)

The DSK comes complete with the DSK-specific Code Composer Studio (CCS). CCS provides an integrated development environment (IDE) for project management, editing, compiling, debugging, and visualization. Both C/C$^{++}$ and assembly language codes can be developed and debugged.

To use CCS, we need to know how to build applications and how to debug or test them using a target such as the DSK board. These two aspects are considered in the following subsections.

### 6.8.1 Building a Project

A new project is built by choosing "New" in the Project menu. The Project Creation window appears, allowing one to specify the project name, location, and type. The project type executable generates an .out extension executable file. Ending the project creation takes you to the Project View window, where files to be used in the project can be added. These files are the source files (both assembly and $C^{++}$), library files, and the command file. Select "Add Files" under the Project menu and specify the file type and its location to add it to the project. The include files are not added; these are automatically added by the CCS after scanning the source files.

A project configuration is selected from the Project toolbar. Two configurations, Debug or Release, are available for different phases of program development. The output generated after the project is built is placed in the configuration-specific subdirectory in the directory for the project.

Figure 6.6 shows a sample project file generated by the CCS in response to selections and the files used. This file contains all the information about the project, such as project settings, source files, compiler settings, and linker settings. The details of the settings for the compiler and linker are given in reference [2].

The project is built by choosing "Rebuild All" in the Project toolbar. The executable file is placed in the appropriate directory, such as the Debug directory. The executable program can be loaded to the board using "Load Program" under the File menu. The program can be executed or debugged using the Debug option in the File menu. The debugging can be done using various controls and options to run the program and view its results. Some of these options are discussed in the next section.

### 6.8.2 The Debug Options

The CCS debugger provides a powerful debugging capability by permitting the execution of a program in many different ways and viewing the results in many different formats. The basic debug capabilities of CCS consist of provisions to download a program to the DSK board, run the program, single-step through instructions, modify registers and memory locations, view registers and locations, and apply reset to the processor. In addition to basic capabilities, there are a number of advanced debugging features provided in CCS. Some of these features are as follows:

```
; Code Composer Project File, Version 2.0 (do not modify or remove this
line)

[Project Settings]
ProjectDir="C:\ti\myprojects\example6p1\"
ProjectType=Executable
CPUFamily=TMS320C54XX
Tool="Compiler"
Tool="DspBiosBuilder"
Tool="Linker"
Config="Debug"
Config="Release"

[Source Files]
Source="..\..\..\WINDOWS\Desktop\DSPBookPgm\ch6pgms\example6p1.asm"
Source="..\..\..\WINDOWS\Desktop\DSPBookPgm\ch6pgms\example6p1.cmd"

["Compiler" Settings: "Debug"]
Options=-g -q -fr"C:\ti\myprojects\example6p1\Debug" -d"_DEBUG"

["Compiler" Settings: "Release"]
Options=-q -o2 -fr"C:\ti\myprojects\example6p1\Release"

["DspBiosBuilder" Settings: "Debug"]
Options=-v54

["DspBiosBuilder" Settings: "Release"]
Options=-v54

["Linker" Settings: "Debug"]
Options=-q -c -o".\Debug\example6p1.out" -x

["Linker" Settings: "Release"]
Options=-q -c -o".\Release\example6p1.out" -x
```

**Figure 6.6** A sample project file created by the CCS

**Breakpoints:** A breakpoint can be set on an instruction. Execution of the program stops at the breakpoint, giving an opportunity to view the results produced by the part of the program that has been executed.

**Watch Window:** This feature allows one to monitor program variables as the execution takes place.

**Probe Points:** By adding a Probe Point on a line of the program, data can be transferred either from a file on the host to the DSK memory or from the DSK memory to a file on the host. The program execution resumes after transferring the data.

**Graphing:** CCS provides a number of ways to graph the data processed by the program. This capability is particularly useful in viewing a signal in the frequency and time domains.

**Profiling:** A profiler can be used to determine the number of cycles a particular function or a program takes to execute or how many times the function is called. This capability can be used to optimize the program performance.

**Real-Time Analysis:** The CCS provides the capability to monitor and analyze a real-time program without interfering with its execution. This capability is provided by way of a DSP/BIOS kernel and RTDX (real-time data exchange) technology. The kernel, which is loaded to the board, uses API functions to implement run-time services. These functions can be linked into an application and allow a user to implement performance monitoring and program tracing. The RTDX provides a link to obtain and monitor target data in real time. This capability allows the user to transfer data between the host and the target without interfering with the target application. RTDX has two components, one of which runs on the target to provide a link to the target data. On the host platform, RTDX runs in conjunction with CCS to provide data visualization and analysis. For more information on this capability, the reader is advised to run the DSP/BIOS and RTDX tutorials available in the CCS environment by invoking the Help function.

## 6.9 DSP Software Development Example

In this section, we will go through the various steps of building and debugging an application for the DSK using the CCS. These steps will be illustrated using the source program of Figure 6.4 and the command file of Figure 6.5. The process illustrated here does not demonstrate the complete power of the tools; it is a simplified version of the tools and illustrates the basic process of application development.

1. We start by creating a new project, as shown in Figure 6.7, by selecting "New" under the Project toolbar. The project name, example6p1, can be entered along with its location. The project type chosen will be Executable (.out). The target is TMS320C54xx.
2. The project window after creating and selecting the project is shown in Figure 6.8.
3. The project files are added to the project by selecting "Add Files to Project" under the Project toolbar. As shown in Figure 6.9, we add the source file example6p1.asm. The process is repeated for the command

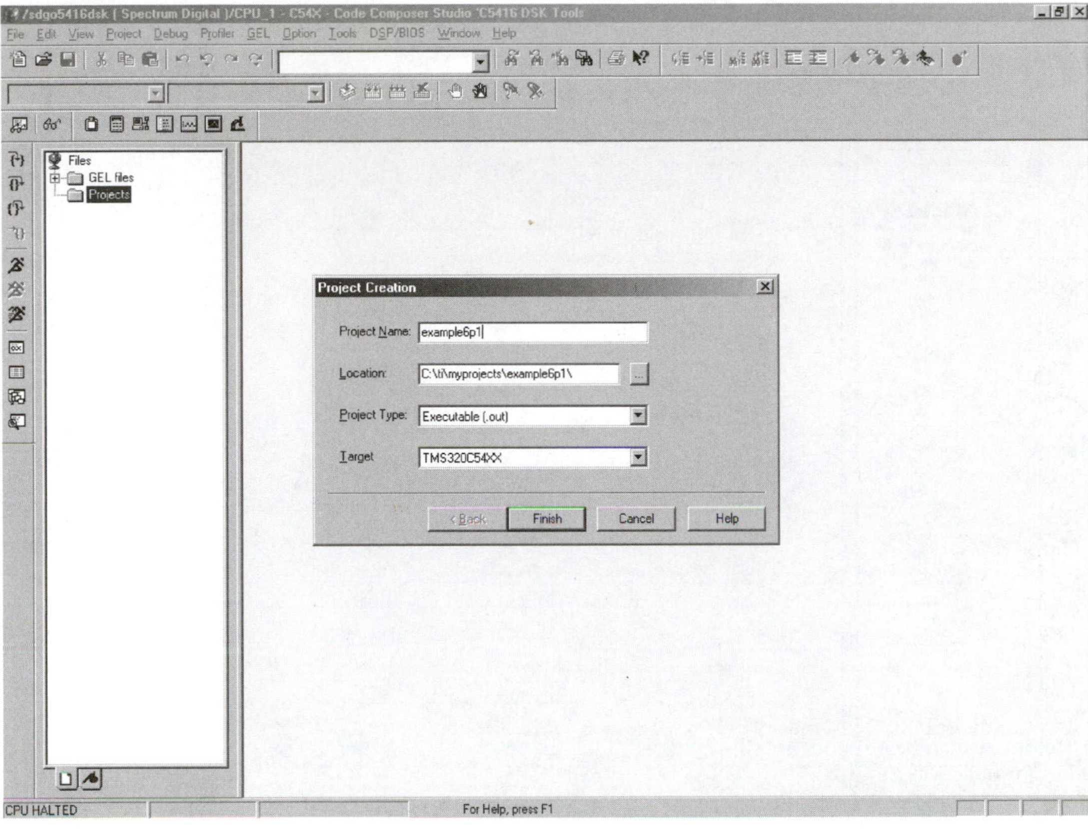


**Figure 6.7** Creating a new project in CCS IDE

file example6p1.cmd. While selecting a file, the file location and its type must be selected to see the file in the window before it can be added.

4. Figure 6.10 shows the Project window after adding and selecting the source and the command files.
5. Figure 6.11(a) shows how project build options can be selected for the assembling, compiling, and linking. The build options are selected from the Project toolbar. Here we can specify options for the assembler, compiler, and the linker. Figure 6.11(b) shows where the place for the object files is specified.
6. The project is built by selecting the "Build" option under the Project toolbar. Figure 6.12 shows the building of the project. The lower window shows any error if it occurs during the build process.
7. The built program can be downloaded to the DSK board by selecting "Load Program" in the File menu. This is shown in Figure 6.13.

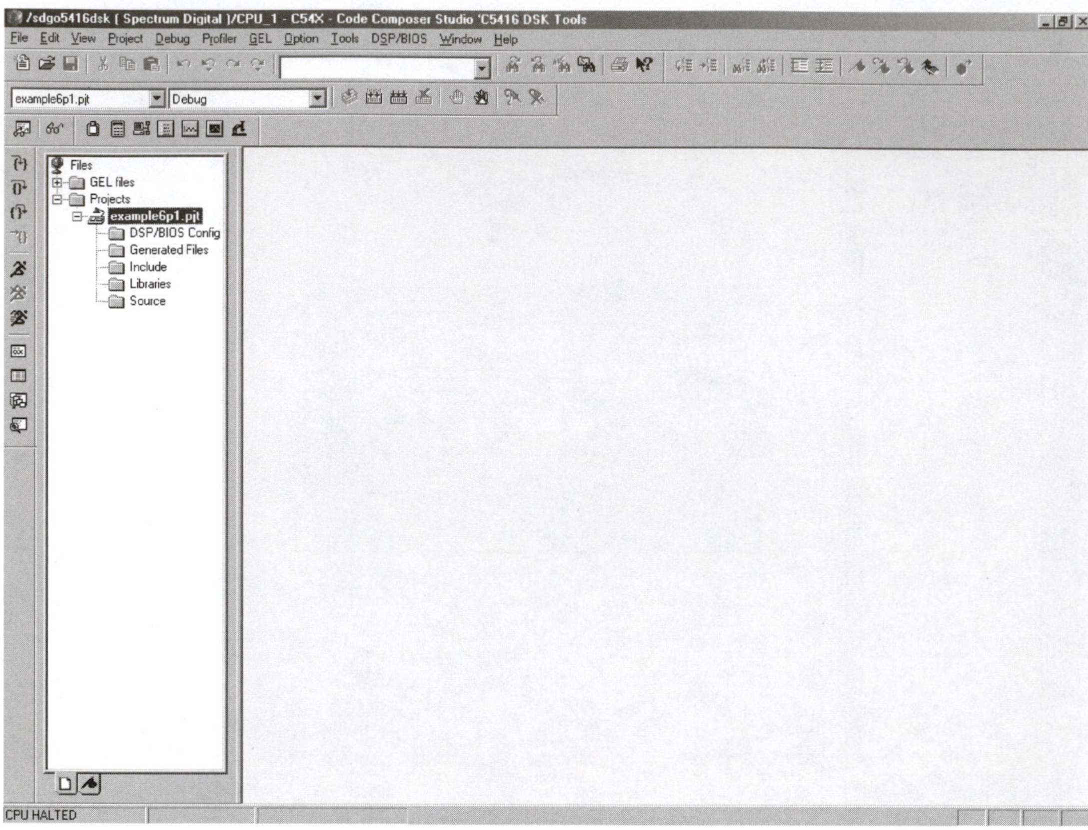


**Figure 6.8** The project window for the project being created

8. After downloading the program, it can be debugged, or simply run, by choosing the debug features. Restarting the program makes it begin from the first instruction, as shown in Figure 6.14. The right arrow shows the start point. In order to execute it to the end, we may set a breakpoint at the last "nop" instruction. The breakpoint is selected from the Debug toolbar. The filled circle on the *nop* instruction shows the breakpoint location. Since the program uses registers, we may view these as shown in Figure 6.14. The register window is selected from the View menu. Notice that A = 0, AR1 = 0, and AR2 = 0. These are the registers used in the program. The program adds the numbers starting at the location Number. Note that the location Number is at address 400h, as the data section is defined in the command file to start at this address.
9. Executing the program with a run command generates the sum of 16 numbers in register A. The result is shown in Figure 6.15. It is easy

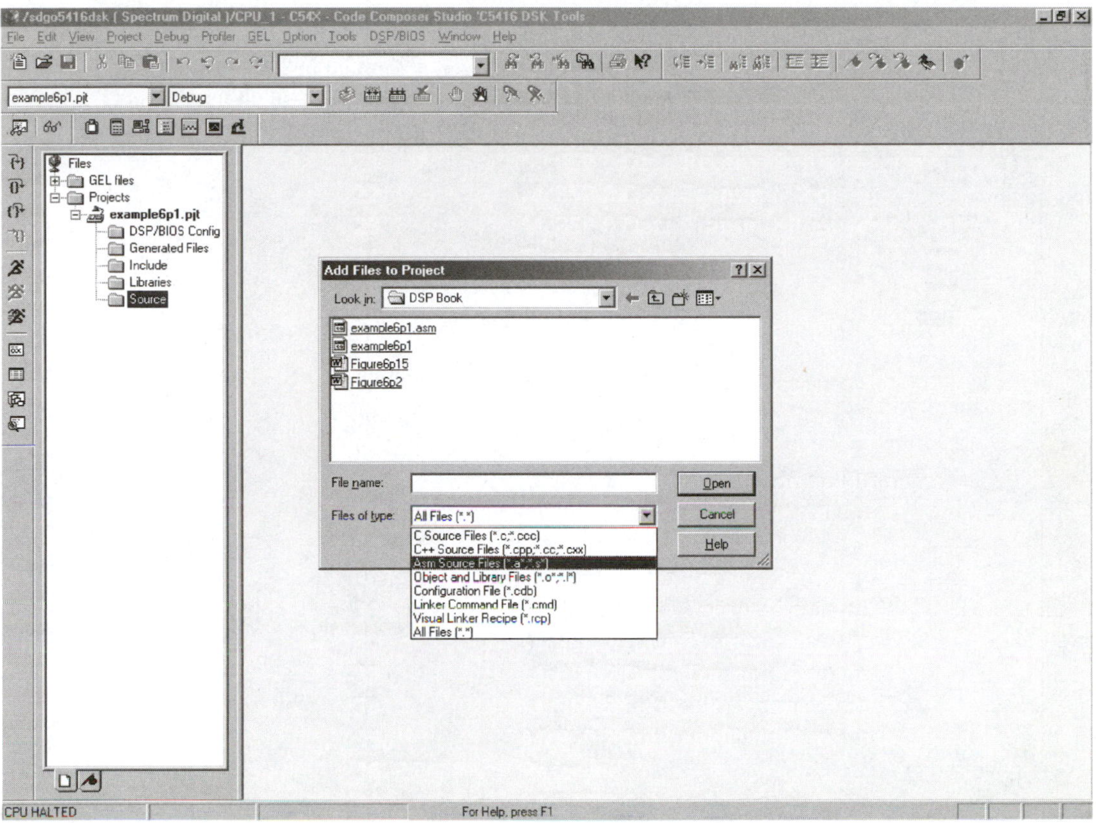


**Figure 6.9** Adding files to the project

to verify that the AR1 and AR2 registers also contain the appropriate numbers after completing the program execution. To debug the program we also can run the program using single-step execution, in which case one instruction at a time is executed. Simultaneously we can view the contents of registers as the instructions are executed.

## 6.10 **Summary**

In this chapter, we introduced an important and inexpensive tool called the DSP System Design Kit (DSK) for the C54xx DSP devices and its associated development software called Code Composer Studio Integrated Development Environment (CCS IDE). The package can be used to develop DSP applications. These tools were illustrated using a simple example.

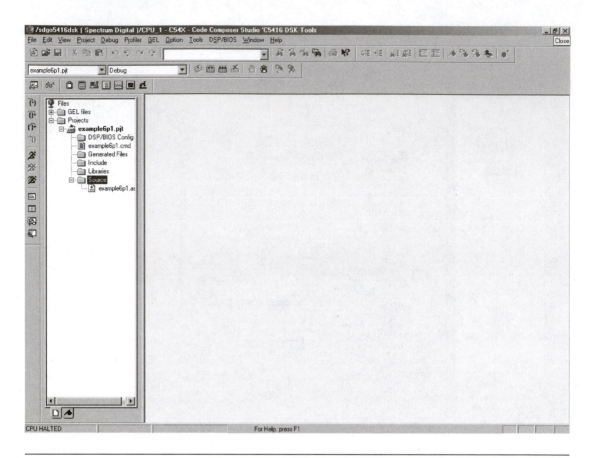


**Figure 6.10** Project window after adding source and command files

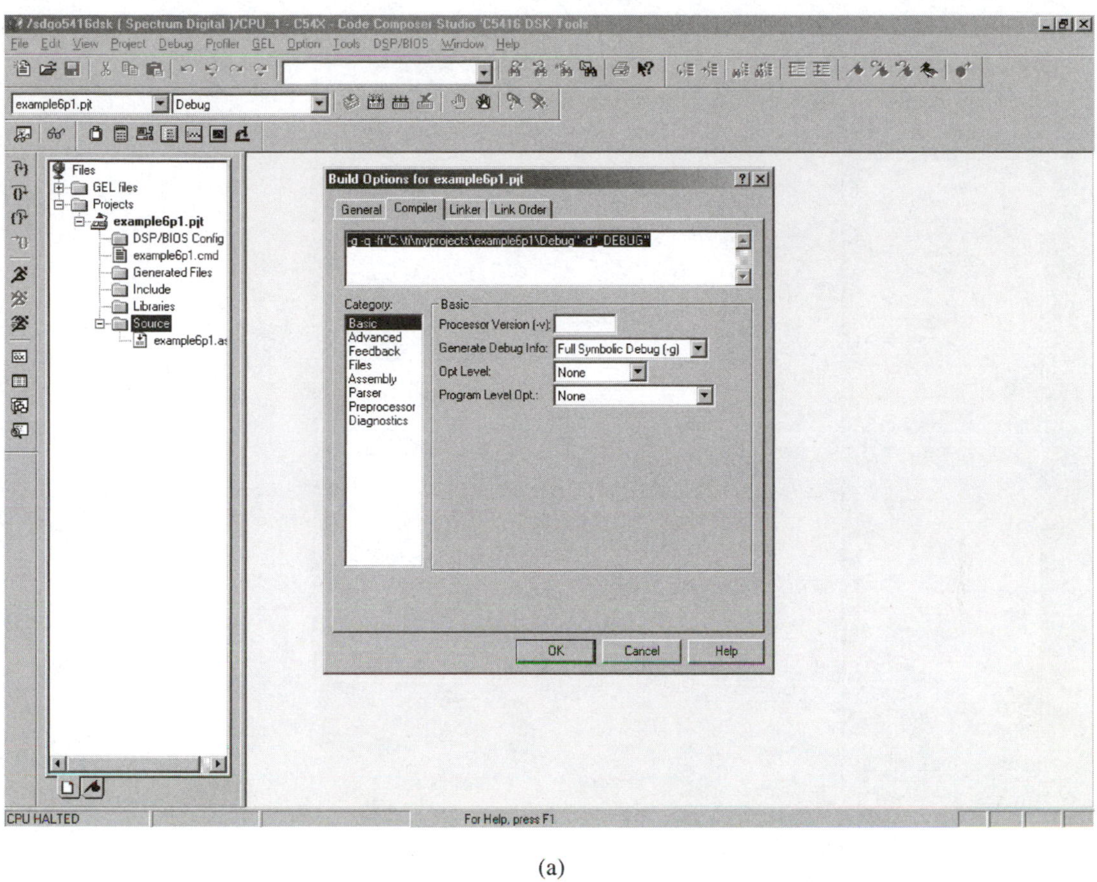


(a)

**Figure 6.11(a)** Selecting project build options

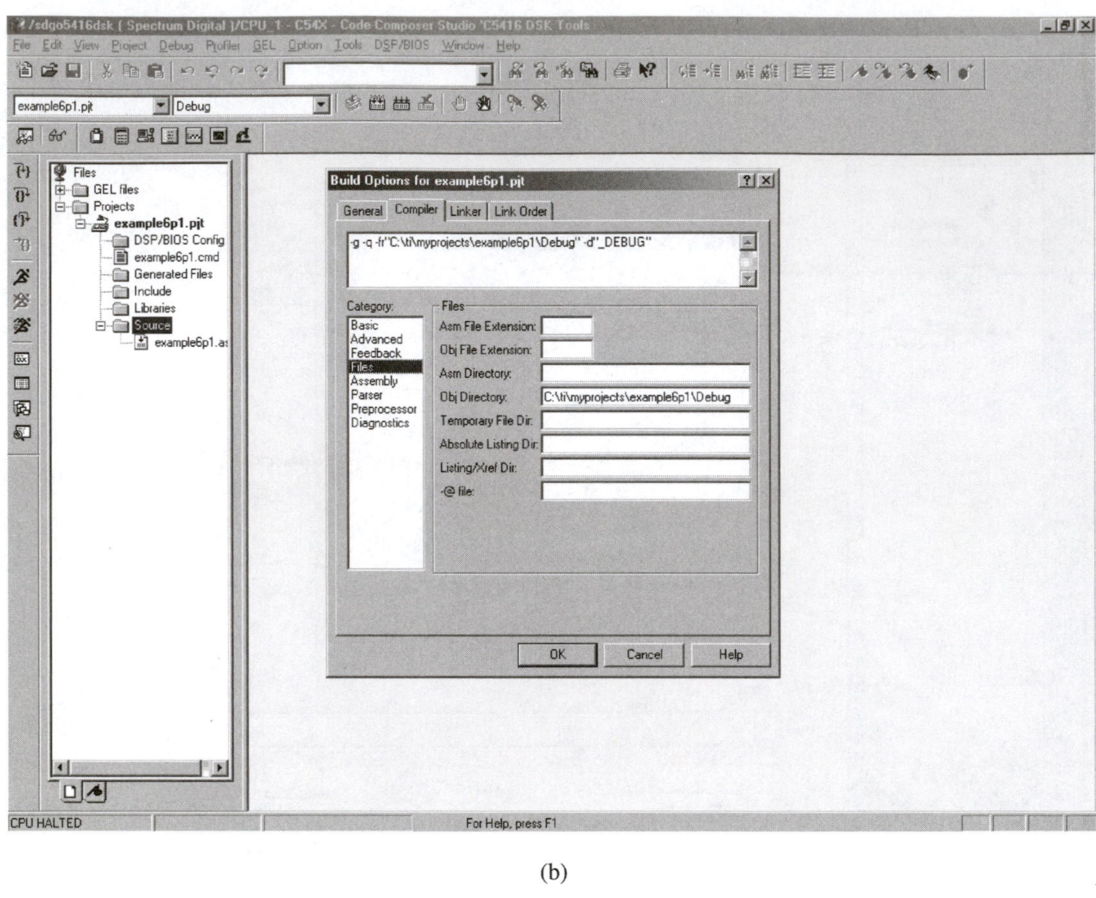


(b)

**Figure 6.11(b)** Selecting project build options

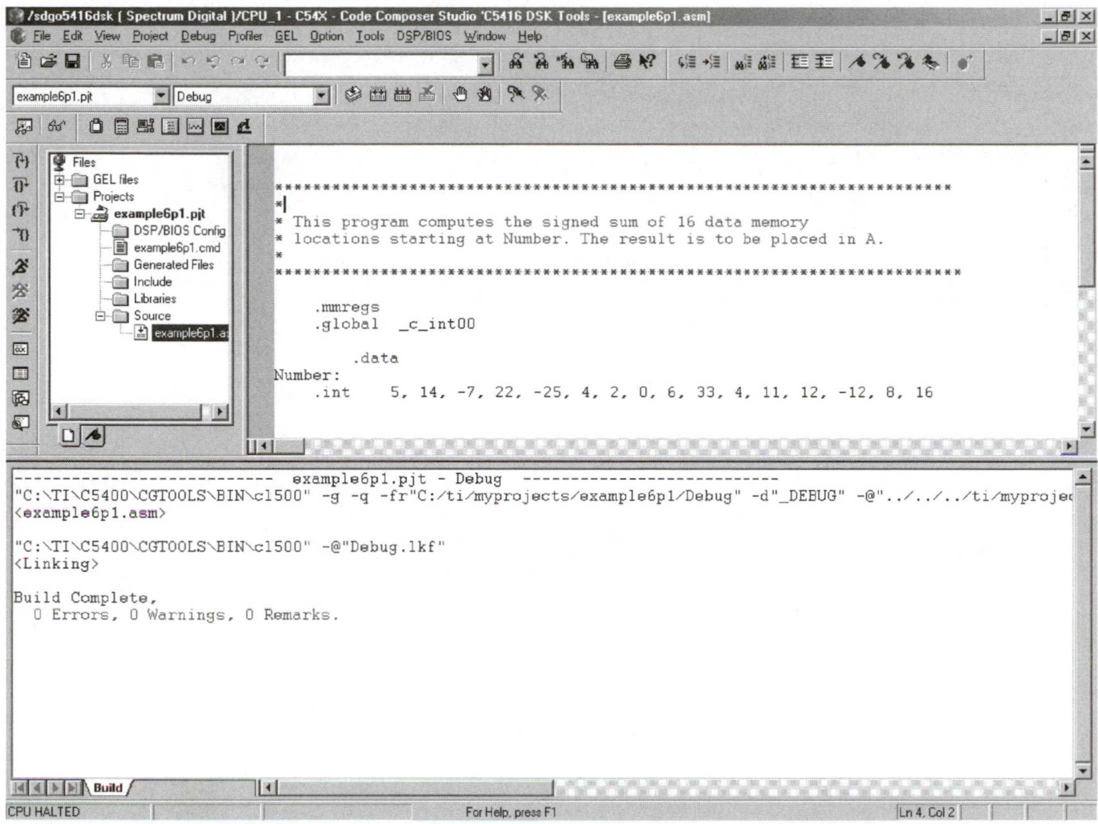


**Figure 6.12** Building the project

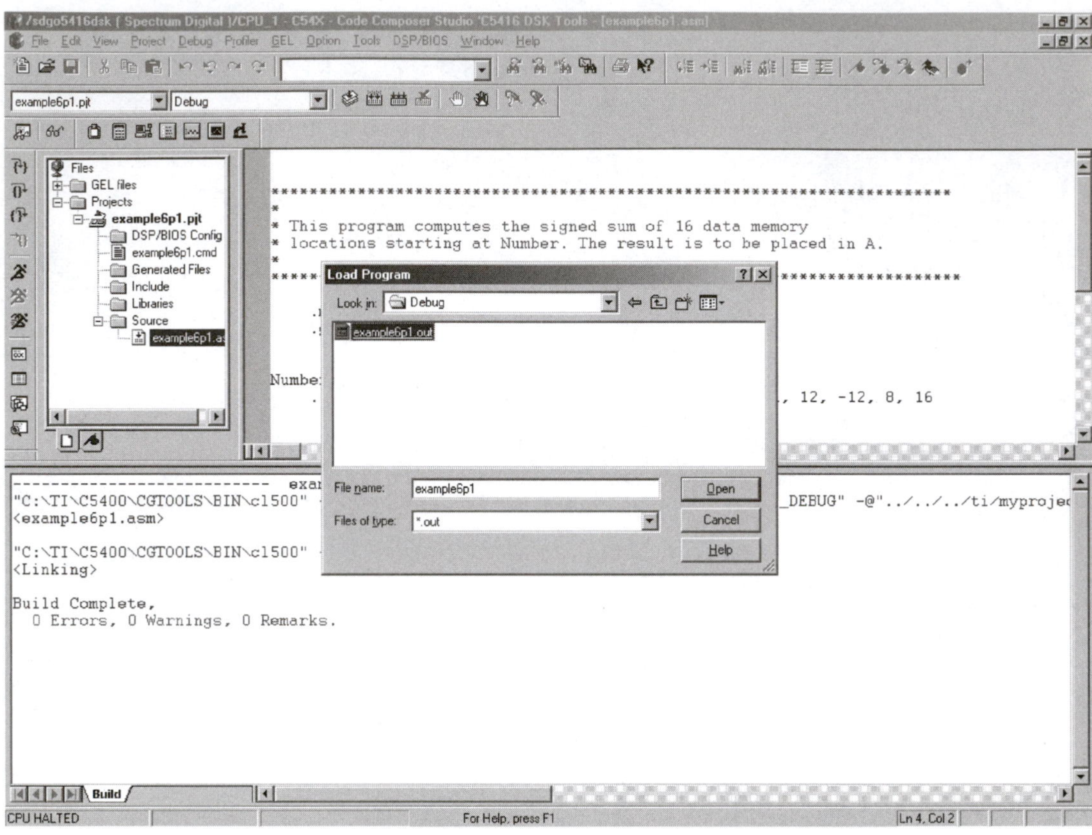


**Figure 6.13** Downloading the project to the DSK

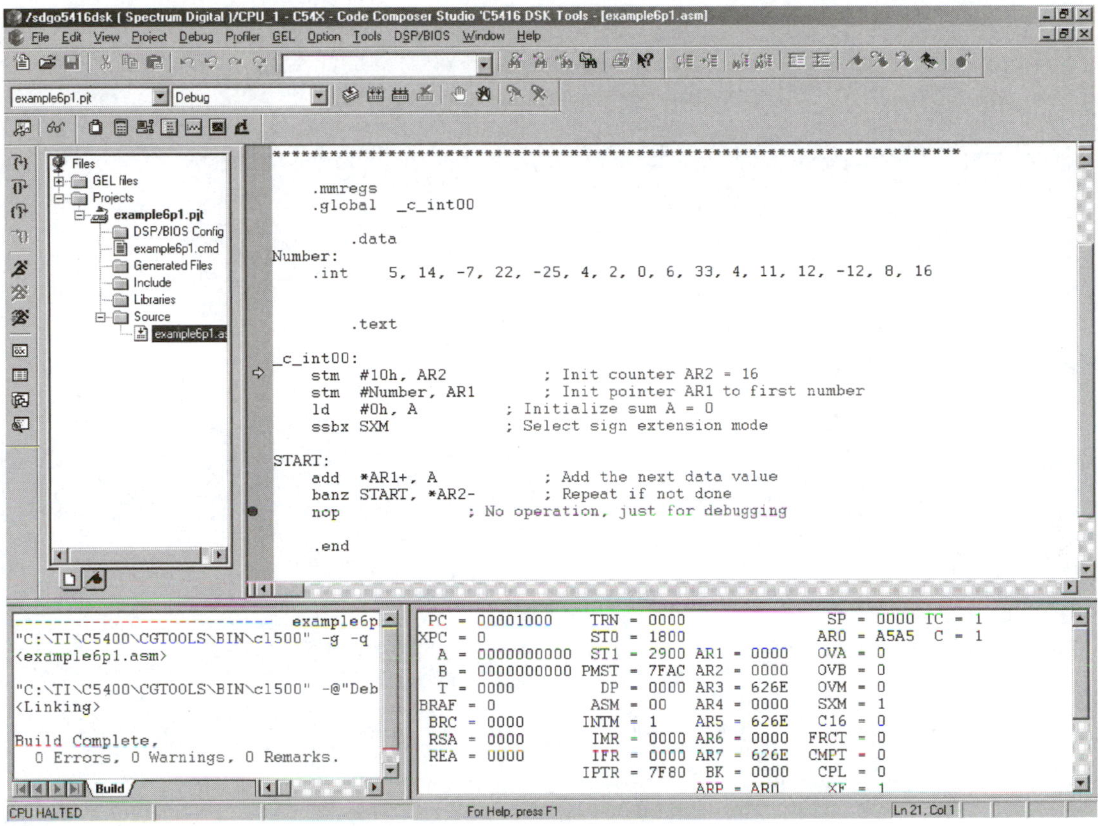


**Figure 6.14** Debugging the project using a breakpoint

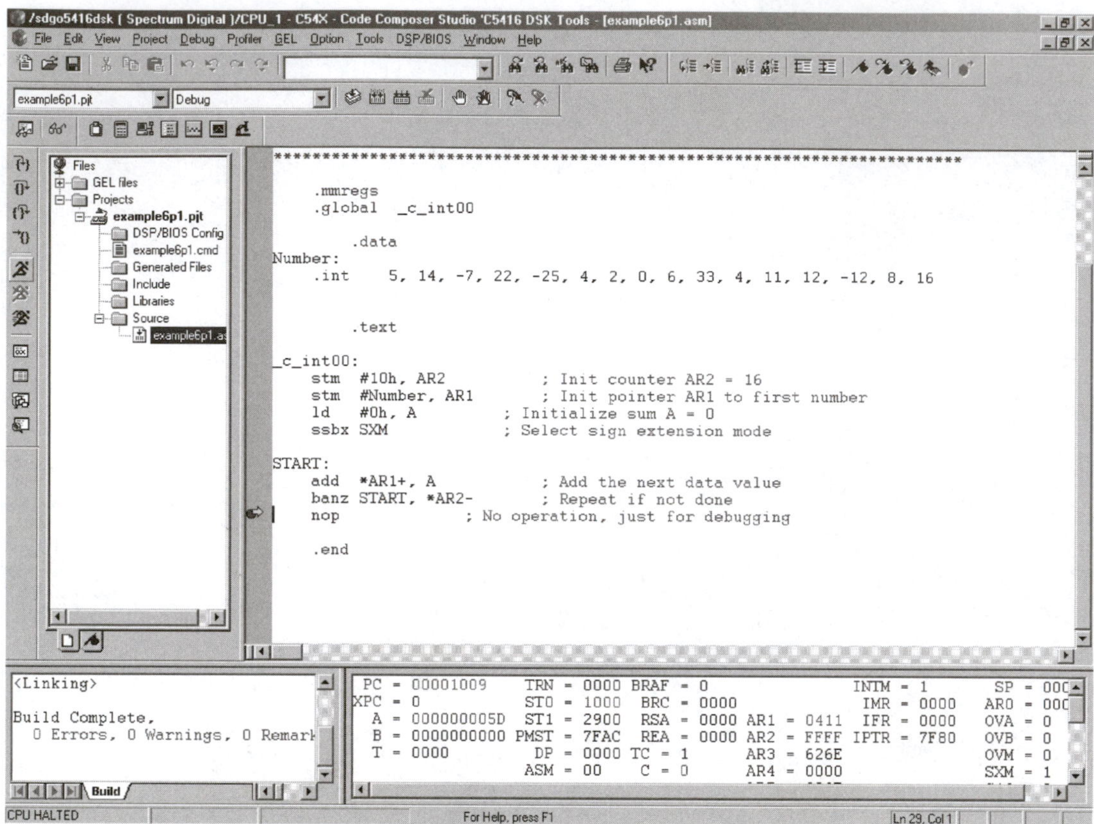

**Figure 6.15** Result of executing the project using a breakpoint

# References


1. TMS320VC5416 DSP Starter Kit (DSK) *Technical Reference*, Spectrum Digital Inc., 50 6005-0001 Rev.A, 2002. (http://www.spectrumdigital.com).
2. *Code Composer Studio Getting Started Guide*, Texas Instruments Literature Number SPRU509.
3. TMS320C54xx Assembly Language Tools *User's Guide*, Texas Instruments Literature Number SPRU102.
4. TMS320C54xx Optimizing C/C$^{++}$ Compiler *User's Guide*, Texas Instruments Literature Number SPRU103.

## Laboratory Assignments

**L6.1** Build a project to verify Example 5.11 using the following data:

$$h(0) = 5, h(1) = 31, h(2) = 13, x(n) = 1, x(n-1) = 5, x(n-2) = -3.$$

**L6.2** Build a project to verify Example 5.12 using the following data:

$$h(0) = 5, h(1) = 31, h(2) = 13, x(n) = 1, x(n-1) = 5, x(n-2) = -3.$$

**L6.3** Build a project to verify Example 5.13 using the following data:

$$h(0) = 5, h(1) = 31, h(2) = 13, x(n) = 1, x(n-1) = 5, x(n-2) = -3.$$

**L6.4** Write a program that computes the square of the distance between the two points with the coordinates $(x_1, y_1)$ and $(x_2, y_2)$. Build a project and verify the program using a set of points.

**L6.5** Use the program in L6.4 to write another program that computes the distance between the points. Build a project and verify the program operation using a set of points. You may use the following algorithm to compute the square root:

Square root of $N$ = Number of sequential odd integers starting at 1 that add to (or whose total approaches) $N$. For instance, $25 = 1 + 3 + 5 + 7 + 9$, or it is the sum of five odd integers and 5 is the squareroot of 25.

# Chapter 7

## Implementations of Basic DSP Algorithms

### 7.1 Introduction

In this chapter, we deal with implementations of DSP algorithms. Here we write programs to implement the core algorithms only. However, these programs can be combined with input/output routines to create applications that work with a specific hardware. Specifically, in this chapter, the following C54xx implementations using assembly language [5, 6, 7] are covered:

- Q-notation
- FIR filters
- IIR filters
- Interpolation filters
- Decimation filters
- PID controller
- Adaptive filters
- 2-D signal processing

### 7.2 The Q-notation

DSP algorithm implementations deal with signals and coefficients. To use a fixed-point DSP device efficiently, one must consider representing filter coefficients and signal samples using fixed-point 2's complement representation. Typically, filter coefficients are fractional numbers. To represent such numbers, the Q-notation has been developed. The Q-notation specifies the number of fractional bits. For instance, Q7 for a 16-bit number means that the most significant 9 bits represent the whole part and the sign of the number and the least significant 7 bits are the fractional part of the number. In other words, the assumed decimal point lies between bit 6 and bit 7.

A commonly used notation for DSP implementations is Q15. In the Q15 representation, the least significant 15 bits represent the fractional part of a number. In a processor where 16 bits are used to represent numbers, the Q15 notation uses the MSB to represent the sign of the number and the rest of the bits represent the value of the number. In general, the value of a 16-bit Q15 number $N$, represented as $b_{15}b_{14}b_{13}\ldots b_1b_0$, can be determined from the equation

$$N = -b_{15} + b_{14}2^{-1} + b_{13}2^{-2} + \cdots + b_1 2^{-14} + b_0 2^{-15} \tag{7.1}$$

Thus, the numbers that can be represented by the Q15 notation, using 16 bits, range from $-1$ to $1 - 2^{-15}$. This range is generally adequate to represent filter coefficients in DSP algorithms.

▷ **Example 7.1** What values are represented by the 16-bit fixed point number $N = 4000\text{h}$ in the Q15 and the Q7 notations?

Solution 4000h = 0100 0000 0000 0000b. In the Q15 notation, it represents 0.100 0000 0000 0000b with the assumed decimal point. Use of Eq. 7.1, to compute its value, yields

$$N = +0.5$$

Similarly, the same number in the Q7 notation represents 0100 0000 0.000 0000b, which, using Eq. 7.1, computes to

$$N = +128.0$$

---

Multiplication of numbers represented using the Q-notation is important for DSP implementations. Figure 7.1(a) shows typical cases encountered in such implementations. For instance, if two 16-bit Q15 numbers are multiplied as integers, the 32-bit result is a number in Q30 representation. In other words, the two MSBs are the sign bits. If this result is to be used as it is, it is important to know where the position of the decimal point is. If the 32-bit result is left shifted one bit position and the 16 MSBs are extracted, we have the final result in Q15 representation. This procedure of dealing with the Q15 numbers can be employed in DSP implementations. Figure 7.1(b) is a TMS320C54xx program that illustrates how to multiply two Q15 numbers and produce a Q15 result. This program also illustrates how to minimize the error due to truncation of the 16 LSBs to obtain a Q15 result. This can be done by rounding off the result before truncation.

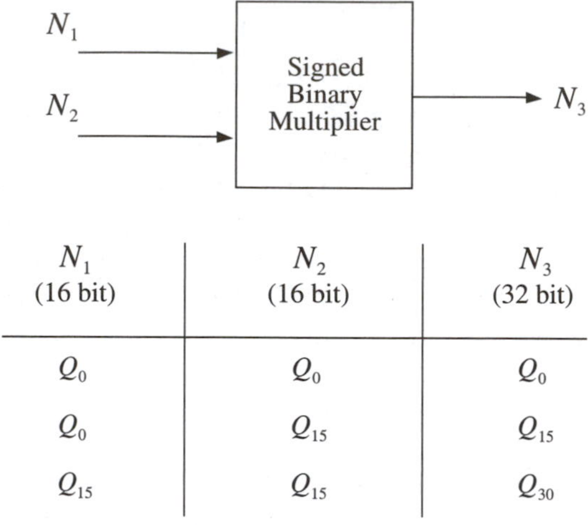


| $N_1$ (16 bit) | $N_2$ (16 bit) | $N_3$ (32 bit) |
|---|---|---|
| $Q_0$ | $Q_0$ | $Q_0$ |
| $Q_0$ | $Q_{15}$ | $Q_{15}$ |
| $Q_{15}$ | $Q_{15}$ | $Q_{30}$ |

**Figure 7.1(a)** Multiplication of numbers represented using Q-notation

## 7.3 FIR Filters

A finite impulse response (FIR) filter of order $N$ can be described by the difference equation

$$y(n) = \sum_{m=0}^{m=N-1} h(m)x(n-m) \tag{7.2}$$

or in expanded form we have

$$y(n) = h(0)x(n) + h(1)x(n-1) + \cdots + h(N-1)x(n-(N-1)) \tag{7.3}$$

For FIR filter implementation, we use Eq. 7.3 to illustrate how the DSP code can be written. Figure 7.2 shows a block diagram for the implementation. To compute $y(n)$, we start with the computation and accumulation of the last product, followed by the one before the last, and so on. The implementation requires signal delay for each sample to compute the next output. The next output, $y(n+1)$, is given as

$$y(n+1) = h(0)x(n+1) + h(1)x(n) + \cdots + h(N-1)x(n-(N-2)) \tag{7.4}$$

Figure 7.3 shows the memory organization for the implementation of the filter. The filter coefficients and the signals samples are stored in two circular buffers each of a size equal to the filter. AR2 is used to point to the samples

```
;---------------------------------------------------------------
; Program Name:  ex7p1Qxx.asm
;
; Description:   This is an example to show how to multiply numbers
;                represented using Q-notation. It implements the
;                following:
;
;                N1xN2 = N1 * N2
;
;                where
;                N1 and N2 are 16-bit numbers in Q15 notation
;                N1xN2 is the 16-bit result in Q15 notation
;
; Author:        Avtar Singh, SJSU
;---------------------------------------------------------------
; Definitions
                 .mmregs                ; memory-mapped registers

                 .data                  ; sequential locations
N1:              .word    4000h         ; N1 = 0.5 (Q15 number)
N2:              .word    2000h         ; N2 = 0.25 (Q15 number)
N1xN2:           .space   10h           ; space for N1 x N2

                 .text
                 .ref     _c_int00

                 .sect    ".vectors"
RESET:           b        _c_int00      ; Reset vector
                 nop
                 nop

_c_int00:
                 stm      #N1, AR2      ; AR2 points to N1
                 ld       *AR2+, T      ; T reg = N1
                 mpy      *AR2+, A      ; A = N1 * N2 in Q30 notation
                 add      #1, 14, A     ; round the result
                 sth      A, 1, *AR2    ; save N1 * N2 as Q15 number
                 nop
                 nop

                 .end
```

**Figure 7.1(b)** TMS320C54xx program to multiply two Q15 numbers

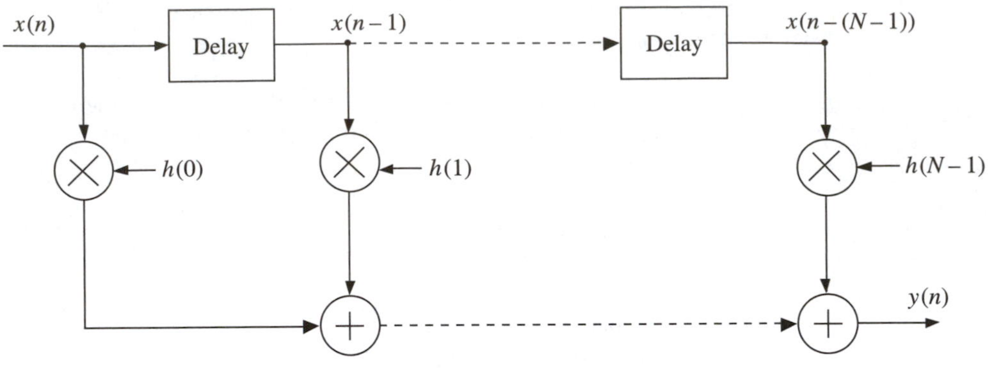


**Figure 7.2** A FIR filter implementation block diagram

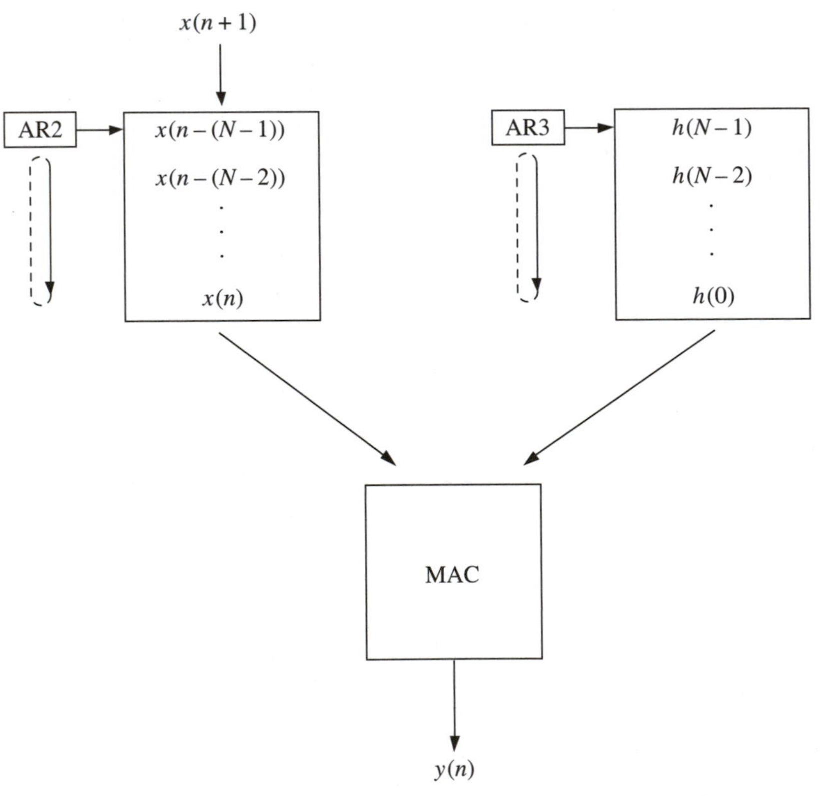


**Figure 7.3** Organization of signal samples and filter coefficients in circular buffers for a FIR filter implementation

and AR3 to the coefficients. In order to start with the last product, the pointer register AR2 must be initialized to access the signal sample $x(n-(N-1))$, and the pointer register AR3 to access the filter coefficient $h(N-1)$. As each product is computed and added to the previous result, the pointers advance circularly. At the end of the computation, the signal sample pointer is at the oldest sample, which is replaced with the newest sample to proceed with the next output computation.

Figure 7.4 shows the TMS320C54xx program to implement the FIR filter. In this implementation, it is assumed that the most recent incoming signal sample is available from a buffer addressed by the pointer AR5. The computed outputs are placed in another buffer using the pointer AR6. In a real-time DSP system, the incoming samples can be from an A/D converter and the outgoing samples can be applied to a D/A converter. Such interfaces are covered in Chapters 9 and 10.

## 7.4 IIR Filters

An infinite impulse response (IIR) filter is represented by a transfer function, which is a ratio of two polynomials in $z$. To implement such a filter, the difference equation representing the transfer function can be derived and implemented using multiply and add operations. To show such an implementation, we consider a second-order transfer function given by

$$H(z) = \frac{Y(z)}{X(z)} = \frac{b_0 + b_1 z^{-1} + b_2 z^{-2}}{1 - a_1 z^{-1} - a_2 z^{-2}} \tag{7.5}$$

A higher-order IIR filter can be constructed by cascading second-order sections [1, 2]. To develop the difference equation for the IIR filter in Eq. 7.5, we rewrite it as

$$\frac{Y(z)}{X(z)} = \frac{Y(z)}{W(z)} \cdot \frac{W(z)}{X(z)} \tag{7.6}$$

where $W(z)$, an intermediate variable, has been introduced to facilitate implementation. Next, we assign the numerator of the transfer function as

$$\frac{Y(z)}{W(z)} = b_0 + b_1 z^{-1} + b_2 z^{-2} \tag{7.7}$$

which can be represented by a difference equation as

$$y(n) = b_0 w(n) + b_1 w(n-1) + b_2 w(n-2) \tag{7.8}$$

```
;-----------------------------------------------------------------------------
; Program:  ex7p2FIR.asm
; Description:  This is an example to show how to implement an FIR filter.
;               It implements the following equation
;
;               y(n)=h(N-1)x(n-(N-1))+h(N-2)x(n-(N-2))+ ...h(1)x(n-1)+h(0)x(n)
;
;               where  N = Number of filter coefficients = 16.
;                      h(N-1), h(N-2),...h(0) etc are filter coeffs (q15 numbers)
;                      The coefficients are available in file: coeff_fir.dat.
;                      x(n-(N-1)),x(n-(N-2)),...x(n) are signal samples(integers).
;                      The input x(n) is received from the data file: data_in.dat.
;                      The computed output y(n) is placed in a data buffer.
;
; Author:       Avtar Singh, SJSU
;-----------------------------------------------------------------------------
; Definitions
                .mmregs
                .def   _c_int00

                .sect "samples"
InSamples       .include "data_in.dat"    ; Allocate space for x(n)s
OutSamples      .bss y,200,1              ; Allocate space for y(n)s
SampleCnt       .set 200                  ; Number of samples to filter

                .bss CoefBuf, 16, 1       ; Memory for coeff circular buffer
                .bss SampleBuf, 16, 1     ; Memory for sample circular buffer

                .sect "FirCoeff"          ; Filter coeff (seq locations)
FirCoeff        .include "coff_fir.dat"

Nm1             .set 15                   ; N - 1

                .text
_c_int00:
                STM #OutSamples, AR6      ; Clear output sample buffer
                RPT #SampleCnt
                ST #0, *AR6+

                STM #InSamples, AR5       ; AR5 points to InSamples buffer
                STM #OutSamples, AR6      ; AR6 points to OutSample buffer
                STM #SampleCnt, AR4       ; AR4 = Number of samples to filter
                CALL fir_init             ; Init for filter calculations
                SSBX SXM                  ; Select sign extension mode
```

**Figure 7.4** TMS320C54xx implementation of a FIR filter *(continued)*

```
loop:
            LD *AR5+,A                  ; A = next input sample (integer)
            CALL fir_filter             ; Call Filter Routine
            STH A,1,*AR6+               ; Store filtered sample (integer)
            BANZ loop,*AR4-             ; Repeat till all samples filtered
            nop
            nop
            nop
;---------------------------------------------------------------------------
; FIR Filter Initialization Routine
; This routine sets AR2 as the pointer for the sample circular buffer, and
; AR3 as the pointer for coefficient circular buffer.
; BK = Number of filter taps - 1.
; AR0 = 1 = circular buffer pointer increment
;---------------------------------------------------------------------------
fir_init:
            ST #CoefBuf,AR3             ; AR3 is the CB Coeff Pointer
            ST #SampleBuf,AR2           ; AR2 is the CB sample pointer
            STM #Nm1,BK                 ; BK = number of filter taps
            RPT #Nm1
            MVPD #FirCoeff, *AR3+%      ; Place coeff in circular buffer
            RPT #Nm1 - 1                ; Clear circular sample buffer
            ST #0h,*AR2+%
            STM #1,AR0                  ; AR0 = 1 = CB pointer increment
            RET
            nop
            nop
            nop
;---------------------------------------------------------------------------
; FIR Filter Routine
; Enter with A = the current sample x(n) - an integer,
;           AR2 pointing to the location for the current sample x(n),
;           and AR3 pointing to the q15 coefficient h(N-1).
; Exit with A = y(n) as q15 number.
;---------------------------------------------------------------------------
fir_filter:
            STL A, *AR2+0%              ; Place x(n)in the sample buffer

            RPTZ A, #Nm1                ; A = 0
            MAC *AR3+0%,*AR2+0%,A       ; A = filtered sum (q15)
            RET
            nop
            nop
            nop
            .end
```

**Figure 7.4** Continued

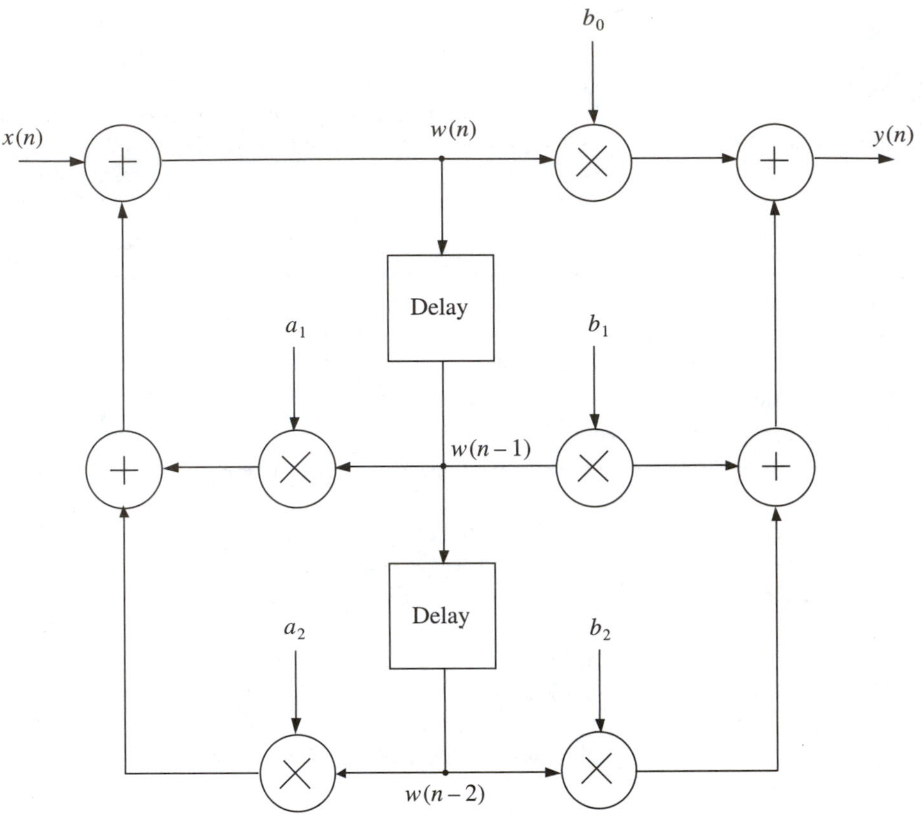


**Figure 7.5** A second-order IIR filter

Similarly, assigning the denominator as

$$\frac{W(z)}{X(z)} = \frac{1}{1 - a_1 z^{-1} - a_2 z^{-2}} \tag{7.9}$$

gives the difference equation

$$w(n) = x(n) + a_1 w(n-1) + a_2 w(n-2) \tag{7.10}$$

Figure 7.5 shows a block diagram of this IIR filter. To compute $y(n)$, we first compute $w(n)$ from $w(n-1)$, $w(n-2)$, and $x(n)$. Next, $w(n)$, $w(n-1)$, and $w(n-2)$ are used to compute $y(n)$. The program in Figure 7.6 shows the TMS320C54xx implementation of the second-order IIR filter. The filter coefficients are stored in memory in the order $b_0$, $b_1$, $b_2$, $a_1$, and $a_2$. The intermediate signals are stored in the order $w(n)$, $w(n-1)$, and $w(n-2)$. Like the

```
;------------------------------------------------------------------------------
; Program Name:  ex7p3IIR.asm
;
; Description:   This is an example to show how to implement an IIR filter. It
;                implements the transfer function
;
;                H(z) = [b0 + b1.z**(-1) + b2.z**(-2)]/[1-a1.z**(-1)-a2.z**(-2)]
;
;                which is equivalent to the equations:
;
;                w(n) = x(n) + a1.w(n-1) + a2.w(n-2)
;                y(n) = b0.w(n) + b1.w(n-1) + b2.w(n-2)
;
;                where
;                w(n), w(n-1), and w(n-2) are the intermediate variables used in
;                computations (integers).
;                a1, a2, b0, b1, and b2 are the filter coefficients (q15 numbers).
;                x(n) is the input sample (integer). Input samples are placed in
;                the buffer, InSamples, from a data file, data_in.dat
;                y(n) is the computed output (integer). The output samples are
;                placed in a buffer, OutSamples.
;
; Author:        Avtar Singh, SJSU
;------------------------------------------------------------------------------
; Definitions
                 .mmregs
                 .def _c_int00

                 .sect "samples"
InSamples        .include "data_in.dat"   ; Allocate space for x(n)s
OutSamples       .bss y,200,1             ; Allocate buffer for y(n)s
SampleCnt        .set 200                 ; Number of samples to filter

; Intermediate variables (sequential locations)
wn               .word 0                  ;initial w(n)
wnm1             .word 0                  ;initial w(n-1) = 0
wnm2             .word 0                  ;initial w(n-2) = 0

                 .sect "coeff"
; Filter coefficients (sequential locations)
b0               .word 3431               ;b0 = 0.104
b1               .word -3356              ;b1 = -0.102
b2               .word 3431               ;b2 = 0.104
a1               .word -32767             ;a1 = -1
a2               .word 20072              ;a2 = 0.612
```

**Figure 7.6** TMS320C54xx implementation of the second-order IIR filter *(continued)*

```
                .text
_c_int00:
                STM #OutSamples, AR6     ; Clear output sample buffer
                RPT #SampleCnt
                ST #0, *AR6+

                STM #InSamples, AR5      ; AR5 points to InSamples buffer
                STM #OutSamples, AR6     ; AR6 points to OutSample buffer
                STM #SampleCnt, AR4      ; AR4 = Number of samples to filter
loop:
                LD *AR5+,15,A            ; A = next input sample (q15)
                CALL iir_filter          ; Call Filter Routine
                STH A,1,*AR6+            ; Store filtered sample (integer)
                BANZ loop,*AR4-          ; Repeat till all samples filtered
                nop
                nop
                nop
;-------------------------------------------------------------------------------
; IIR Filter Subroutine
; Enter with A = x(n) as q15 number
; Exit with A = y(n) as q15 number
; Uses AR2 and AR3
;-------------------------------------------------------------------------------
iir_filter:
                SSBX SXM                 ; Select sign extension mode

                ;w(n)=x(n)+a1.w(n-1)+a2.w(n-2)

                STM #a2,AR2              ; AR2 points to a2
                STM #wnm2, AR3           ; AR3 points to w(n-2)
                MAC *AR2-,*AR3-,A        ; A = x(n)+a2.w(n-2)
                                         ; AR2 points to a1 & AR3 to w(n-1)
                MAC *AR2-,*AR3-,A        ; A = x(n)+a1.w(n-1)+a2.w(n-2)
                                         ; AR2 points to b2 & AR3 to w(n)
                STH A,1,*AR3             ; Save w(n)

                ;y(n)=b0.w(n)+b1.w(n-1)+ b2.w(n-2)

                LD #0,A                  ; A = 0
                STM #wnm2,AR3            ; AR3 points to w(n-2)

                MAC *AR2-,*AR3-,A        ; A = b2.w(n-2)
                                         ; AR2 points to b1 & AR3 to w(n-1)
                DELAY *AR3               ; w(n-1) -> w(n-2)
```

**Figure 7.6** Continued

```
MAC *AR2-,*AR3-,A        ; A = b1.w(n-1)+b2.w(n-2)
                         ; AR2 points to b0 & AR3 to w(n)
DELAY *AR3               ; w(n) -> w(n-1)
MAC *AR2,*AR3,A          ; A = b0.w(n)+b1.w(n-1)+b2.w(n-2)

RET                      ; Return
nop
nop
nop

.end
```

**Figure 7.6** Continued

FIR filter implementation, the incoming sample $x(n)$ is obtained from the buffer InSamples. This buffer is set up using samples in the data file data_in. The filtered signal sample is placed in another buffer called OutSamples. The program uses linearly addressed buffers and the delay instruction in implementation.

## 7.5 Interpolation Filters

An *interpolation filter* is used to increase the sampling rate. The interpolation process involves inserting samples between the incoming samples to create additional samples to increase the sampling rate for the output.

One way to implement an interpolation filter is to first insert zeros between samples of the original sample sequence. The zero-inserted sequence is then passed through an appropriate lowpass digital FIR filter to generate the interpolated sequence [4]. The interpolation process is depicted in Figure 7.7.

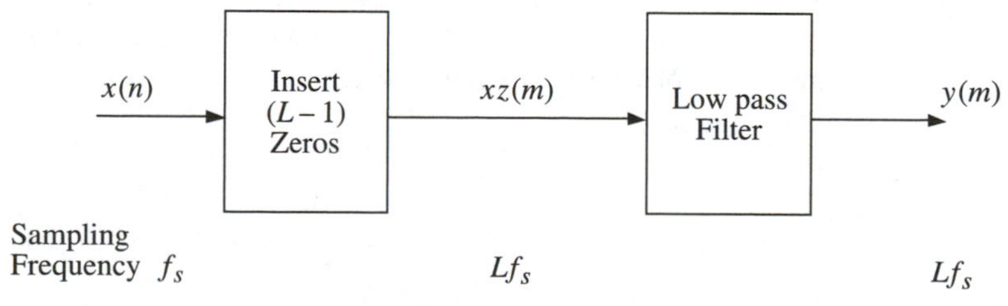


**Figure 7.7** Digital interpolation *with* interpolation factor = $L$

▷ **Example 7.2** Consider the sample sequence $x(n)$ given by

$$x(n) = [0\ \ 2\ \ 4\ \ 6\ \ 8\ \ 10]$$

Let us insert a zero between each two samples to generate the zero-inserted sequence $xz(n)$ as

$$xz(n) = [0\ \ 0\ \ 2\ \ 0\ \ 4\ \ 0\ \ 6\ \ 0\ \ 8\ \ 0\ \ 10\ \ 0]$$

Now, if this sequence is convolved with the sequence $h(n)$, given as

$$h(n) = [0.5\ \ 1\ \ 0.5]$$

the result is a linearly interpolated sequence $y(n)$, given by

$$y(n) = [0\ \ 0\ \ 1\ \ 2\ \ 3\ \ 4\ \ 5\ \ 6\ \ 7\ \ 8\ \ 9\ \ 10\ \ 5\ \ 0]$$

---

The kind of interpolation carried out in the example is called *linear interpolation* because the convolving sequence $h(n)$ is derived based on linear interpolation of samples. Further, in this case, the $h(n)$ selected is just a second-order filter and therefore uses just two adjacent samples to interpolate a sample. A higher-order filter can be used to base interpolation on more input samples. To implement an ideal interpolation, it is shown in the literature that a filter based on samples of an appropriate *sinc* function can be used.

If we assume that the unit sample response of such a filter is available, we need to consider only the implementation technique. Figure 7.8 shows how an interpolating filter using a 15-tap FIR filter and an interpolation factor of 5 can be implemented. In this example, each incoming sample is followed by four zeros to increase the number of samples by a factor of 5. The interpolated samples are computed using a program similar to the one used for a FIR filter implementation.

One drawback of using the implementation strategy depicted in Figure 7.8 is that there are many multiplies in which one of the multiplying elements is zero. Such multiplies need not be included in computation if the computation is rearranged to take advantage of this fact. One such scheme, based on generating what are called *polyphase subfilters*, is available for reducing the computation. For a case where the number of filter coefficients $N$ is a multiple of the interpolating factor $L$, the scheme implements the interpolation filter using the equation

$$y(m+i) = \sum_{k=0}^{N/L-1} h(kL+i)x(n-k) \tag{7.11}$$

where $i = 0, 1, 2, (L-1)$ and $m = nL$.

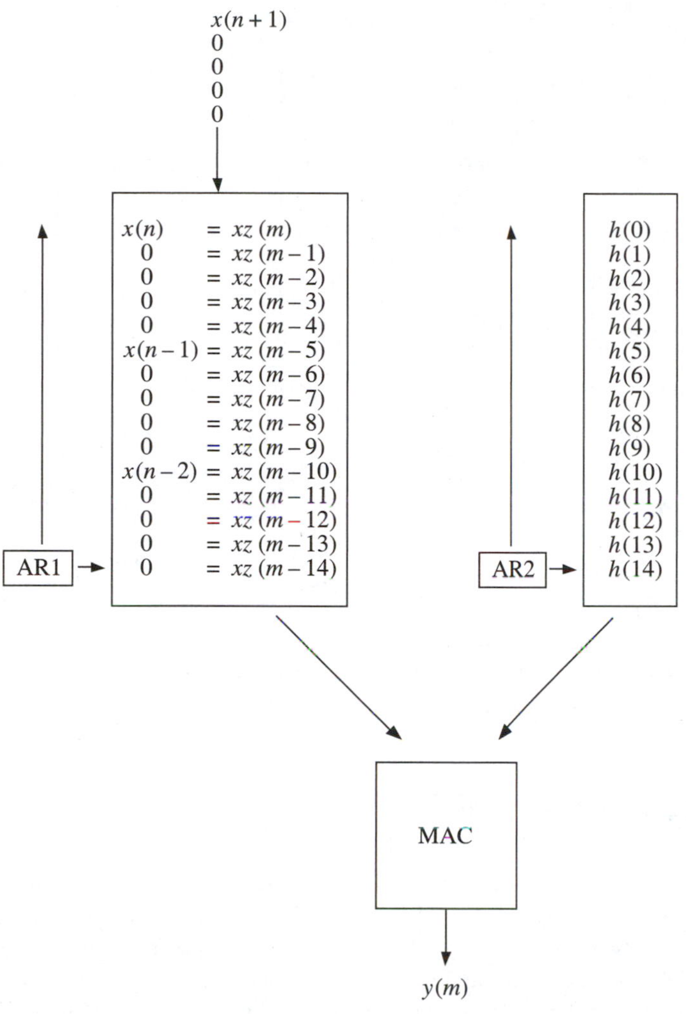


**Figure 7.8** Digital interpolation using a FIR filter with interpolation factor = 5

Figure 7.9 shows a scheme that uses polyphase subfilters to implement the interpolating filter using the 15-tap FIR filter and an interpolation factor of 5. In this implementation, the 15 filter taps are arranged as shown and divided into five 3-tap subfilters. The input samples $x(n)$, $x(n-1)$, and $x(n-2)$ are used five times to generate the five output samples. This implementation requires 15 multiplies as opposed to 75 in the direct implementation of Figure 7.8. The TMS320C54xx implementation for the interpolating scheme of Figure 7.9 is shown in Figure 7.10.

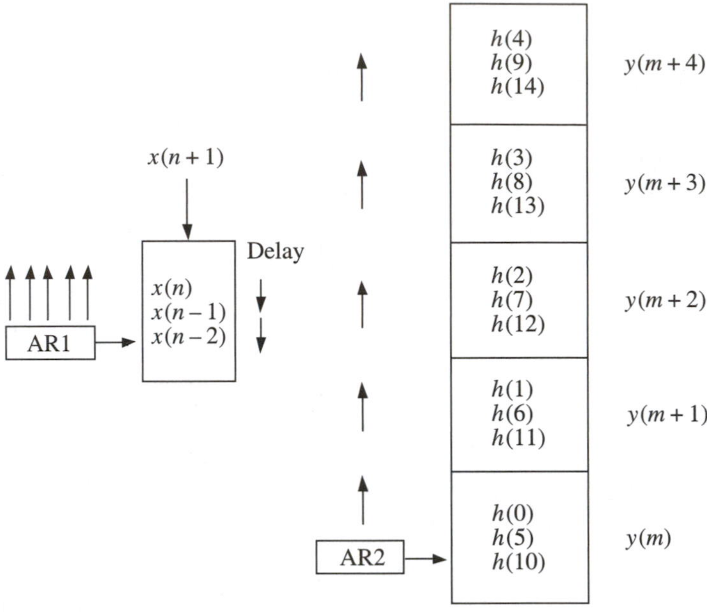


**Figure 7.9** Digital interpolation implementation using five polyphase subfilters; interpolation factor = 5, i.e., $m = 5n$

# 7.6 Decimation Filters

A *decimation filter* is used to decrease the sampling rate. The decrease in sampling rate can be achieved by simply dropping samples. For instance, if every other sample of a sampled sequence is dropped, the sampling rate of the resulting sequence will be half that of the original sequence. The problem with dropping samples is that the new sequence may violate the sampling theorem, which requires that the sampling frequency must be greater than two times the highest frequency contents of the signal.

To circumvent the problem of violating the sampling theorem, the signal to be decimated is first filtered using a lowpass filter. The cutoff frequency of the filter is chosen so that it is less than half the final sampling frequency. The filtered signal can be decimated by dropping samples. In fact, the samples that are to be dropped need not be computed at all. Thus, the implementation of a decimator is just a FIR filter implementation in which some of the outputs are not calculated. This process can be described by the following equation [4]:

$$y(m) = y(nL) = \sum_{k=0}^{N-1} h(k)x(nL - k); \quad n = 0, 1, 2, \ldots \tag{7.12}$$

where $L$ is the decimation factor and $N$ is the filter size.

```
;-------------------------------------------------------------------------------
;
; Program Name:   ex7p4INT.asm
;
; Description:    This is an example to show how to implement an interpolating FIR
;                 filter. The filter length is 15 and the interpolating factor
;                 is 5. It implements the equations
;
;                 y(m)   = h(10)x(n-2) + h(5)x(n-1) + h(0)x(n)
;                 y(m+1) = h(11)x(n-2) + h(6)x(n-1) + h(1)x(n)
;                 y(m+2) = h(12)x(n-2) + h(7)x(n-1) + h(2)x(n)
;                 y(m+3) = h(13)x(n-2) + h(8)x(n-1) + h(3)x(n)
;                 y(m+4) = h(14)x(n-2) + h(9)x(n-1) + h(4)x(n)
;
;                 where,
;                 m = 5n.
;                 h(0), h(1),...etc. are the filter coefficients (q15 numbers)
;                 stored in data memory in the order: h(4), h(9), h(14), h(3), h(8),
;                 h(13), h(2), h(7), h(12), h(1), h(6), h(11), h(0), h(5), h(10).
;                 x(n), x(n-1), and x(n-2) are signal samples (integers) used in
;                 computing the next five output samples.
;                 The input samples are obtained from a file and placed in memory
;                 starting at address InSamples.
;                 The computed output samples are placed starting at data memory
;                 location OutSamples.
;
; Author:         Avtar Singh, SJSU
;
;-------------------------------------------------------------------------------
; Definitions
                  .mmregs
                  .def _c_int00

                  .sect "samples"
InSamples         .include "data_in.dat"  ; Incoming data (from a file)
InSampCnt         .set 50                 ; Input sample count
                  .bss sample,3,1         ; Input samples: x(n),x(n-1),x(n-2)

OutSamples        .bss y,250,1            ; Allocate space for y(n)s
SampleCnt         .set 250                ; Number of samples
```

**Figure 7.10** TMS320C54xx program for an interpolation filter implementation *(continued)*

```
Coeff          .sect "Coeff"
               .word 2560, 3072, 512   ; Filter coeffs h(4), h(9), h(14)
               .word 2048, 3584, 1024  ; Filter coeffs h(3), h(8), h(13)
               .word 1536, 4096, 1536  ; Filter coeffs h(2), h(7), h(12)
               .word 1024, 3584, 2048  ; Filter coeffs h(1), h(6), h(11)
               .word 512, 3072, 2560   ; Filter coeffs h(0), h(5), h(10)

CoeffEnd

Nm1            .set 2                  ; # of coeff/interp factor-1
IFm1           .set 4                  ; Interpolating factor-1

               .text
_c_int00:
               ssbx SXM                ; Select sign extension mode
               rsbx FRCT
               stm #InSamples,ar6      ; ar6 points to the input samples
               stm #InSampCnt-1,ar7    ; ar7 = input sample count - 1
               stm #OutSamples,ar5     ; ar5 points to the output samples
               rpt #SampleCnt-1        ; Reset output samples memory
               st #0,*ar5+

               stm #OutSamples,ar5     ; ar5 points to the output samples
               stm #sample,ar3         ; ar3 points to current input samples
               rpt #Nm1                ; Reset the input samples
               st #0, *ar3+
INTloop1:
               stm #CoeffEnd-1,ar2     ; ar2 points to the last coeff
               stm #IFm1,ar4           ; ar4 = Interpolation factor -1
INTloop2:
               stm #sample+Nm1,ar3     ; ar3 points to last sample in use
               stm #Nm1,ar1            ; ar1 = samples for use
               ld #0,A                 ; A = 0
NXTcoeff:
               mac *ar2-,*ar3-,A       ; Compute interpolated sample
               banz NXTcoeff,*ar1-
               banz INTloop2,*ar4-
               sth A,1,*ar5+           ; Store the interpolated sample

               stm #sample+Nm1-1, ar3  ; Delay the sample array
               rpt #Nm1-1
               delay *ar3-
```

**Figure 7.10** Continued

```
ld *ar6+,A              ; Get the next sample
stm #sample,ar2
stl A,*ar2              ; Place it in the sample buffer

banz INTloop1,*ar7-     ; Repeat for all input samples

nop
nop
nop

.end
```

**Figure 7.10** Continued

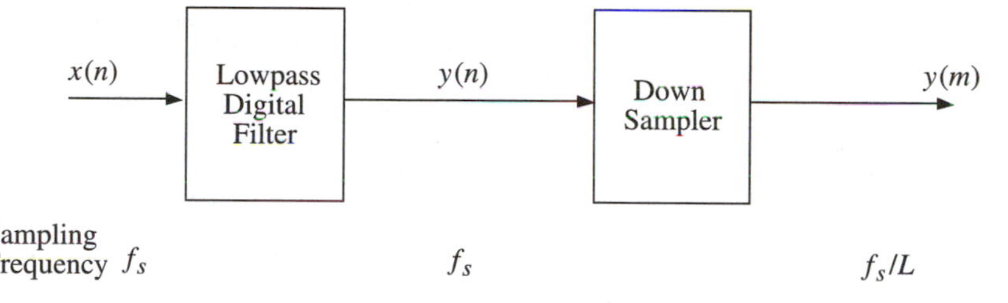


**Figure 7.11** Digital decimation with decimation factor = $L$

Figure 7.11 shows a block diagram of a decimation filter. Digital decimation can be implemented as depicted in Figure 7.12 for an example of a decimation filter with decimation factor of 3. It uses a lowpass FIR filter with 5 taps. The computation is similar to that of a FIR filter. However, after computing each output sample, the signal array is delayed by three sample intervals by bringing in the next three samples into the circular buffer to replace the three oldest samples. The TMS320C54xx implementation of the decimation filter is shown in Figure 7.13.

## 7.7 PID Controller

A basic feedback control system is shown in Figure 7.14. The signal $x(n)$ is the desired plant output and $y(n)$ is the actual response. The error, $e(n)$, is the difference between $x(n)$ and $y(n)$. The PID controller uses the error to generate input to the plant. In a continuous-time system the PID control output is

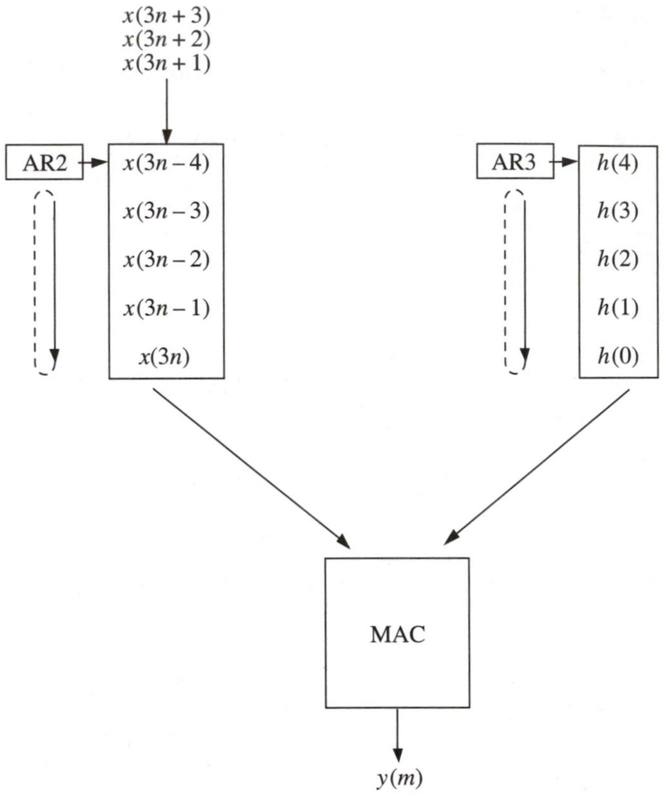

**Figure 7.12** Digital decimation filter implementation or a decimation factor = 3 and a lowpass filter of length 5

generated from the equation

$$u(t) = K_p e(t) + K_i \int e(t)\, dt + K_d \frac{de}{dt} \tag{7.13}$$

where $K_p$, $K_i$, and $K_d$ are constants that depend upon how the plant is to be controlled. The control is based on the error, error integral, and error derivative, giving it the name PID.

The continuous-time equation can be digitized using approximations for the derivative and the integral. The digital equivalent of Eq. 7.13 can be shown to be

$$u(n) = u(n-1) + K_0 e(n) + K_1 e(n-1) + K_2 e(n-2) \tag{7.14}$$

where $K_0$, $K_1$, and $K_2$ are new constants that are related to the constants $K_p$, $K_i$, $K_d$, and the sampling interval.

```
;------------------------------------------------------------------------------
; Program Name:  ex7p5DEC.asm
;
; Description:   This is an example to show how to implement a decimation filter.
;                It implements the following equation
;
;                y(m)   = h(4)x(3n-4) + h(3)x(3n-3) + h(2)x(3n-2) + h(1)x(3n-1) +
;                         h(0)x(3n)
;
;                followed by the equation
;
;                y(m+1) = h(4)x(3n-1) + h(3)x(3n) + h(2)x(3n+1) + h(1)x(3n+2) +
;                         h(0)x(3n+3)
;
;                and so on for a decimation factor of 3 and a filter length of 5.
;
;                Where
;                h(0), h(1), h(2), h(3), and h(4) are the filter coefficients.
;                x(3n), x(3n-1), x(3n-2), x(3n-3), and x(3n-4) are signal samples.
;                x(3n+1), x(3n+2), x(3n+3) are incoming signal samples.
;                y(m), y(m+1) ... etc. are the output signal samples.
;                Signal samples are integers and the filter coefficients are
;                q15 numbers.
;
; Author:        Avtar Singh, SJSU
;------------------------------------------------------------------------------
; Definitions
                 .mmregs
                 .def _c_int00

                 .sect "samples"
InSamples        .include "data_in.dat"     ; Allocate space for x(n)s
OutSamples       .bss y,80,1                ; Allocate space for y(n)s
SampleCnt        .set 240                   ; Number of samples to decimate

                 .sect "FirCoeff"           ; Filter coeff (sequential locations)
FirCoeff         .include "coeff_dec.dat"
Nm1              .set 4                     ; Number of filter taps - 1

                 .bss CoefBuf, 5, 1         ; Memory for coeff circular buffer
                 .bss SampleBuf, 5, 1       ; Memory for sample circular buffer
```

**Figure 7.13** The TMS320C54xx implementation of the decimation filter *(continued)*

```
                .text
_c_int00:
                STM #OutSamples, AR6      ; Clear output sample buffer
                RPT #SampleCnt
                ST #0, *AR6+

                STM #InSamples, AR5       ; AR5 points to InSamples buffer
                STM #OutSamples, AR6      ; AR6 points to OutSample buffer
                STM #SampleCnt, AR4       ; AR4 = Number of samples to filter
                CALL dec_init             ; Init for filter calculations
loop:
                CALL dec_filter           ; Call Filter Routine
                STH A,1,*AR6+             ; Store filtered sample (integer)
                BANZ loop,*AR4-           ; Repeat till all samples filtered
                nop
                nop
                nop
;-----------------------------------------------------------------------------
; Decimation Filter Initialization Routine
; This routine sets AR2 as the pointer for the sample circular buffer, and
; AR3 as the pointer for coefficient circular buffer.
; BK = Number of filter taps.
; AR0 = 1 = circular buffer pointer increment.
;-----------------------------------------------------------------------------
dec_init:
                ST #CoefBuf,AR3           ; AR3 is the CB Coeff Pointer
                ST #SampleBuf,AR2         ; AR2 is the CB sample pointer
                STM #Nm1,BK               ; BK = number of filter taps
                RPT #Nm1
                MVPD #FirCoeff, *AR3+%    ; Place coeff in circular buffer
                RPT #Nm1                  ; Clear circular sample buffer
                ST #0h,*AR2+%
                STM #1,AR0;               ; AR0 = 1 = CB pointer increment
                RET                       ; Return
                nop
                nop
                nop
;-----------------------------------------------------------------------------
; FIR Filter Routine
; Enter with A = x(n), AR2 pointing to the
; circular sample buffer, and AR3 to the
; circular coeff buffer. AR0 = 1.
; Exit with A = y(n) as q15 number.
;-----------------------------------------------------------------------------
```

**Figure 7.13** Continued

```
dec_filter:
            LD *AR5+,A                ; Place next 3 input samples
            STL A, *AR2+0%            ; into the signal buffer
            LD *AR5+,A
            STL A, *AR2+0%
            LD *AR5+,A
            STL A, *AR2+0%

            RPTZ A, #Nm1              ; A = 0
            MAC *AR3+0%,*AR2+0%,A     ; A = filtered signal
            RET                       ; Return
            nop
            nop
            nop

            .end
```

**Figure 7.13** Continued

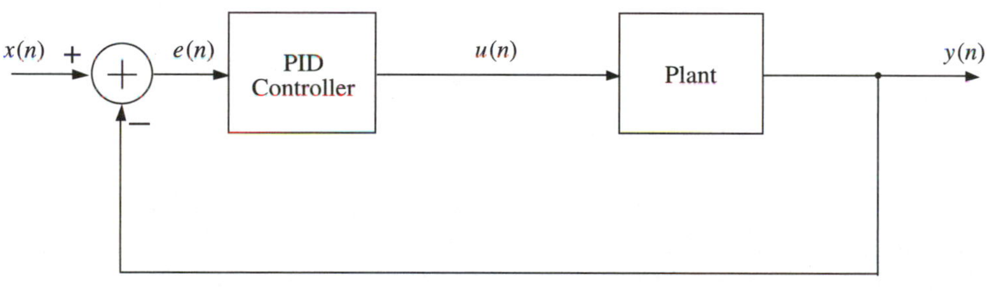


**Figure 7.14** A PID controller for a plant

The implementation of the PID controller requires programming the difference equation 7.14. Figure 7.15 shows the block diagram that can be used to write the code to realize the controller. The program for the TMS320C54xx is shown in Figure 7.16. Note that to actually use the program, we need to generate the error signal outside the signal processor. Alternatively, we need to have desired input and actual output samples that can be subtracted to generate the error signal. For the real-time implementation, these signals are received from A/D converters and the computed control is applied to a D/A converter.

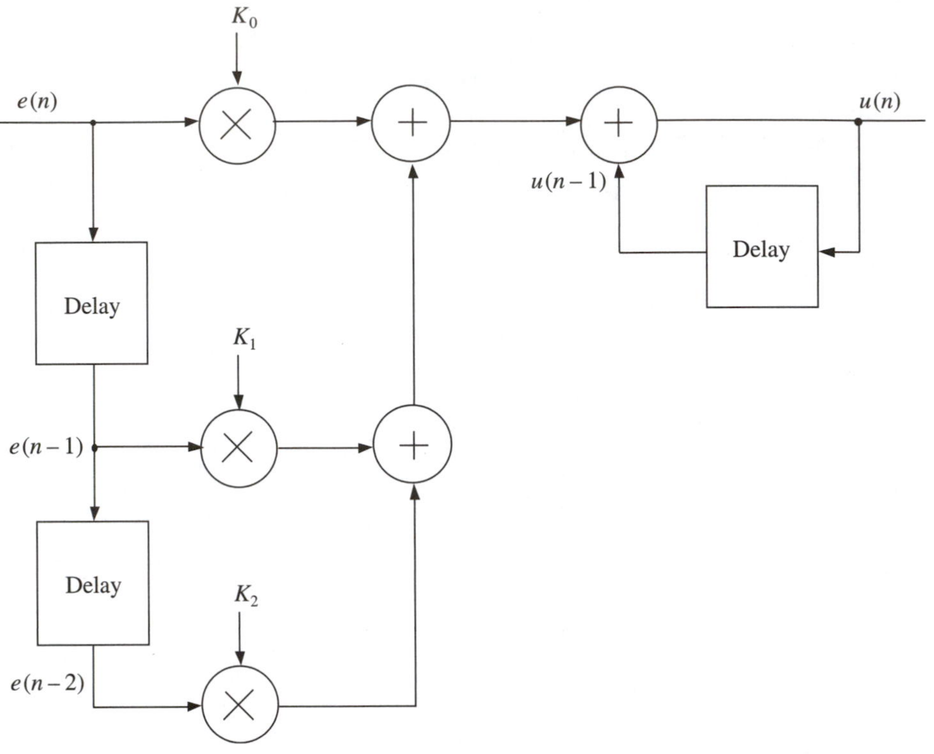


**Figure 7.15** PID controller implementation

## 7.8 Adaptive Filters

An *adaptive filter* is a filter whose coefficients can be updated on-line to counter varying signal distortions. Figure 7.17 is a block diagram of an adaptive filter. The filter in the diagram is typically a FIR filter whose coefficients can be adjusted to minimize some measure of the error signal. The error signal $e(n)$ is generated by subtracting the actual filter output $y(n)$ from the desired output $d(n)$. The desired output is application dependent. A technique used extensively to design an adaptive filter is based on minimizing the mean square error (MSE) [3]. The following equations can be derived using the MSE technique:

$$y(n) = \sum_{k=0}^{N-1} b_k(n)x(n-k) \tag{7.15}$$

$$e(n) = d(n) - y(n) \tag{7.16}$$

```
;-----------------------------------------------------------------------------
;
; Program Name:   ex7p6PID.asm
;
; Description:    This is an example to show how to implement a PID controller.
;                 It implements the following equation
;
;                 u(n) = u(n-1) + K0.e(n) + K1.e(n-1) + K2.e(n-2)
;
;                 where
;                 K0, K1, and K2 are controller coefficients (q15 numbers).
;                 e(n), e(n-1), and e(n-2) are error signal samples (integers).
;
;                 The error samples are the stored values and the computed control
;                 values are also stored in a buffer.
;
;                 The program can be modified for a realtime control system using an
;                 interrupt invoked at the sampling interval, reading the next incoming
;                 error sample from an input port, and applying the computed control
;                 through an output port.
;
; Author:         Avtar Singh, SJSU
;
;-----------------------------------------------------------------------------
                .mmregs                 ; memory-mapped registers
                .def _c_int00

ErrSamples      .bss e, 200, 1          ; Allocate space for e(n)s
ContSamples     .bss u, 200, 1          ; Allocate space for u(n)s
SampleCnt       .set 200                ; Sample count

                .data
; Control and error signals (sequential locations)
un:             .word 0                 ; computed control u(n) as integer
en:             .word 1                 ; error samples e(n) as integer
enm1:           .word 2                 ; error samples e(n-1) as integer
enm2:           .word 1                 ; error samples e(n-2) as integer

                .sect "coeff"
; PID Controller coefficients (sequential locations)
K0:             .word 2000h             ; 1/4 in q15
K1:             .word 0400h             ; 1/32 in q15
K2:             .word 0040h             ; 1/512 in q15
```

**Figure 7.16** TMS320C54xx implementation of a PID controller *(continued)*

```
                .text
_c_int00:
                STM #ContSamples,AR6   ; Clear control sample buffer
                RPT #SampleCnt
                ST #0, *AR6+

                STM #ErrSamples, AR5   ; AR5 points to InSamples buffer start
                STM #ContSamples,AR6   ; AR6 points to OutSample buffer start
                STM #SampleCnt, AR4    ; AR4 = Number of samples to filter
loop:
                LD *AR5+,B             ; B = next error sample
                CALL PID               ; Call PID Control Routine
                STH B,*AR6+            ; Store computed control
                BANZ loop,*AR4-        ; Repeat till all samples done
                nop
                nop
                nop
;---------------------------------------------------------------------------------
; PID Controller Subroutine
; Enter with B = e(n) as integer
; Exit with B = u(n) as integer
; Uses A, AR2, and AR3
;---------------------------------------------------------------------------------
PID:
                SSBX SXM               ; Select sign extension mode
                STM #enm2, AR2         ; AR2 points to current e(n-2)
                STM #K2, AR3           ; AR3 points to current K2
                LD #0, A               ; A = 0
                MAC *AR2-, *AR3-, A    ; A = K2.e(n-2)
                DELAY *AR2             ; e(n-1) -> e(n-2)
                MAC *AR2-, *AR3-, A    ; A = K1.e(n-1) + K2.e(n-2)
                DELAY *AR2             ; e(n) -> e(n-1)
                STL B, *AR2            ; new e(n)
                MAC *AR2-, *AR3, A     ; A = K0.e(n) + K1.e(n-1) + K2.e(n-2)
                ADD *AR2, 15, A        ; A = u(n-1) + K0.e(n) + K1.e(n-1) + K2.e(n-2)
                ADD #1, 14, A          ; Round the result
                STH A, 1, *AR2         ; new u(n)
                LD *AR2, B             ; B = new control
                RET                    ; Return
                nop
                nop
                nop

                .end
```

**Figure 7.16** Continued

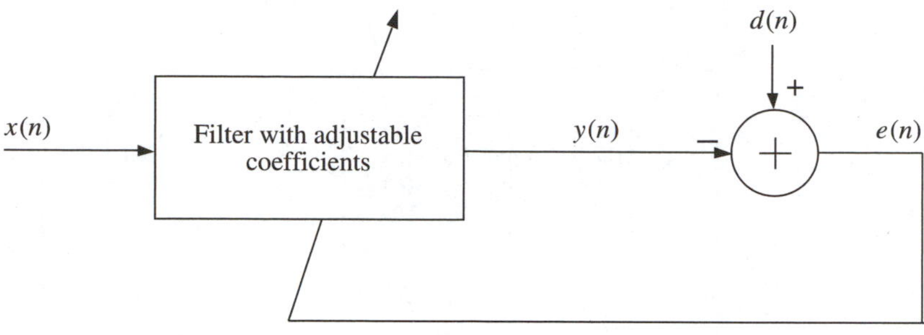


**Figure 7.17** An adaptive filter

$$b_k(n+1) = b_k(n) + 2\mu e(n)x(n-k)$$
$$= b_k(n) + \mathrm{erf}(n)x(n-k) \tag{7.17}$$

where

$$\mathrm{erf}(n) = 2\mu e(n) \tag{7.18}$$

Equation 7.15 is that of a FIR filter. Here, $b_k(n)$ is the $k$th filter coefficient at instant $n$. $N$ represents the number of filter coefficients. The $\mu$ in Eq. 7.17 is called the *coefficient of adaptation*. The adaptation speed and accuracy depend upon $\mu$.

The updating scheme for the coefficients is shown in Figure 7.18. Each coefficient is updated using the erf($n$) which can be computed in advance using Eq. 7.18. The program in Figure 7.19 shows the implementation of a 9-tap adaptive filter for the TMS320C54xx.

## 7.9 2-D Signal Processing

Consider the example of the $N$-tap FIR filter discussed earlier. If the values of the samples $\{x(n), x(n-1), x(n-2), \ldots, x(n-N+1)\}$ are considered as a vector $\mathbf{X}_n$ and the values of the coefficients $\{h(0), h(1), h(2), \ldots, h(N-1)\}$ are considered as another vector **H**, the value of the output sample given by

$$y(n) = x(n)h(0) + x(n-1)h(1) + x(n-2)h(2) + \cdots + x(n-N+1)h(N-1) \tag{7.19}$$

can be considered as the dot product of the two vectors $\mathbf{X}_n$ and **H**. In other words,

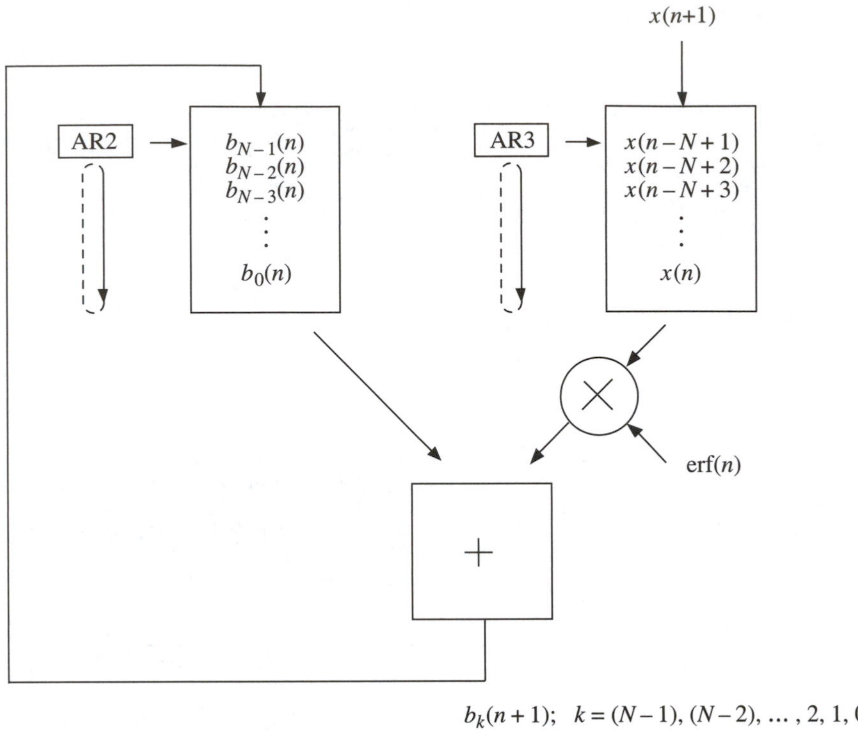


**Figure 7.18** Updating filter coefficients in the adaptive filter implementation

$$\mathbf{Y}_n = \mathbf{X}_n \cdot \mathbf{H} \tag{7.20}$$

where · denotes the dot product. Many times in digital signal processing, one or both the operands **X** and **H** may be two-dimensional, i.e., matrices instead of vectors. A typical application is in image processing. In such a case, **X** may represent intensity values of pixels (picture elements) in the horizontal and vertical directions of a two-dimensional image and **H** may represent coefficients in the horizontal and vertical directions of a two-dimensional filter. One of the most frequently used operations in image processing involves sliding the two-dimensional window of filter coefficients (usually much smaller in size compared to the size of the image) on the image to perform an operation such as filtering out an unwanted feature or enhancing a desirable feature. All these operations can be basically reduced to multiplication of two matrices. Therefore, it becomes essential to know how to write a program to multiply two matrices in order to be able to use the device in two-dimensional signal-processing applications.

```
;--------------------------------------------------------------------------------
; Program Name:   ex7p7ADP.asm
;
; Description:    This is an example to show how to implement an adaptive filter. It
;                 implements a 9-tap adaptive filter using the following equations
;
;                 y(n) =  b0(n)x(n) + b1(n)x(n-1) + b2(n)x(n-2) +
;                         b3(n)x(n-3) + b4(n)x(n-4) + b5(n)x(n-4)+
;                         b7(n)x(n-7) + b7(n)x(n-7) + b8(n)x(n-8)
;
;                 b0(n+1) = b0(n) + erf(n).x(n)
;                 b1(n+1) = b1(n) + erf(n).x(n-1)
;                         .
;                         .
;                 b8(n+1) = b8(n) + erf(n).x(n-8)
;
;                 where
;                 b0(n), b1(n), ... etc. are filter coeff at n, and b0(n+1),
;                 b1(n+1), ... etc. are same filter coeff at n+1.
;                 These coefficients are q15 numbers and are stored in a circular buffer
;                 (CoefBuf).
;
;                 x(n), x(n-1), ... etc. are input samples (integers) stored in a signal
;                 circular buffer (SampleBuf).
;
;                 y(n) is the filtered output (integer).
;                 d(n) is the desired output (integer).
;                 e(n) = d(n) - y(n) (integer)
;                 erfn = e(n) * mu (integer)
;                 mu is the adaptation coefficient (q15 number).
;
; Author:         Avtar Singh, SJSU
;--------------------------------------------------------------------------------
; Definitions
                  .mmregs
                  .def _c_int00

                  .sect "samples"
InSamples         .include "data_in.dat"    ; Input samples to be filtered
OutSamples        .bss y,400,1              ; Output samples
SampleCnt         .set 400                  ; Input sample buffer size

                  .bss CoefBuf,9,1          ; Coeff circular buffer
                  .bss SampleBuf,9,1        ; Sample circular buffer
FilterSize        .set 9                    ; Filter size
```

**Figure 7.19** The TMS320C54xx implementation of an adaptive filter *(continued)*

```
mu              .set 328                ; mu = 0.01 (as q15 number)

dn              .word 0                 ; Desired Signal d(n)
en              .word 0                 ; Error Signal e(n)
yn              .word 0                 ; Filtered Signal y(n)
erfn            .word 0                 ; erfn = e(n).mu

                .text
_c_int00:
                SSBX SXM                ; select sign extension mode
                STM #OutSamples, AR6    ; AR6 points to out sample buffer
                RPT #SampleCnt-1
                ST #0, *AR6+            ; Reset the output sample buffer
                STM #OutSamples, AR6    ; Reset output sample buffer pointer

                STM #InSamples, AR5     ; AR5 points to input sample buffer
                STM #SampleCnt-1, AR4   ; AR4 = the sample count
                STM #FilterSize-1, BK   ; BK = filter size
                STM #SampleBuf, AR3     ; AR3 points to the sample CB
                STM #CoefBuf, AR2       ; AR2 points to the coeff CB

                RPT #FilterSize-1       ; Reset coeff buffer (CoefBuf)
                ST #0h,*AR2+%

                RPT #FilterSize-1       ; Reset sample buffer (SampleBuf)
                ST #0h,*AR3+%

loop:
                CALL adaptive_filter    ; Do adaptive filtering
                BANZ loop, *AR4-        ; Repeat for all samples
                nop
                nop
                nop

adaptive_filter:
; Compute y(n) using current filter coefficients
                STM #1, AR0             ; AR0 = 1 for increment
                RPTZ A, #FilterSize-1   ; y(n) = b0(n)x(n) ... b8(n)x(n-8)
                MAC *AR3+0%, *AR2+0%, A
                STM #yn, AR1
                STH A, 1, *AR1          ; Save y(n) as an integer
                STH A, 1, *AR6+         ; Save filtered signal
```

**Figure 7.19** Continued

```
; Generates d(n) from the computed y(n).
; d(n) can be generated (or obtained) in many different ways.
; Generation of d(n) depends on the problem at hand.

                LD *AR1, B               ; B = y(n)
                STM #dn, AR1             ; AR1 points to the d(n)
                BC high, bgt             ; Branch to high if y(n) > 0
low:            ST  #0c000h, *AR1        ; d(n) = c000h if y(n) < 0
                B end_dn
high:           ST #4000, *AR1           ; d(n) = 4000h if y(n) > 0
end_dn:

; Compute the error e(n)
                STM #dn, AR1             ; e(n) = d(n) - y(n)
                LD *AR1, A
                STM #yn, AR1
                SUB *AR1, A
                STM #en, AR1
                STL A, *AR1

; Update coefficients
                STM #mu, T               ; erfn = mu.e(n)
                MPY *AR1, A
                STM #erfn, AR1
                STH A, 1, *AR1

                STM #FilterSize-1, BRC   ; BRC = No of Taps - 1
                LD *AR1, T               ; T = erfn
                RPTB end_update          ; Update coefficients
                MPY *AR3+0%, A           ; A = erfn*x(n)
                ADD *AR2,15, A           ; Update coefficient
end_update:
                STH A,1, *AR2+0%         ; Save the updated coefficient

; Obtain new input sample
                LD *AR5+, B              ; Get the new Input sample
                STL B, *AR3+0%           ; Put new sample in sample buffer

                RET                      ; Return
                nop
                nop

                .end
```

**Figure 7.19** Continued

$$\begin{bmatrix} a_{11} & a_{12} \cdots a_{1J} \\ a_{21} & a_{22} \cdots a_{2J} \\ \vdots & \\ a_{I1} & a_{I2} \cdots a_{IJ} \end{bmatrix} \bullet \begin{bmatrix} b_{11} & b_{12} \cdots b_{1L} \\ b_{21} & b_{22} \cdots b_{2L} \\ \vdots & \\ b_{K1} & b_{K2} \cdots b_{KL} \end{bmatrix} = \begin{bmatrix} c_{11} & c_{12} \cdots c_{1N} \\ c_{21} & c_{22} \cdots c_{2N} \\ \vdots & \\ c_{M1} & c_{M2} \cdots c_{MN} \end{bmatrix}$$

where J = K, I = M, and L = N

**Figure 7.20** Organization of matrices **A**, **B**, and **C**

### 7.9.1 Matrix Multiplication

Let $\mathbf{A}(i, j)$ be an I × J matrix and $\mathbf{B}(k, l)$, a K × L matrix. In order to be able to multiply the matrix **A** by the matrix **B**, J should be equal to K. We call the product matrix $\mathbf{C}(m, n)$, with M rows and N columns. Since we have multiplied an I × J matrix by a K × L matrix (J being equal to K), the resulting matrix will have I rows and L columns, i.e., M = I, and N = L. Figure 7.20 shows the organization of the matrices **A**, **B**, and **C**.

Each element of the matrix **C** is the dot product of a vector representing a row of the matrix **A** with a vector representing a column of matrix **B**. For example,

$$\begin{aligned}
c_{11} &= a_{11}b_{11} + a_{12}b_{21} + a_{13}b_{31} + \cdots + a_{1J}b_{K1} \\
c_{12} &= a_{11}b_{12} + a_{12}b_{22} + a_{13}b_{32} + \cdots + a_{1J}b_{K2} \\
&\vdots \\
c_{1N} &= a_{11}b_{1l} + a_{12}b_{2l} + a_{13}b_{3l} + \cdots + a_{1J}b_{Kl} \\
c_{21} &= a_{21}b_{11} + a_{22}b_{21} + a_{23}b_{31} + \cdots + a_{2J}b_{K1} \\
c_{22} &= a_{21}b_{12} + a_{22}b_{21} + a_{23}b_{32} + \cdots + a_{2J}b_{K2} \\
&\vdots \\
c_{2N} &= a_{21}b_{1l} + a_{22}b_{2l} + a_{23}b_{3l} + \cdots + a_{2J}b_{Kl} \\
&\vdots \\
c_{M1} &= a_{I1}b_{11} + a_{I2}b_{21} + a_{I3}b_{31} + \cdots + a_{IJ}b_{K1} \\
c_{M2} &= a_{I1}b_{12} + a_{I2}b_{22} + a_{I3}b_{32} + \cdots + a_{IJ}b_{K2} \\
&\vdots \\
c_{MN} &= a_{I1}b_{I1} + a_{I2}b_{2l} + a_{I3}b_{3l} + \cdots + a_{IJ}b_{Kl}
\end{aligned}$$

In other words, in order to obtain the element $c_{11}$, row 1 of **A** is multiplied with column 1 of **B**; to get $c_{12}$, row 1 of **A** is multiplied by column 2 of **B**, etc., until all the elements of row 1 of **C** are computed. Then the operation is repeated with row 2 of **A** to get the elements of row 2 of **C** and so on until all the elements of **C** are computed.

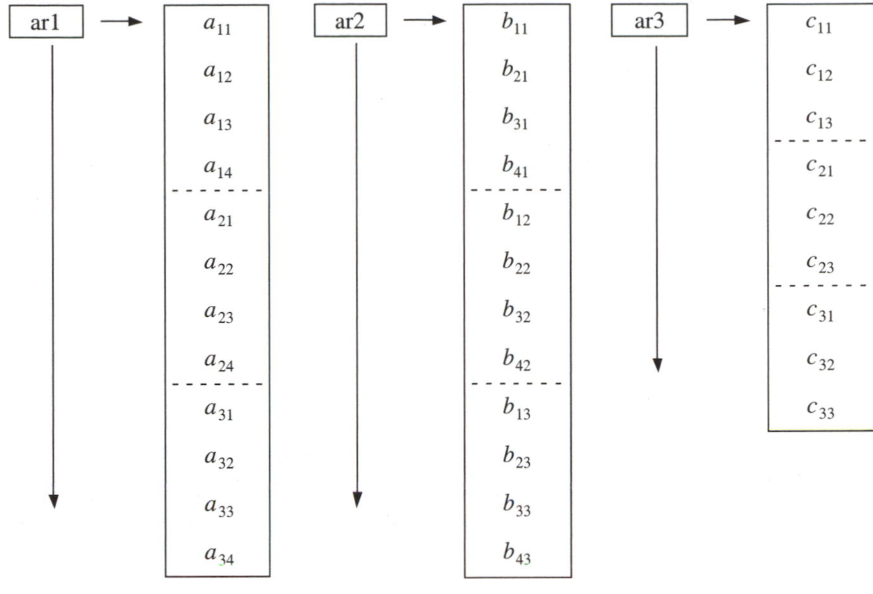


**Figure 7.21** Memory organization for matrix multiplication of a $3 \times 4$ matrix with a $4 \times 3$ matrix

In order to implement the matrix multiplication algorithm on the TMS320C54xx, the data corresponding to matrices **A** and **B** and the resulting matrix **C** should be organized in the DSP memory as shown in Figure 7.21. The elements of matrix **A** are ordered row by row, those of matrix **B** are ordered column by column, and the elements of the product matrix **C** are stored row by row. Note that Figure 7.21 is an example in which **A** is a $3 \times 4$ matrix, **B** is a $4 \times 3$ matrix, and therefore, **C** is a $3 \times 3$ matrix.

Three pointers are required to keep track of the elements in the matrices **A**, **B**, and **C**. Let these pointers be ar1, ar2, and ar3, respectively. All the pointers are initialized to the starting addresses of the respective matrices. To compute $c_{11}$, ar1 has to advance from $a_{11}$ to $a_{14}$ and ar2 from $b_{11}$ to $b_{41}$. To compute $c_{12}$, ar1 has to be reset to $a_{11}$ and has to go again from $a_{11}$ to $a_{14}$. On the other hand, ar2 continues from $b_{12}$ to $b_{42}$. This is repeated for all the elements of the first row of **C**, i.e., ar1 goes from $a_{11}$ to $a_{14}$ three times while ar2 goes all the way from $b_{11}$ to $b_{34}$. After computing all the elements of the first row of **C**, ar1 is set to $a_{21}$ while ar2 is reset to $b_{11}$ to compute the elements of row 2 of **C** in the same way as was done for row 1. This process is repeated for all the rows of **C**. ar3 starts at $c_{11}$ and moves to $c_{33}$ with the computation of each element of **C**. The memory organization for the matrix elements is shown in Figure 7.21. The TMS320C54xx program for matrix multiplication is shown in Figure 7.22. Note that the program uses the repeat-block instruction to compute the dot product used in the matrix multiplication.

```
;________________________________________________________________________________
;
; Program Name:  ex7p8MAT.asm
;
; Description:   This is an example to show how to implement matrix mutiplication.
;                It implements the following equation
;
;                C = A.B
;
;                where
;                A is a 3 x 4 matrix,
;                B is a 4 x 3 matrix, and
;                C is a 3 x 3 matrix
;
;                Matrix A elements are stored in data memory row after row.
;                Matrix B elements are stored in data memory column after column.
;                Matrix C elements are stored in data memory row after row.
;
;                All elements are q15 numbers.
;
; Author:        Avtar Singh, SJSU
;
;________________________________________________________________________________
                .mmregs                          ; memory-mapped registers

                .ref   _c_int00

                .sect  ".vectors"

RESET:          B _c_int00                       ; Reset vector
                NOP
                NOP

                .data

matArow1:       .word 1000h,2000h,3000h,4000h   ; row 1 of matrix A
matArow2:       .word 1000h,2000h,3000h,4000h   ; row 2 of matrix A
matArow3:       .word 1000h,2000h,3000h,4000h   ; row 3 of matrix A

matBcol1:       .word 1000h,2000h,3000h,4000h   ; column 1 of matrix B
matBcol2:       .word 1000h,2000h,3000h,4000h   ; column 2 of matrix B
matBcol3:       .word 1000h,2000h,3000h,4000h   ; column 3 of matrix B
```

**Figure 7.22** The TMS320C54xx implementation of the matrix multiplication *(continued)*

```
matC:           .word 0,0,0                 ; row 1 of matrix C
                .word 0,0,0                 ; row 2 of matrix C
                .word 0,0,0                 ; row 3 of matrix C

Nm1             .set 3                      ; columns of matrix A - 1

                .text

_c_int00:
                ssbx sxm                    ; select sign extension mode
                stm #matC, ar3              ; ar3 = matrix C start address
                stm #matArow1, ar1          ; ar1 = matrix A row 1 start address
                stm #matBcol1, ar2          ; ar2 = matrix B col 1 start address
                stm #Nm1, BRC               ; BRC = row/col elements - 1
                call DOTPROD                ; find dot product
                sth a,1,*ar3+               ; save the result as matrix C element

                stm #matArow1, ar1          ; ar1 = matrix A row 1 start address
                stm #matBcol2, ar2          ; ar2 = matrix B col 2 start address
                stm #Nm1, BRC               ; BRC = row/col elements - 1
                call DOTPROD                ; find dot product
                sth a,1,*ar3+               ; save the result as matrix C element

                stm #matArow1, ar1          ; ar1 = matrix A row 1 start address
                stm #matBcol3, ar2          ; ar2 = matrix B col 3 start address
                stm #Nm1, BRC               ; BRC = row/col elements - 1
                call DOTPROD                ; find dot product
                sth a,1,*ar3+               ; save the result as matrix C element

                stm #matArow2, ar1          ; ar1 = matrix A row 2 start address
                stm #matBcol1, ar2          ; ar2 = matrix B col 1 start address
                stm #Nm1, BRC               ; BRC = row/col elements - 1
                call DOTPROD                ; find dot product
                sth a,1,*ar3+               ; save the result as matrix C element

                stm #matArow2, ar1          ; ar1 = matrix A row 2 start address
                stm #matBcol2, ar2          ; ar2 = matrix B col 2 start address
                stm #Nm1, BRC               ; BRC = row/col elements - 1
                call DOTPROD                ; find dot product
                sth a,1,*ar3+               ; save the result as matrix C element

                stm #matArow2, ar1          ; ar1 = matrix A row 2 start address
                stm #matBcol3, ar2          ; ar2 = matrix B col 3 start address
                stm #Nm1, BRC               ; BRC = row/col elements - 1
                call DOTPROD                ; find dot product
                sth a,1,*ar3+               ; save the result as matrix C element
```

**Figure 7.22** Continued

```
            stm #matArow3, ar1              ; ar1 = matrix A row 3 start address
            stm #matBcol1, ar2              ; ar2 = matrix B col 1 start address
            stm #Nm1, BRC                   ; BRC = row/col elements - 1
            call DOTPROD                    ; find dot product
            sth a,1,*ar3+                   ; save the result as matrix C element

            stm #matArow3, ar1              ; ar1 = matrix A row 3 start address
            stm #matBcol2, ar2              ; ar2 = matrix B col 2 start address
            stm #Nm1, BRC                   ; BRC = row/col elements - 1
            call DOTPROD                    ; find dot product
            sth a,1,*ar3+                   ; save the result as matrix C element

            stm #matArow3, ar1              ; ar1 = matrix A row 3 start address
            stm #matBcol3, ar2              ; ar2 = matrix B col 3 start address
            stm #Nm1, BRC                   ; BRC = row/col elements - 1
            call DOTPROD                    ; find dot product
            sth a,1,*ar3+                   ; save the result as matrix C element
            nop
            nop
            nop
;--------------------------------------------------------------------------
;
;       Dot Product Routine
;
;       This routine determines the dot product of two vectors
;
;       Input:  ar1 = pointer to the first element of vector 1
;               ar2 = pointer to the first element of vector 2
;               BRC = size - 1 for either vector
;
;               All elements are q15 numbers
;
;       Output: A = dot product as q30 number
;
;--------------------------------------------------------------------------
DOTPROD:
            ld #0, a                        ; A = 0
NXTeleofA:
            rptb end_dotp-1                 ; A = sum of ar1(i)*ar2(i) for all i
            ld *ar2+, t
            mac *ar1+, a
end_dotp:   ret                             ; return
            nop
            nop

            .end
```

**Figure 7.22** Continued

## 7.10 Summary

In this chapter, we have covered some basic DSP implementations with the view of using a fixed-point programmable DSP device such as the TMS320C54xx. All these implementations require some sort of multiply and accumulate operation on two arrays, typically an array of samples and an array of coefficients. In all these implementations, memory organization is important, as it leads to the specific programming strategy to do the computations. Another important aspect of these implementations is how signal samples and coefficients are represented. The Q-notation is handy when representing fractional filter coefficients. However, care must be exercised in using the multiply operation on numbers represented in the Q-notation.

The implementations covered in this chapter include FIR filters, IIR filters, interpolation filters, decimation filters, PID controller, adaptive filters, and 2-D signal processing. In these implementations, it is assumed that the input signal samples are available in a memory buffer or in a data file. The computed output samples are also placed in a memory buffer. However, to design a real-time application requires inclusion of A/D and D/A interfacing along with the appropriate software to control them for data acquisition. Real-time signal processing is considered in Chapters 9 and 10.

## References


1. Strum, R. D., and Kirk, D. E. *First Principles of Discrete Systems and Digital Signal Processing*, Addison-Wesley, 1988.
2. Peled, A., and Liu, B. *Digital Signal Processing*, John Wiley, 1976.
3. Stearns, S. D., and Ruth, D. A. *Signal Processing Algorithms*, Prentice-Hall, 1988.
4. Orfanidis, S. J. *Introduction to Signal Processing*, Prentice-Hall, 1996.
5. *TMS320C54x DSP Reference Set, Volume 1*, Texas Instruments, 2001.
6. *TMS320C54x DSP Reference Set, Volume 2*, Texas Instruments, 1999.
7. *TMS320C54x Assembly Language Tools, User's Guide, SPRU102D*, Texas Instruments, December 1999.


## Assignments

**7.1.** Determine the value of each of the following 16-bit numbers represented using the given Q-notation:

a. 4400h as a Q0 number

b. 4400h as a Q15 number

c. 4400h as a Q7 number

**7.2.** Represent each of the following as 16-bit numbers in the desired Q-notation:

a. 0.3125 as a Q15 number

b. −0.3125 as a Q15 number

c. 3.125 as a Q7 number

d. −352 as a Q0 number

**7.3.** Modify the TMS320C54xx program in Figure 7.1(b) so that it can be used to multiply a Q15 number with a Q0 number to obtain the result in Q0 notation.

**7.4.** Modify the TMS320C54xx program in Figure 7.1(b) so that the rounding is done as follows: Use ordinary rounding as in the program except when the part to be truncated is exactly equal to half the largest value represented by the dropped bits, in which case the part to be kept is incremented only if, as a binary number, it represents an odd integer.

**7.5.** Analyze the following program to answer the questions at the end. Assume that all specified data locations are on the same page starting at a0.

```
.data
ao      .word  6000h
b1      .word  2000h
xn      .word  4000h
yn      .word  0h
ynm1    .word  3000h

.text
ld      #a0, dp
ld      a0, t
mpy     xn, a
ld      b1, t
mac     ynm1, a
sth     a, 1, yn
```

Assuming that all memory contents for constants and signals are in Q15 notation, determine the

a. decimal values represented by ao, b1, xn, and ynm1,

b. decimal value of the computed yn and that of the error due to truncation.

c. equation for yn implemented by the above program.

**7.6.** For the following program determine (a) the difference equation, and (b) the transfer function for the implemented filter.

```
AGAIN:
   Ld     #yn, dp      ; Set the data page
   portr  inport, xn   ; Get the new input x(n) sample
   ld     #0, a
   ld     xnm2, t
   mpy    a2, a
   ld     xnm1, t
   delay  xnm1
   mac    a1, a
   ld     xn, t
   delay  xn
   mac    a0, a
   ld     ynm2, T
   delay  ynm2
   mac    b2, a
   ld     ynm1, t
   delay  ynm1
   mac    b1, a
   ld     yn, t
   delay  yn
   mac    b0, a
   sth    a, yn        ; Replace y(n) with the computed y(n)
   b      AGAIN
```

Assume that all signals are integers and stored in the order $y(n)$, $y(n-1)$, $y(n-2)$, $x(n)$, $x(n-1)$, $x(n-2)$ starting at the lowest address and proceeding to the higher addresses on the same page. Note that ynm1 in the code stands for $y(n-1)$ and similarly other signals are denoted. All coefficients such as a0, a1, . . . , etc. are also stored as integers on the same data page.

**7.7.** An $N$-tap FIR filter has

$$h(i) = h(N-1-i)$$

where $i = 0, 1, \ldots, (N/2) - 1$, for an even value of $N$. Use the coefficient symmetry to rewrite Eq. 7.2 so that the number of multiplies is minimized. Show an implementation scheme similar to Figure 7.3 for the filter.

**7.8.** An $N$-tap FIR filter has

$$h(i) = h(N-1-i)$$

where $i = 0, 1, \ldots, (N-1)/2$, for an odd value of $N$. Use the coefficient symmetry to rewrite Eq. 7.2 so that number of multiplies is minimized. Show an implementation scheme similar to Figure 7.3 for the filter.

**7.9.** Modify the TMS320C54xx program for the FIR filter implementation shown in Figure 7.4 to implement the symmetrical tap filter in Problem 4 with $N = 30$. Test the filter implementation using an appropriate set of tap values.

**7.10.** Modify the TMS320C54xx program for the FIR filter implementation shown in Figure 7.4 to implement the symmetrical tap filter in Problem 5 with $N = 31$. Test the filter implementation using an appropriate set of tap values.

**7.11.** Implement the IIR filter represented by the following difference equation on the TMS320C54xx. Assume that Q15 notation is used to represent the values of coefficients and Q0 to represent the signal samples.

$$y(n) = b(0)x(n) + b(1)x(n-1) + a(0)y(n-1) + a(1)y(n-2) + a(2)y(n-3)$$

**7.12.** Using the program of Figure 7.6, develop a TMS320C54xx program to implement the following FIR filter:

$$H(z) = \frac{(0.1 + 0.2z^{-1} + 0.1z^{-2})(0.5 - 0.2z^{-2})}{(1 + 0.25z^{-1})(1 - 0.15z^{-1} - 0.5z^{-2})}$$

**7.13.** Determine the linearly interpolated sequence from the given sequence

$$x(n) = [0\ \ 4\ \ 8\ \ 12\ \ 16\ \ 12\ \ 8\ \ 4\ \ 0]$$

for an interpolation factor of 3. What interpolating sequence $h(n)$ can achieve the specified interpolation?

**7.14.** Modify the interpolation filter implementation scheme of Figure 7.9 so as to avoid going over the sample sequence five times. This can be done using more memory locations.

**7.15.** Use the scheme of Problem 11 to write a TMS320C54xx program for the interpolation filter. Use appropriate data to test the program.

**7.16.** If decimation by a factor of 8 is achieved by decimating by a factor of 2 followed by another factor of 4, determine the cutoff frequencies of the two lowpass filters that should be used in the decimation scheme.

**7.17.** Develop a decimation filter program that can be used to decimate by a factor of $2^5$ using a subroutine to decimate by a factor of 2 in conjunction with appropriate filters.

**7.18.** In the PID controller of Figure 7.14, $K_3 = K_1/64$, $K_2 = K_1/8$. Modify Eq. 7.14 so that a minimum number of multiplies are used for its implementation. What processor operation can be used to achieve this?

**7.19.** Develop a TMS320C54xx program for the PID controller of Problem 5.

**7.20.** Modify the adaptive filter implementation scheme of Figure 7.18 so that the adaptive filter is also an interpolation filter with an interpolation factor of 2.

**7.21.** Develop a TMS320C54xx program for the scheme of the adaptive and interpolation filter in Problem 17.

**7.22.** Develop a TMS320C54xx subroutine to multiply two $3 \times 3$ matrices.

**7.23.** Use the subroutine developed in Problem 19 to develop a TMS320C54xx program to implement 2-D convolution. Assume appropriate values for the 2-D signal samples and the convolution coefficients.

# Chapter 8

# Implementation of FFT Algorithms

## 8.1 Introduction

In this chapter, we cover the implementation of FFT algorithms for DFT computation and related issues. As an example, an 8-point DIT FFT algorithm is implemented with considerations for computational structure and scaling to avoid overflow. The following topics are covered in this chapter:

An FFT algorithm for DFT computation
A butterfly computation
Overflow and scaling
Bit-reversed index generation
An 8-point FFT implementation on the TMS320C54xx
Computation of the signal spectrum

## 8.2 An FFT Algorithm for DFT Computation

Here we consider the DFT computation using FFT algorithms. We discuss these algorithms from the implementation point of view. For a detailed treatment of the FFT, we refer the reader to the many available excellent books on the subject [1].

The discrete Fourier transform (DFT) pair is given as

$$X(k) = \sum_{n=0}^{n=N-1} x(n)e^{-j2\pi nk/N}; \quad k = 0, 1, 2, \ldots, (N-1) \tag{8.1}$$

and

$$x(n) = 1/N \sum_{k=0}^{k=N-1} X(k)e^{j2\pi nk/N}; \quad n = 0, 1, 2, \ldots, (N-1) \tag{8.2}$$

where $x(n)$ is the time-domain sequence, $X(k)$ is the corresponding frequency-domain sequence, and $N$ is the number of elements of each sequence.

Equation (8.1) is known as the forward transform, or DFT, and (8.2) as the inverse transform, or IDFT. Replacing $e^{-j2\pi/N}$ by $W_N$, we get

$$X(k) = \sum_{n=0}^{n=N-1} x(n){W_N}^{nk}; \quad k = 0, 1, 2, \ldots, (N-1) \tag{8.3}$$

and

$$x(n) = (1/N) \sum_{k=0}^{k=N-1} X(k){W_N}^{-nk}; \quad n = 0, 1, 2, \ldots, (N-1) \tag{8.4}$$

where ${W_N}^{nk}$ is known as the twiddle factor.

Note that the direct DFT computation of (8.1) or (8.2) requires $N^2$ complex multiplies and $N(N-1)$ complex additions. That is, it requires approximately $N^2$ complex operations. Let us now consider a few specific cases starting with the 2-point DFT. The objective is to derive an algorithm for efficient computation of the DFT and IDFT.

## 8.2.1 2-Point DFT Computation

For $N = 2$, Equation 8.3 written explicitly for $k = 0$ and 1 gives

$$X(0) = x(0){W_2}^0 + x(1){W_2}^0 \tag{8.5}$$

$$X(1) = x(0){W_2}^0 + x(1){W_2}^{-1} \tag{8.6}$$

Note that the twiddle factor ${W_2}^0 = e^0 = 1$ and ${W_2}^{-1} = e^{-j\pi} = -1$.

Substituting for twiddle factors in Equations 8.5 and 8.6 gives

$$X(0) = x(0) + x(1) \tag{8.7}$$

$$X(1) = x(0) - x(1) \tag{8.8}$$

The computation represented by these equations is shown in the signal flow graph of Figure 8.1. This computation is called an in-place computation if the computed values $X(0)$, $X(1)$ replace $x(0)$ and $x(1)$, respectively. Note that the 2-point DFT computation requires only add and subtract operations to implement. The structure in Figure 8.1 is called a *butterfly*.

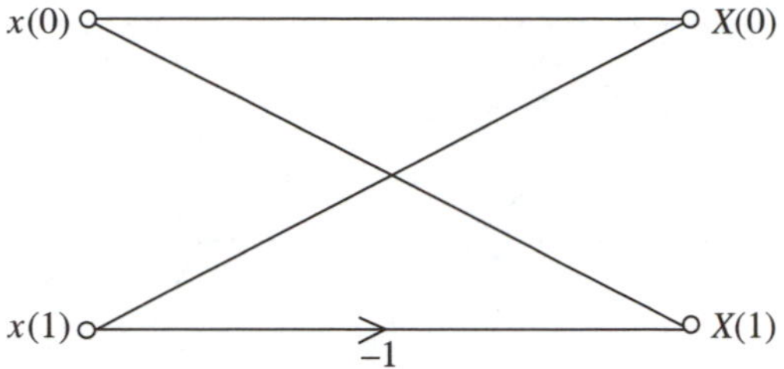


**Figure 8.1** Signal flow graph for a 2-point DFT computation

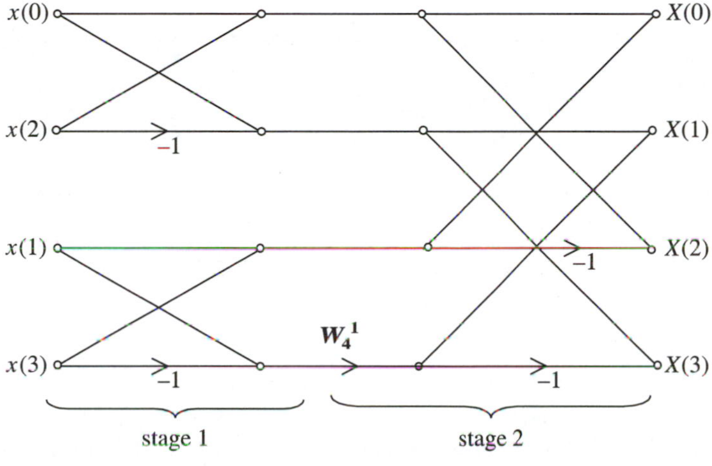


**Figure 8.2** Signal flow graph for a 4-point DFT computation

### 8.2.2 4-Point DFT Computation

Computation of a 4-point DFT can be shown to yield the structure shown in Figure 8.2. Note that now we require a total of four butterflies in two stages of computation. The first stage has two butterflies, one operating on $x(0)$ and $x(2)$ and the second operating on $x(1)$ and $x(3)$. In the second stage, the first butterfly operates on upper outputs of the first-stage butterflies and the second one operates on the lower outputs of the first-stage butterflies. Also, note that the lower output of the second butterfly of the first stage needs to be multiplied with the twiddle factor $W_4{}^1$.

Further, note that the input samples $x(0)$ through $x(3)$ are required to be rearranged in the order $x(0)$, $x(2)$, $x(1)$, $x(3)$ to implement the computation depicted in Figure 8.2. Now, if the naturally occurring input sample indices 0, 1, 2, 3 are represented by their binary equivalents 00, 01, 10, 11 and these binary numbers are reversed, we get 00, 10, 01, 11, which are 0, 2, 1, 3, the

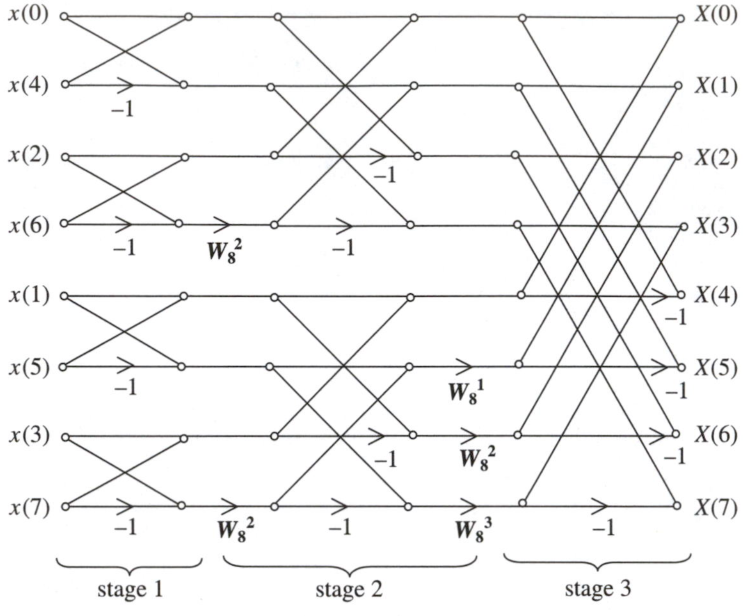


**Figure 8.3** Signal flow graph for an 8-point DFT computation

indices for the sequence in which the signals must be processed by the computational structure of Figure 8.2. This process of rearrangement of indices for DFT computation is called *bit reversing* and is further considered in a subsequent section.

### 8.2.3 8-Point DFT Computation

When 8 points are used to compute DFT, the result is the computational structure of Figure 8.3. Now we have three computational stages, each stage requiring 4 butterflies for a total of 12 butterflies. Note that the input is rearranged following bit-reversed indices of eight input samples. The relationship between the input indices and the bit-reversed indices required for DFT computation will be explored further in a subsequent section. Further, note that now more twiddle factors are needed to compute the DFT.

### 8.2.4 $N = 2^M$ Point FFT Computation

The above approach to DFT computation extended to a case of $N$ points, where $N$ is a power of 2, yields $\log_2 N$ stages of computation, with each stage requiring $N/2$ butterflies. This computational structure is the *fast Fourier transform*, or FFT.

### Another FFT Algorithm

Two types of commonly used FFT algorithms are available, decimation-in-time (DIT) and decimation-in-frequency (DIF). If the naturally occurring input time-sequence sample indices are bit reversed and processed by the above algorithm, the frequency domain output is in the natural order. Such a computation is called a *DIT FFT algorithm.* Another algorithm results if a time-domain sample sequence is used without bit-reversing the indices. The latter algorithm is similar to the former, with small changes in the butterfly computational structure. The output generated by the latter algorithm has bit-reversed indices. This second approach is called the *DIF FFT algorithm.* The details of the DIF FFT algorithm can be found in most books on DSP fundamentals [1] and are left for the reader to explore.

### Zero-Padding

At times, the sequence to be transformed is appended with zeros before computing the DFT. This can be done to satisfy the condition that the FFT algorithm requires that the number of points be a power of 2. Another objective of zero-padding is to increase the transformed points to decrease the frequency interval between adjacent points represented by the $X(k)$ sequence. This leads to improvement in frequency resolution for representing signals in the frequency domain.

## 8.3 A Butterfly Computation

A general DIT FFT butterfly in-place computation structure is shown in Figure 8.4. Its implementation requires the following computation:

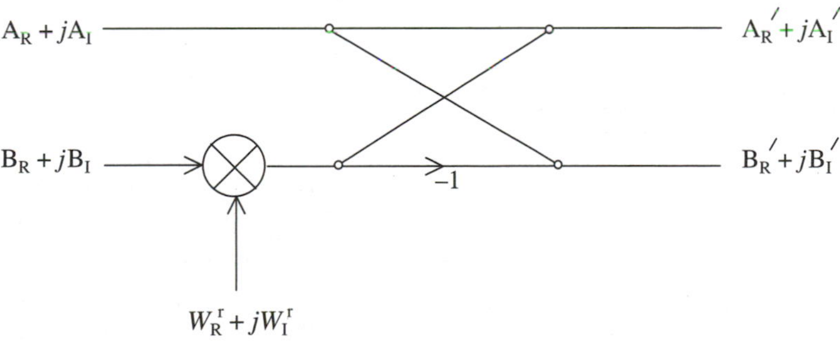


**Figure 8.4** A general butterfly computation structure

$$\begin{aligned} A_R{}' + jA_I{}' &= (A_R + jA_I) + (B_R + jB_I).(W_R{}^r + jW_I{}^r) \\ &= A_R + B_R W_R{}^r - B_I W_I{}^r + j(A_I + B_I W_R{}^r + B_R W_I{}^r) \qquad (8.9) \\ B_R{}' + jB_I{}' &= (A_R + jA_I) - (B_R + jB_I)(W_R{}^r + jW_I{}^r) \\ &= A_R - B_R W_R{}^r + B_I W_I{}^r + j(A_I - B_I W_R{}^r - B_R W_I{}^r) \qquad (8.10) \end{aligned}$$

Equating real and imaginary parts yields

$$\begin{aligned} A_R{}' &= A_R + B_R W_R{}^r - B_I W_I{}^r \\ A_I{}' &= A_I + B_I W_R{}^r + B_R W_I{}^r \\ B_R{}' &= A_R - B_R W_R{}^r + B_I W_I{}^r \\ B_I{}' &= A_I - B_I W_R{}^r - B_R W_I{}^r \end{aligned} \qquad (8.11)$$

or

$$\begin{aligned} A_R{}' &= A_R + \mathrm{TMP1} \\ A_I{}' &= A_I + \mathrm{TMP2} \\ B_R{}' &= A_R - \mathrm{TMP1} \\ B_I{}' &= A_I - \mathrm{TMP2} \end{aligned} \qquad (8.12)$$

where

$$\mathrm{TMP1} = B_R W_R{}^r - B_I W_I{}^r \qquad (8.13)$$

and

$$\mathrm{TMP2} = B_I W_R{}^r + B_R W_I{}^r \qquad (8.14)$$

Thus, to compute the butterfly one can use Equations 8.13 and 8.14 to first compute TMP1 and TMP2 and then use these in Equation 8.12.

## 8.4 Overflow and Scaling

The data must be properly scaled down before or during a butterfly computation to avoid overflow at any stage of calculations. Overflow leads to a useless transformed result. However, excessive scaling leads to precision problems due to dropping of the least significant bits. Thus, one needs to have an idea about the magnitudes of signal values so that scaling is applied only when needed. In essence, the purpose of scaling should be to avoid overflow without sacrificing precision.

Consider the following equation in the butterfly computation

$$A_I{}' = A_I + B_I W_R{}^r + B_R W_I{}^r \tag{8.15}$$

where $W_R{}^r = \cos\theta$, $W_I{}^r = \sin\theta$, $\theta = 2\pi nk/N$. Substituting for the twiddle factor gives

$$A_I{}' = A_I + B_I \cos\theta + B_R \sin\theta \tag{8.16}$$

The maximum value of $A_I{}'$ occurs when $\partial A_I{}'/\partial\theta = 0$. This yields

$$\partial A_I{}'/\partial\theta = -B_I \sin\theta + B_R \cos\theta = 0$$

That is,

$$\tan\theta = B_R/B_I \tag{8.17}$$

which yields

$$\sin\theta = \frac{B_R}{\sqrt{B_R{}^2 + B_I{}^2}}$$

$$\cos\theta = \frac{B_I}{\sqrt{B_R{}^2 + B_I{}^2}} \tag{8.18}$$

Substituting $\sin\theta$ and $\cos\theta$ in Equation 8.16 yields

$$AI'_{max} = A_I + \sqrt{B_R{}^2 + B_I{}^2} \tag{8.19}$$

If we assume that the maximum value of each variable in Equation 8.19 is 1, then the maximum possible value that $AI'$ can attain is given as

$$AI'_{max} = 1 + \sqrt{2} = 2.414$$

Similarly it can be shown that the maxima for other computed variables in Equation 8.12 in the butterfly computations are also equal to 2.414. Therefore, to avoid overflow each input variable can be multiplied by $1/2.414 = 0.414$ before computing the butterfly. The butterfly computation is modified, by including this scale factor, as shown in Figure 8.5(a). If a shift operation, which is simpler to implement, is used to scale the variables, the scale factor to avoid overflow should be 0.25. Figure 8.5(b) shows the butterfly computation that uses 0.25 as the scale factor. Use of the shift operation is preferred in programmable signal processors where it is implemented as part of data transfer and requires no additional execution time. However, in such a case we may be scaling more than what is absolutely needed and thus compromising the computational accuracy. For simplicity we will use shift in the FFT implementation example considered in a subsequent section. Figure 8.5(c) is the

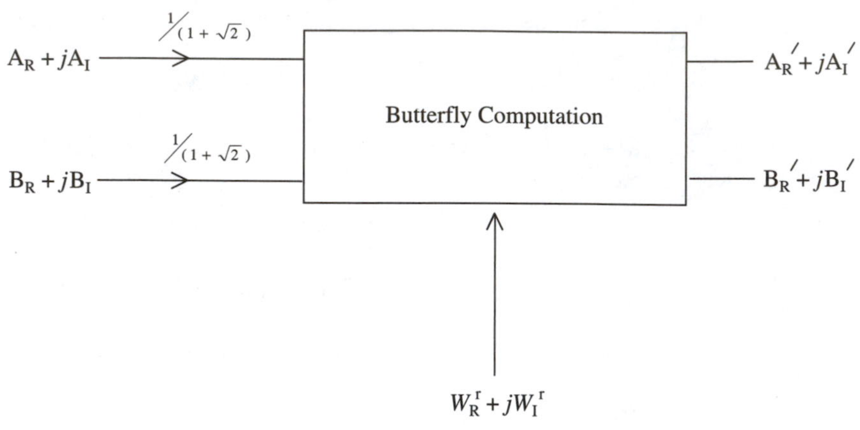


**Figure 8.5(a)** Butterfly computation, where the magnitude of all numbers is limited to less than 1, using a scale factor = $1/(1 + \sqrt{2})$

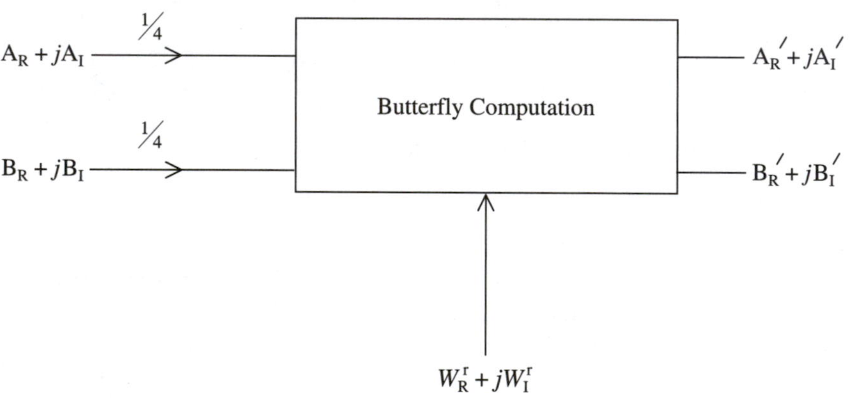


**Figure 8.5(b)** Butterfly computation where all magnitudes must be less than 1 and the scale factor is a power of 2

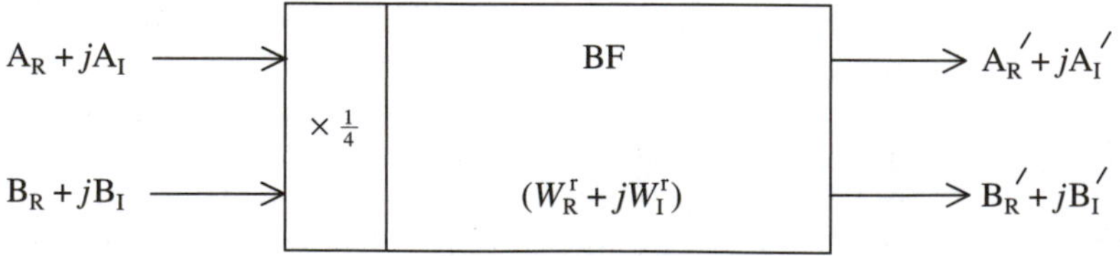


**Figure 8.5(c)** A representation for the butterfly computation using a scale factor of 1/4

representation for the butterfly structure that will be used in the DIT FFT implementation. This representation includes the scale factor as well as the twiddle factor.

## 8.5 Bit-Reversed Index Generation

The table in Figure 8.6(a) shows the relationship between the naturally occurring original input indices and the indices with reference to which DIT DFT is computed. The bit-reversed indices, needed for the DIT FFT implementation, can be generated using a reverse carry add operation, as shown in the example of Figure 8.6(b). For instance, if the current bit-reversed index is $0100_2$ in an 8-point DFT, then, the next bit-reversed index is obtained by adding $0100_2$ (half the DFT size) using reverse carry propagation (carry moving to the right).

As discussed in Chapter 5 TMS320C54xx has an addressing mode that allows one to implement bit reversing in a very convenient manner. As the naturally sequenced input data is obtained, it is bit reversed before placing in memory for FFT computation.

| Original Index | Bit-Reversed Index |
|---|---|
| 000 | 000 |
| 001 | 100 |
| 010 | 010 |
| 011 | 110 |
| 100 | 001 |
| 101 | 101 |
| 110 | 011 |
| 111 | 111 |

**Figure 8.6(a)** Bit-reversed indices in an 8-point DFT computation

| | |
|---|---|
| 0010 | (Carry in) |
| 0100 | (Current bit-reversed index) |
| + 0100 | (Half the number of DFT points) |
| 0010 | (Next bit-reversed index) |
| 0100 | (Carry out) |

**Figure 8.6(b)** Bit-reversed index generation example

## 8.6 An 8-Point FFT Implementation on the TMS320C54xx

An 8-point DIT FFT implementation structure based on the butterfly of Figure 8.5(c) is shown in Figure 8.7. The TMS320C54xx program that implements the algorithm is shown in Figure 8.8. The program uses subroutines for bit

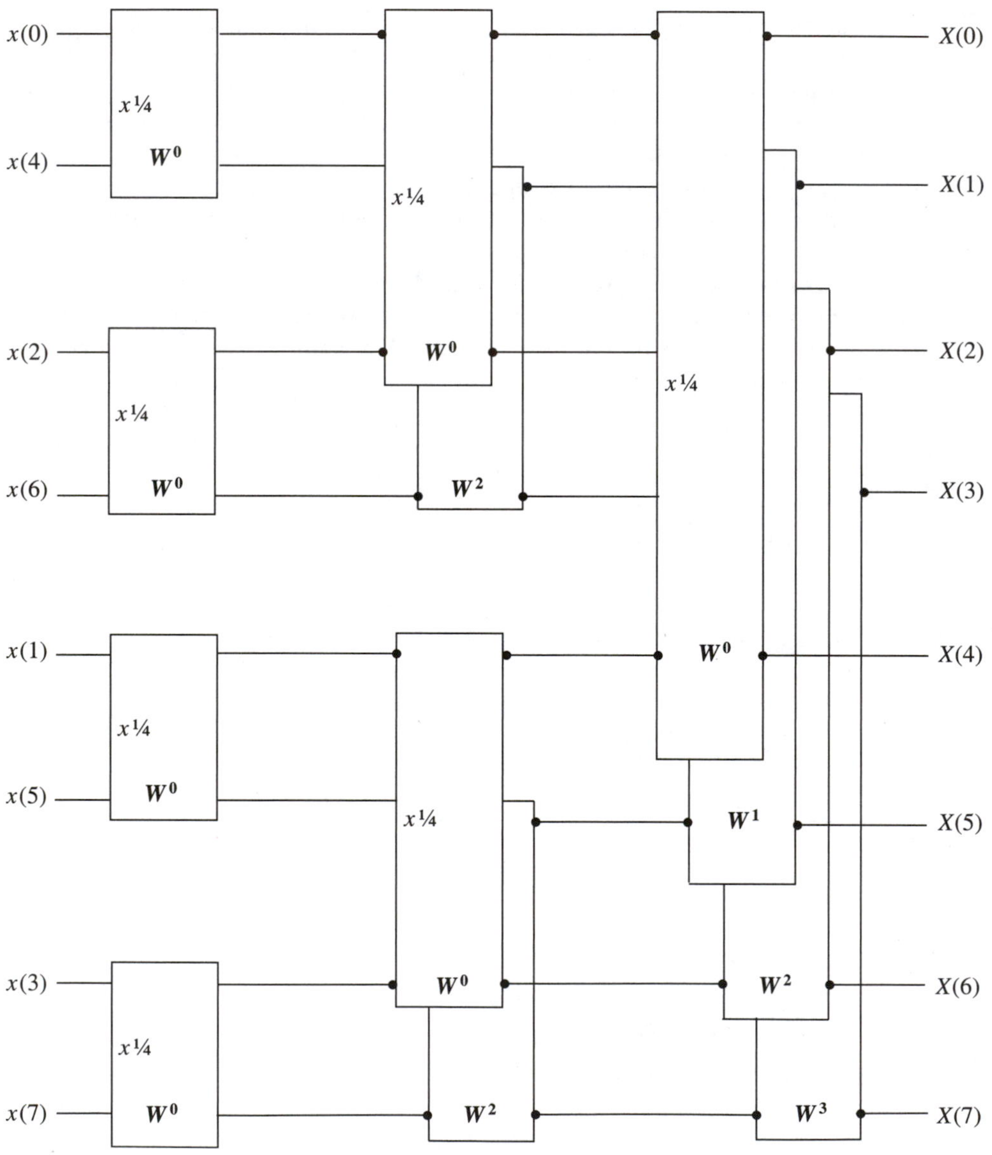


**Figure 8.7** An 8-point FFT implementation structure; scale factor for all butterflies = 1/4

```
;------------------------------------------------------------------------
;
; Program Name:  FFT8.asm
;
; Description:   This program implements an 8-point DIT FFT algorithm.
;
; Author:        Avtar Singh, SJSU
;
;------------------------------------------------------------------------
       .mmregs
       .def _c_int00

       .data
;------------------------------------------------------------------------
; Transformed data
;------------------------------------------------------------------------
X0R    .word 0                  ; Real part of X0
X0I    .word 0                  ; Imag part of X0
X1R    .word 0                  ; Real part of X1
X1I    .word 0                  ; Imag part of X1
X2R    .word 0                  ; Real part of X2
X2I    .word 0                  ; Imag part of X2
X3R    .word 0                  ; Real part of X3
X3I    .word 0                  ; Imag part of X3
X4R    .word 0                  ; Real part of X4
X4I    .word 0                  ; Imag part of X4
X5R    .word 0                  ; Real part of X5
X5I    .word 0                  ; Imag part of X5
X6R    .word 0                  ; Real part of X6
X6I    .word 0                  ; Imag part of X6
X7R    .word 0                  ; Real part of X6
X7I    .word 0                  ; Imag part of X7
;------------------------------------------------------------------------
; Input data. It should be replaced with the actual data for which the
; FFT is to be computed
;------------------------------------------------------------------------
x0     .word 0
x1     .word 23170
x2     .word 32767
x3     .word 23170
x4     .word 0
x5     .word -23170
x6     .word -32767
x7     .word -23170
```

**Figure 8.8** FFT implementation program for the TMS320C54xx *(continued)*

```
;-----------------------------------------------------------------------
; Twiddle Factors (q15 numbers)
;-----------------------------------------------------------------------
W08R  .word 32767             ; cos(0)
W08I  .word 0                 ; -sin(0)
W18R  .word 23170             ; cos(pi/4)
W18I  .word -23170            ; -sin(pi/4)
W28R  .word 0                 ; cos(pi/2)
W28I  .word -32767            ; -sin(pi/2)
W38R  .word -23170            ; cos(3pi/4)
W38I  .word -23170            ; -sin(3pi/4)
;-----------------------------------------------------------------------
; Spectrum Data
;-----------------------------------------------------------------------
S0    .word 0                 ; S0 = Freq 0.fs/8 contents
S1    .word 0                 ; S1 = Freq 1.fs/8 contents
S2    .word 0                 ; S2 = Freq 2.fs/8 contents
S3    .word 0                 ; S3 = Freq 3.fs/8 contents
S4    .word 0                 ; S4 = Freq 4.fs/8 contents
S5    .word 0                 ; S5 = Freq 5.fs/8 contents
S6    .word 0                 ; S6 = Freq 6.fs/8 contents
S7    .word 0                 ; S7 = Freq 7.fs/8 contents
;-----------------------------------------------------------------------
; Butterfly scratch-pad locations
;-----------------------------------------------------------------------
TMP1  .word 0
TMP2  .word 0

      .text
;-----------------------------------------------------------------------
;
; Main Program
; This program computes 8-point DFT using DIT FFT algorithm.
; It also computes signal spectrum using the transformed data.
;
;-----------------------------------------------------------------------
_c_int00:
        SSBX SXM              ; Select sign extension mode
        CALL _clear           ; Clear FFT data locations
        CALL _bitrev          ; Get bit-reversed input data

; STAGE 1 Butterflies:
```

**Figure 8.8** Continued

```
; Call BUTTERFLY with AR = X0R, AI = X0I, BR = X1R, BI = X1I
; Replace X0R, X0I, X1R, X1I

        STM  #X0R, AR1
        STM  #X1R, AR2
        STM  #W08R, AR3
        CALL _butterfly

; Call BUTTERFLY with AR = X2R, AI = X2I, BR = X3R, BI = X3I
; Replace X2R, X2I, X3R, X3I

        STM  #X2R, AR1
        STM  #X3R, AR2
        STM  #W08R, AR3
        CALL _butterfly

; Call BUTTERFLY with AR = X4R, AI = X4I, BR = X5R, BI = X5I
; Replace X4R, X4I, X5R, X5I

        STM  #X4R, AR1
        STM  #X5R, AR2
        STM  #W08R, AR3
        CALL _butterfly

; Call BUTTERFLY with AR = X6R, AI = X6I, BR = X7R, BI = X7I
; Replace X6R, X6I, X7R, X7I

        STM  #X6R, AR1
        STM  #X7R, AR2
        STM  #W08R, AR3
        CALL _butterfly

; STAGE 2 Butterflies:

; Call BUTTERFLY with AR = X0R, AI = X0I, BR = X21R, BI = X2I
; Replace X0R, X0I, X2R, X2I

        STM  #X0R, AR1
        STM  #X2R, AR2
        STM  #W08R, AR3
        CALL _butterfly
```

**Figure 8.8** Continued

```
; Apply Twiddle Factor W28 to X3R, X3I
; Call BUTTERFLY with AR = X1R, AI = X1I, BR = X3R, BI = X3I
; Replace X1R, X1I, X3R, X3I

        STM  #X1R, AR1
        STM  #X3R, AR2
        STM  #W28R, AR3
        CALL _butterfly

; Call BUTTERFLY with AR = X4R, AI = X4I, BR = X6R, BI = X6I
; Replace X4R, X4I, X6R, X6I

        STM  #X4R, AR1
        STM  #X6R, AR2
        STM  #W08R, AR3
        CALL _butterfly

; Apply Twiddle Factor W28 to X7R, X7I
; Call BUTTERFLY with AR = X5R, AI = X5I, BR = X7R, BI = X7I
; Replace X5R, X5I, X7R, X7I

        STM  #X5R, AR1
        STM  #X7R, AR2
        STM  #W28R, AR3
        CALL _butterfly

; STAGE 3 Butterflies:

; Call BUTTERFLY with AR = X0R, AI = X0I, BR = X4R, BI = X4I
; Replace X0R, X0I, X4R, X4I

        STM  #X0R, AR1
        STM  #X4R, AR2
        STM  #W08R, AR3
        CALL _butterfly

; Apply Twiddle Factor W18 to X5R, X5I
; Call BUTTERFLY with AR = X1R, AI = X1I, BR = X5R, BI = X5I
; Replace X1R, X1I, X5R, X5I

        STM  #X1R, AR1
        STM  #X5R, AR2
        STM  #W18R, AR3
        CALL _butterfly
```

**Figure 8.8** Continued

```
; Apply Twiddle Factor W28 to X6R, X6I
; Call BUTTERFLY with AR = X2R, AI = X2I, BR = X6R, BI = X6I
; Replace X2R, X2I, X6R, X6I

        STM  #X2R, AR1
        STM  #X6R, AR2
        STM  #W28R, AR3
        CALL _butterfly

; Apply Twiddle Factor W38 to X7R, X7I
; Call BUTTERFLY with AR = X3R, AI = X3I, BR = X7R, BI = X7I
; Replace X3R, X3I, X7R, X7I

        STM  #X3R, AR1
        STM  #X7R, AR2
        STM  #W38R, AR3
        CALL _butterfly

; Spectrum computation

        STM  #X0R, AR1          ; AR1 points to transformed X0R
        STM  #S0, AR2           ; AR2 points to spectrum S0
        STM  #7, AR3            ; AR3 = # of spectrum points-1

        CALL _spectrum          ; Compute signal spectrum
        nop
        nop
;-----------------------------------------------------------------------
; This subroutine moves the data to the FFT memory.
; The data is written in bit-reversed order.
;-----------------------------------------------------------------------
_bitrev:
        STM #x0, AR1            ; AR1 points to input sample x0
        STM #X0R, AR2           ; AR2 points to FFT data memory start
        STM #8, AR0             ; AR0 = FFT order = 8
        STM #7, AR3             ; AR3 = FFT order-1 = 7
loop:
        LD *AR1+, A             ; Get next input data sample
        STL A, *AR2+0B          ; Store bit-reversed in FFT memory
        BANZ loop, *AR3-        ; Repeat for all input samples
        RET
        nop
        nop
```

**Figure 8.8** Continued

```
;-----------------------------------------------------------------------
; Clear FFT data memory routine
;-----------------------------------------------------------------------
_clear:
        STM #XOR, AR2           ; AR2 points to FFT data memory
        RPT #15                 ; Clear FFT memory
        ST #0, *AR2+
        RET
        nop
        nop
;-----------------------------------------------------------------------
; This subroutine implements the butterfly computation
;
;       Use AR1 as pointer to first complex number.
;       Use AR2 as pointer to second complex number.
;       Use AR3 as pointer to twiddle factor.
;
;       AR <= AR + BR*WR - BI*WI
;       AI <= AI + BR*WI + BI*WR
;       BR <= AR - BR*WR + BI*WI
;       BI <= AI - BR*WR - BI*WI
;
;       Scale Factor = 1/4
;
;-----------------------------------------------------------------------
_butterfly:
        MVMM AR1, AR5           ; AR5 = AR1
        STM  #TMP1, AR4         ; AR4 points to TMP1

        LD *AR5, -2, A
        STL A, *AR5+            ; Replace AR with AR/4
        LD *AR5, -2, A
        STL A, *AR5-            ; Replace AI with AI/4

        LD *AR2, -2, A
        STL A, *AR2+            ; Replace BR with BR/4
        LD *AR2, -2, A
        STL A, *AR2-            ; Replace BI with BI/4

        LD *AR5+, A
        STL A, 1, *AR4+         ; Store AR in TMP1
        LD *AR5-, A
        STL A, 1, *AR4-         ; Store AI in TMP2
```

**Figure 8.8** Continued

```
        ;AR <= AR + BR*WR - BI*WI
        LD #0, A                ; A = 0
        MPY *AR2+, *AR3+, A     ; A = BR*WR
        MAS *AR2-, *AR3, A      ; A = (BR*WR) - BI*WI
        ADD *AR5, 15, A         ; A = (BR*WR - BI*WI) + AR
        ADD #1,14,A             ; Round the result
        STH A, 1, *AR5+         ; Save computed AR

        ;AI <= AI + BR*WI + BI*WR
        LD #0, A                ; A = 0
        MPY *AR2+, *AR3-, A     ; A = BR*WI
        MAC *AR2-, *AR3, A      ; A = (BR*WI) + BI*WR
        ADD *AR5, 15, A         ; A = (BR*WI + BI*WR) + AI
        ADD #1,14,A             ; Round the result
        STH A, 1, *AR5-         ; Save computed AI

        ;BR <= AR - (BR*WR - BI*WI)
        LD *AR4+, A             ; A = AR
        SUB *AR5+, A            ; A = AR-(BR*WR - BI*WI)
        STL A, *AR2+            ; Save computed BR

        ;BI <= AI - (BR*WI + BI*WR)
        LD *AR4-, A             ; A = AI
        SUB *AR5-, A            ; A = AI-(BR*WI + BI*WR)
        STL A, *AR2-            ; Save computed BI
        RET
        nop
        nop
;---------------------------------------------------------------------
; This subroutine computes the spectrum of the transformed data.
;
;       Use AR1 as pointer to the transformed data.
;       Use AR2 as pointer to the spectrum data.
;
;       S(k) = (1/N)*|X(k)|*|Conj(X(k))|
;
;---------------------------------------------------------------------
_spectrum:
        LD #0,A                 ; A = 0
        LD #0,B                 ; B = 0
        SQUR *AR1+,A            ; Square X(k) real
        SQUR *AR1+,B            ; Square X(k) imaginary
        ADD B,A                 ; A = |X(k)|.|Conj(X(k))|
```

**Figure 8.8** Continued

```
STH A,1,*AR2
LD *AR2,13,A            ; divide by 8
STH A, *AR2+            ; Store the spectrum result
BANZ _spectrum, *AR3-
RET
nop
nop

.end
```

**Figure 8.8** Continued

reversing and butterfly computation, as described earlier. For programming details see references [2, 3]. The program is written to carry out computation stage by stage, starting from the left and proceeding to the right. For simplicity, the implementation uses the butterfly routine including the scaling within the butterfly. More accurate implementations are possible that exploit scaling only when needed. For instance, the scale factor of 0.25 in an 8-point FFT computation results in overall scaling of $0.25^3 = .015625$. However, the required scaling is $= 0.414^3 = 0.071$. If we apply a scaling of 0.25 to the first two stages and none to the third, the overall scaling will be 0.0625, which is adequate to avoid an overflow. Similarly, other scaling strategies can be developed, and these are left as exercises to explore. The implemented scale factor can be accounted for in the interpretation of the transformed data or it can be used to scale the result back to obtain the true transformed result.

The program in Figure 8.8 can be extended to transform any $x(n)$ sequence with numbers that are powers of 2. A sequence that does not satisfy this condition can be extended to the next power-of-2 number by appending it with zeros. The zero-appended sequence can then be processed to compute the transform. These extensions are left as exercises. In order to extend the program to a higher number of points, such as 16 or 32, we need to include more calls to additional butterflies. A simple extension of the program based on adding more calls makes it unmanageable. In such a case, the program should be restructured to incorporate nested loops. In such an implementation the computation will proceed similarly for each stage, computing the butterfly in the innermost loop. This, however, requires storing all the twiddle factors, including $W_N{}^0$, in sequential memory locations. Such an implementation is left as an exercise for the reader.

## 8.7 Computation of the Signal Spectrum

The *spectrum* of a signal describes the power associated with each frequency content of the signal. The spectrum estimate for an $N$-point transform is given by [1]

$$S(k) = (1/N)X^2(k) = (1/N)X(k)X^*(k) \tag{8.20}$$

where $k = 0, 1, 2, \ldots, (N-1)$. If $(1/N)$ is absorbed in a scale factor, then Equation 8.20 can be computed from

$$S(k) = (\text{Real}(X(k)))^2 + (\text{Imag}(X(k)))^2 \tag{8.21}$$

Figure 8.8 includes a subroutine to compute the signal spectrum using the result of the 8-point FFT.

## 8.8 Summary

This chapter is about the implementation of an FFT algorithm on the fixed-point signal processor TMS320C54xx. The FFT computation structure is described. The butterfly and bit-reversing aspects are covered from an implementation point of view. The implementation issues, such as overflow and scaling, are discussed. The chapter includes an implementation example for an 8-point DIT FFT algorithm. The example also includes spectrum computation using the FFT result.

## References

1. Strum, R. D., and Kirk, D. E. *First Principles of Discrete Systems and Digital Signal Processing*, Addison Wesley, 1988.
2. *TMS320C54x DSP Programmer's Guide* (spru538.pdf, 231 KB), 2001.
3. *MS320C54x DSP Mnemonic Instruction Set, Reference Set, Volume 2 (Rev. C)*, (spru172c.pdf, 1096 KB), 2001.

## Assignments

1. Determine the following for a 128-point FFT computation:
   a. number of stages
   b. number of butterflies in each stage
   c. number of butterflies needed for the entire computation
   d. number of butterflies that need no twiddle factors

e. number of butterflies that require real twiddle factors

f. number of butterflies that require complex twiddle factors.

**2.** What minimum size FFT must be used to compute a DFT of 40 points? What must be done to the samples before the chosen FFT is applied?

**3.** How many add/subtract (A) and multiply (M) operations are needed to implement a general butterfly similar to the one described in Section 8.3?

**4.** Show that the butterfly computation of Section 8.3 can also be implemented using the following equations:

$$A_R' = A_R + B_R W_R^r - B_I W_I^r$$

$$A_I' = A_I + B_I W_R^r + B_R W_I^r$$

$$B_R' = 2A_R - A_R'$$

$$B_I' = 2A_I - A_I'$$

**5.** Compare the butterfly implementation in Problem 3, with that in Problem 4 in terms of multiply, add, and shift operations.

**6.** Compare the following specific cases of butterfly implementation using the equations in Section 8.3:

a. $A_I = B_I = 0,\ W_R^r + jW_I^r = 1$

b. $W_R^r + jW_I^r = 1$

c. $W_R^r + jW_I^r = j$

**7.** Derive equations, similar to the ones in Section 8.3, to implement a butterfly encountered in a DIF FFT implementation. Such a butterfly is represented by the following equations:

$$A_R' + jA_I' = (A_R + jA_I) + (B_R + jB_I)$$

$$B_R' + jB_I' = ((A_R + jA_I) - (B_R + jB_I))(W_R^r + jW_I^r)$$

**8.** Derive the optimum scaling factor for the DIF FFT butterfly.

**9.** How can the program of Figure 8.8 be modified so that scaling is done only when needed?

**10.** Rewrite the program in Figure 8.8 using nested loops so that there is just one CALL statement to call a butterfly routine.

**11.** Modify the program in Problem 10 so that it can be used to compute a FFT for any number of points that are powers of 2.

**12.** Modify the program in Problem 11 so that it can be used to compute a FFT for points that are not powers of 2.

**13.** A time-domain sequence of 73 elements is to be convolved with another time-domain sequence of 50 elements using DFT to transform the two sequences, multiplying them, and then doing IDFT to obtain the resulting time-domain sequence. To implement DFT or IDFT, the DIT-FFT algorithm is to be used.

Determine the total number of complex multiplies needed to implement the convolution. Assume that each butterfly computation requires one complex multiplication.

**14.** The computation in Problem 13 is to be implemented on a fixed-point signal processor that takes 10 ns to do a real integer multiplication. Determine the convolution computation time. If the convolution is to be implemented for a real-time signal and each time a new sample is received the transform is to be calculated; determine the highest frequency signal that can be handled by the signal processor.

# Chapter 9

## Interfacing Memory and Parallel I/O Peripherals to Programmable DSP Devices

## 9.1 Introduction

In previous chapters, we studied the architectures of digital signal processors and learned about their instruction set and programming techniques. In a complex DSP system, in addition to the processor, there are also external peripherals, such as memory and input/output devices. In order to interface such peripherals, we need to understand various interfacing DSP signals and the techniques for using them. Peripherals can be interfaced to a processor either in serial or in parallel mode. In the serial mode, data transfer takes place bit by bit; in the parallel mode transfer takes place word by word. The choice is based on the nature of the peripheral and the desired data transfer rate.

In this chapter, we consider the interfacing signals of the TMS320C54xx processors and use of these signals for parallel interfacing of memory and peripherals. These topics are covered under the following headings:

Memory space organization
External bus interfacing signals
Memory interface
Parallel I/O interface
Programmed I/O
Interrupts and I/O
Direct memory access

## 9.2 Memory Space Organization

The TMS320C54xx devices each support a basic memory space (internal and external) of 192K 16-bit words. This consists of 64K words of program mem-

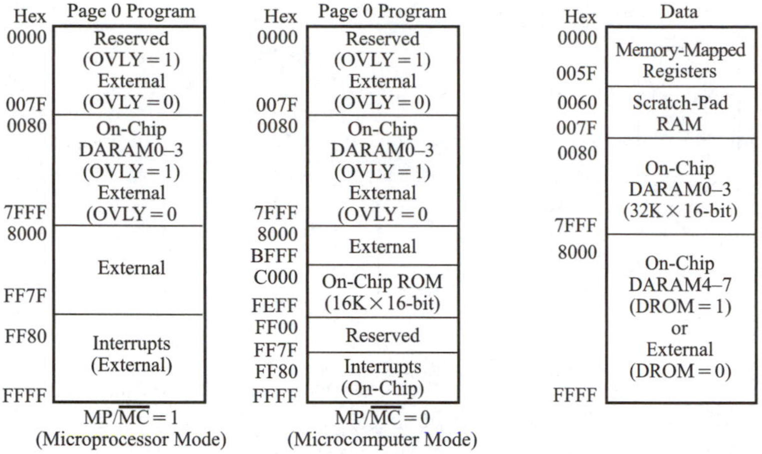

Page 0 Program — MP/$\overline{\text{MC}}$ = 1 (Microprocessor Mode)

| Hex | Page 0 Program |
|---|---|
| 0000–007F | Reserved (OVLY = 1) External (OVLY = 0) |
| 0080–7FFF | On-Chip DARAM0–3 (OVLY = 1) External (OVLY = 0 |
| 8000–FF7F | External |
| FF80–FFFF | Interrupts (External) |

Page 0 Program — MP/$\overline{\text{MC}}$ = 0 (Microcomputer Mode)

| Hex | Page 0 Program |
|---|---|
| 0000–007F | Reserved (OVLY = 1) External (OVLY = 0) |
| 0080–7FFF | On-Chip DARAM0–3 (OVLY = 1) External (OVLY = 0 |
| 8000–BFFF | External |
| C000–FEFF | On-Chip ROM (16K × 16-bit) |
| FF00–FF7F | Reserved |
| FF80–FFFF | Interrupts (On-Chip) |

| Hex | Data |
|---|---|
| 0000–005F | Memory-Mapped Registers |
| 0060–007F | Scratch-Pad RAM |
| 0080–7FFF | On-Chip DARAM0–3 (32K × 16-bit) |
| 8000–FFFF | On-Chip DARAM4–7 (DROM = 1) or External (DROM = 0) |

Address ranges for on-chip DARAM in data memory are: DARAM0: 0080h–1FFFh; DARAM1: 2000h–3FFFh; DARAM2: 4000h–5FFFh; DARAM3: 6000h–7FFFh; DARAM4: 8000h–9FFFh; DARAM5: A000h–BFFFh; DARAM6: C000h–DFFFh; DARAM7: E000h–FFFFh

**Figure 9.1** Memory map of TMS320C5416

(Courtesy of Texas Instruments Inc.)

ory, 64K words of data memory, and 64K words of I/O space. Program and data memories can comprise of both internal (on-chip) and external (off-chip) memories. The actual amount of memory depends upon the particular DSP device of the family.

Depending on a specific C54xx device, the on-chip program memory can be ROM, DARAM, SARAM, or combinations of these types. The on-chip memory of a device is mapped to the space by three CPU status register bits—MP/$\overline{\text{MC}}$, OVLY, and DROM. As shown in Figures 9.1 and 9.2, the on-chip memory of the TMS320VC5416 processor consists of 16K ROM, 64K DARAM, and 64K SARAM [1].

Devices with boot loader ROM, lookup tables such as a sine table, and an interrupt vector table are also available for applications that need these capabilities. In some of the C54xx devices, the program memory can be extended up to 8192K words by providing external memory-addressing capability. For the implementation of external memory systems these devices may be provided with up to 23 address lines to access the memory. For example, the C5416 provides 23 address lines that provide the capability of addressing up to 8192K of memory space in 128 64K word pages, as shown in Figure 9.2.

Data memory can also be both on-chip and off-chip. As shown in Figure 9.1, the on-chip DARAM of the C5416 can be mapped as on-chip program and/or data memory. The on-chip ROM can be mapped as on-chip program

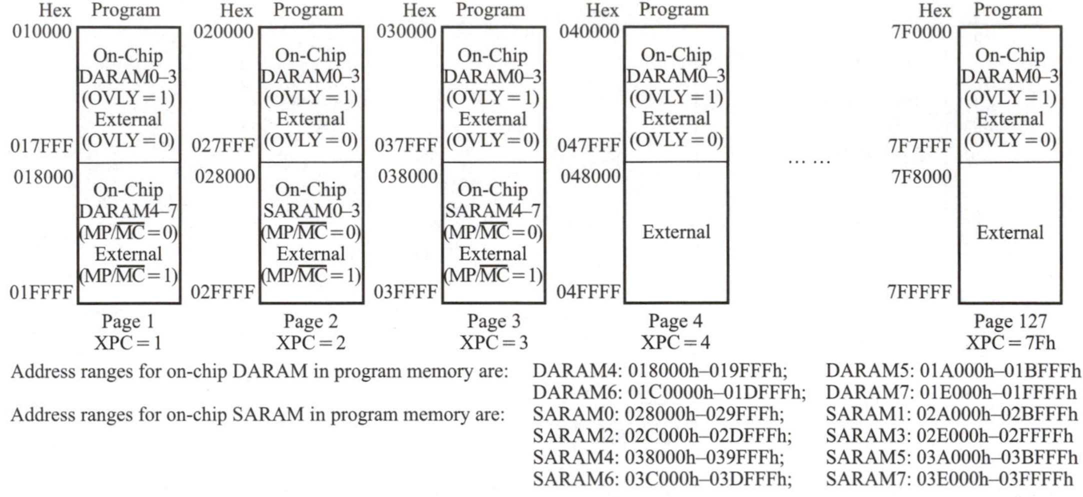


**Figure 9.2** Extended memory map of TMS320C5416

(Courtesy of Texas Instruments Inc.)

memory, or this space can be in the external memory. These flexibilities are provided to support applications with different types of needs.

On-chip memory is faster than external memory and has no interfacing requirements because it is within the chip. It consumes less power compared to external memory and enables higher performance of the DSP because of better flow within the pipeline of the central arithmetic logic unit. However, external memory provides a large memory space and hence is used when large memory size is required.

## 9.3 External Bus Interfacing Signals

A DSP device can be interfaced to a wide variety of peripherals by means of its address bus, data bus, and a set of control signals. Important external interfacing signals of TMS320C5416 devices are given in Table 9.1. The use of many of these signals should become evident when we discuss memory and I/O interfacing later in this chapter.

## 9.4 Memory Interface

In the processor architecture, separate on-chip data and program memories are provided to enhance the speed of program execution by using parallelism.

**Table 9.1** Memory and I/O Interfacing Signals of the TMS320C5416 Device

| Signal | Description |
|---|---|
| A0–A19 | 20-bit Address Bus |
| D0–D15 | 16-bit Data Bus |
| $\overline{\text{DS}}$ | Data Space Select |
| $\overline{\text{PS}}$ | Program Space Select |
| $\overline{\text{IS}}$ | I/O Space Select |
| R/$\overline{\text{W}}$ | Read/Write Signal |
| $\overline{\text{MSTRB}}$ | Memory Strobe |
| $\overline{\text{IOSTB}}$ | I/O Strobe |
| READY | Data Ready Signal |
| $\overline{\text{HOLD}}$ | Hold Request |
| $\overline{\text{HOLDA}}$ | Hold Acknowledge |
| $\overline{\text{MSC}}$ | Micro State Complete |
| $\overline{\text{IRQ}}$ | Interrupt Request |
| $\overline{\text{IACK}}$ | Interrupt Acknowledge |
| XF | External Flag Output |
| $\overline{\text{BIO}}$ | Branch Control Input |

Due to this parallel configuration and their dual-access capability, up to four concurrent memory operations can be performed in one cycle. These include three reads and one write operation. In spite of the advantages of on-chip memory, size constraints may require the designer to use external memory.

The external memory interface of the C54xx processors consists of a 16- to 23-bit address bus (depending on the device), a 16-bit data bus, and interfacing control signals. The interfacing signals are used to generate chip select ($\overline{\text{CS}}$), output enable ($\overline{\text{OE}}$), and write enable ($\overline{\text{WE}}$) signals required for accessing the memory for data transfer [3]. Figure 9.3 shows a block diagram for the memory interface of the C5416 processor. Notice that the job of the interface is to use the processor signals and generate the appropriate signals for setting up communication with the memory.

## 9.4.1 Timing Sequence for External Memory Access

The timing reference for the external memory access is provided by the CLKOUT signal of the C54xx devices. Depending on the operation performed, the external memory requires a number of clock cycles. During the entire memory read and write operations, $\overline{\text{MSTRB}}$ remains low and the $\overline{\text{PS}}$ and $\overline{\text{DS}}$ are active while program memory and data memory, respectively, are

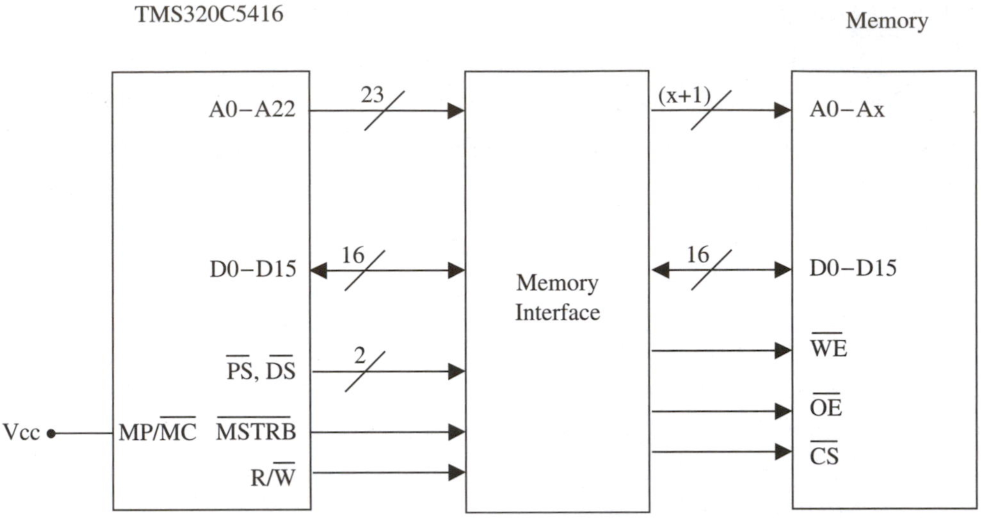


**Figure 9.3** Memory interface block diagram for the TMS320C5416 processor

accessed. The R/$\overline{\text{W}}$ signal is used to specify the direction of data transfer. Figure 9.4 shows the TMS320C54xx signals during two memory reads and a memory write operation. The strobe signal, $\overline{\text{MSTRB}}$ remains low for both read and write operations. R/$\overline{\text{W}}$ is high for the read operations and becomes low for the write operation. Note that the write operation requires two cycles. This is because, in the example, the write operation is for an external memory location. Also note that during the read operation, $\overline{\text{PS}}$ is low since the read locations are in the program space. Likewise, during the write operation, $\overline{\text{DS}}$ is low, indicating a write operation with the data memory.

### 9.4.2 **Wait States**

The TMS320C54xx DSP can be interfaced to slower off-chip memories and I/O devices by introducing wait states. Software programmable wait states are easily incorporated without any external hardware. The user-accessible memory-mapped software wait state register (SWWSR) controls the internal software wait state generator. Program and data memory spaces have two pages each of 32K, and for the I/O, a single page of 64K that can be programmed to have software-generated wait states. This is done by means of a three-bit field, for the corresponding space and address range, in the SWWSR: 000 corresponds to no wait state and 111 to seven wait states. Memory devices that require more than seven wait states have to be interfaced using the hardware READY signal. An external device uses the READY signal to indicate its

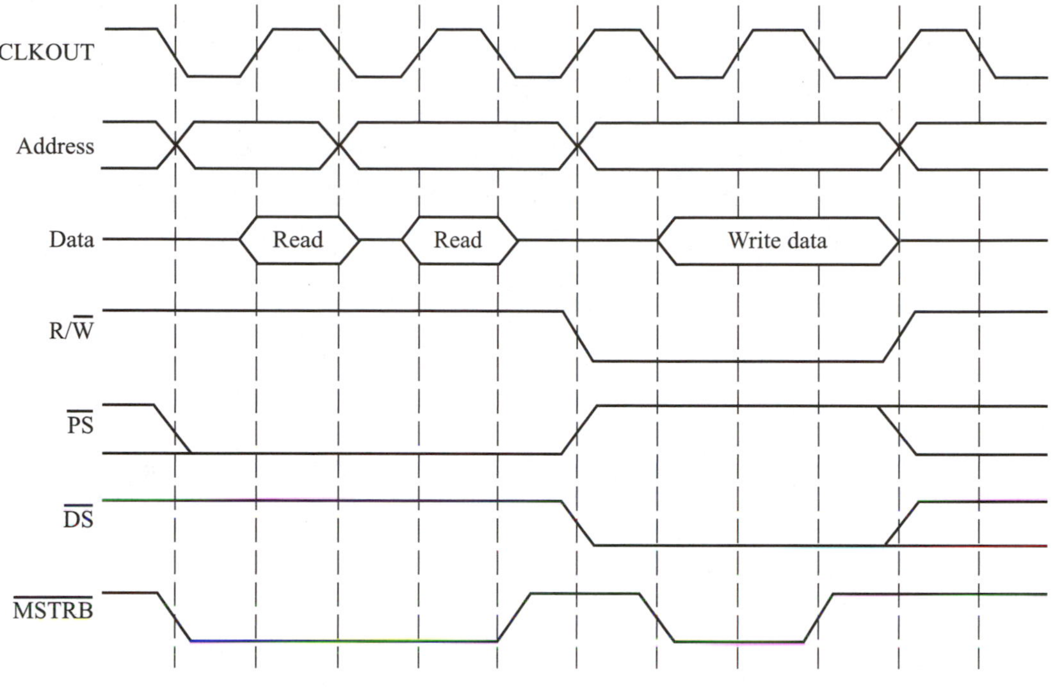


**Figure 9.4** Memory interface signals for a read–read–write sequence of operations
(Courtesy of Texas Instruments Inc.)

readiness for the bus transaction. The processor checks the READY signal during each bus cycle and completes the bus cycle only when the signal becomes logic 1.

The primary goal of a DSP is to make external memory access as fast as possible. The interface hardware introduces signal delay and thus slows the memory accessing. One solution is to design the interface without any device. Such an example is shown in Figure 9.5. There is no address decoding to generate chip-select signals. This means that the entire addressing space is used by just one 8K × 16 SRAM device. For instance, the memory not only responds to the address range 0000–1FFFh, it also responds to all the ranges generated by all possible combinations of the unused address bits A13–A19. The $\overline{PS}$ and $\overline{DS}$ signals are not combined with the R/$\overline{W}$ signal to generate the $\overline{WE}$ signal—only R/$\overline{W}$ is connected to $\overline{WE}$. This means that the SRAM is indistinguishable as a program or data RAM.

One subtle point to remember is that only the program memory can be paged in TMS320C54xx processors. The $\overline{DS}$ pin will never go low above the 0FFFFh address. Paging in program memory is controlled by the XPC register. It allows paging of seven extra address lines in program space. For example, if

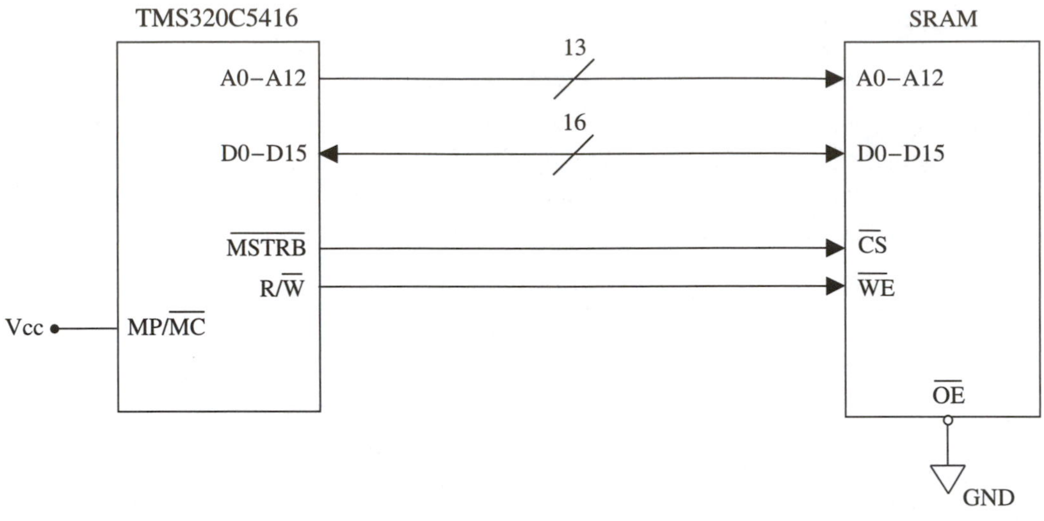


**Figure 9.5** An example of a no-decode external memory interface

A17 of the memory device with storage of 256K × 16 RAM is connected to $\overline{\text{PS}}$, only 192K words can be accessed. This is because 64K words of data memory (corresponding to $\overline{\text{PS}}$ high) will be lost, since data memory cannot be paged beyond 64K words; program memory can be paged for the full 128K words.

A disadvantage of external memory is that it may be slower than the processor and may not be able to keep pace with the processor. However, it can be accessed using wait states to slow the processor for transactions with the slow memory.

A way to access the program memory with zero wait state is to run the code from the internal memory. For this, the OVLY bit in the PMST register has to be set. This, however, causes the internal data memory (SRAM or DARAM) to overlap the program memory region, thus reducing the available memory space.

▷ **Example 9.1** Assuming that the SRAM in Figure 9.5 is to be used to hold a program, how many address ranges exist for the TMS320C5416 processor to access this memory?

Solution The address lines A13–A22 for the C5416 can take any binary value from 0000000000 to 1111111111. Any of these values combined with the specific value of A0–A12 generates the address for the same specific location. Since there are 10 bits that are don't cares, there exist $2^{10}$ or 1024 valid addresses for each location. For instance, the first location in the memory can be accessed using address 00000h, or address 12000h, or address 24000h, etc. Thus, 1024 address ranges exist for the memory in Figure 9.5.

## 9.4.3 Memory Design Examples

We now consider some simple examples to illustrate interfacing external memory devices with the TMS320C54xx signal processors.

▷ **Example 9.2** Design an interface to connect a 64K × 16 flash memory to a TMS320C54xx device. The processor address bus is A0–A15.

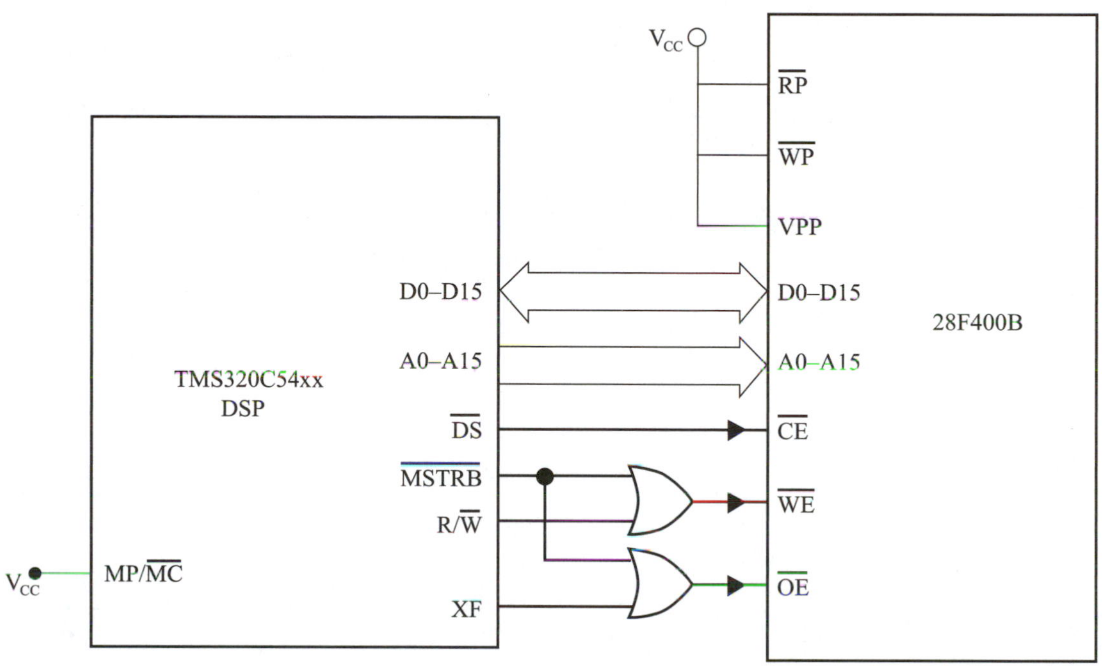


**Figure 9.6** An example of a flash memory interface for the TMS320C54xx DSP

(Courtesy of Texas Instruments Inc.)

Solution Figure 9.6 shows an interface between the TMS320C54xx device and the 64K × 16 flash memory [4]. The 16 address lines (A0–A15) are used to address the 64K flash memory. Writing into flash memory for programming requires wait states, while reading from it does not. Under program control, XF is driven low in the read mode and high in the write mode. In this example, external memory does not use the READY signal to interface with the DSP. Wait states may be introduced by appropriately loading the SWWSR register. The R/$\overline{\text{W}}$ signal is used along with $\overline{\text{MSTRB}}$ to provide the write-enable signal to the memory for programming purposes. For reading the memory, $\overline{\text{MSTRB}}$ is used along with the XF signal to enable the output of the chip.

▷ **Example 9.3** Design a data memory system with address range 000800h–000FFFh for a C5416 processor. Use 2K × 8 SRAM memory chips.

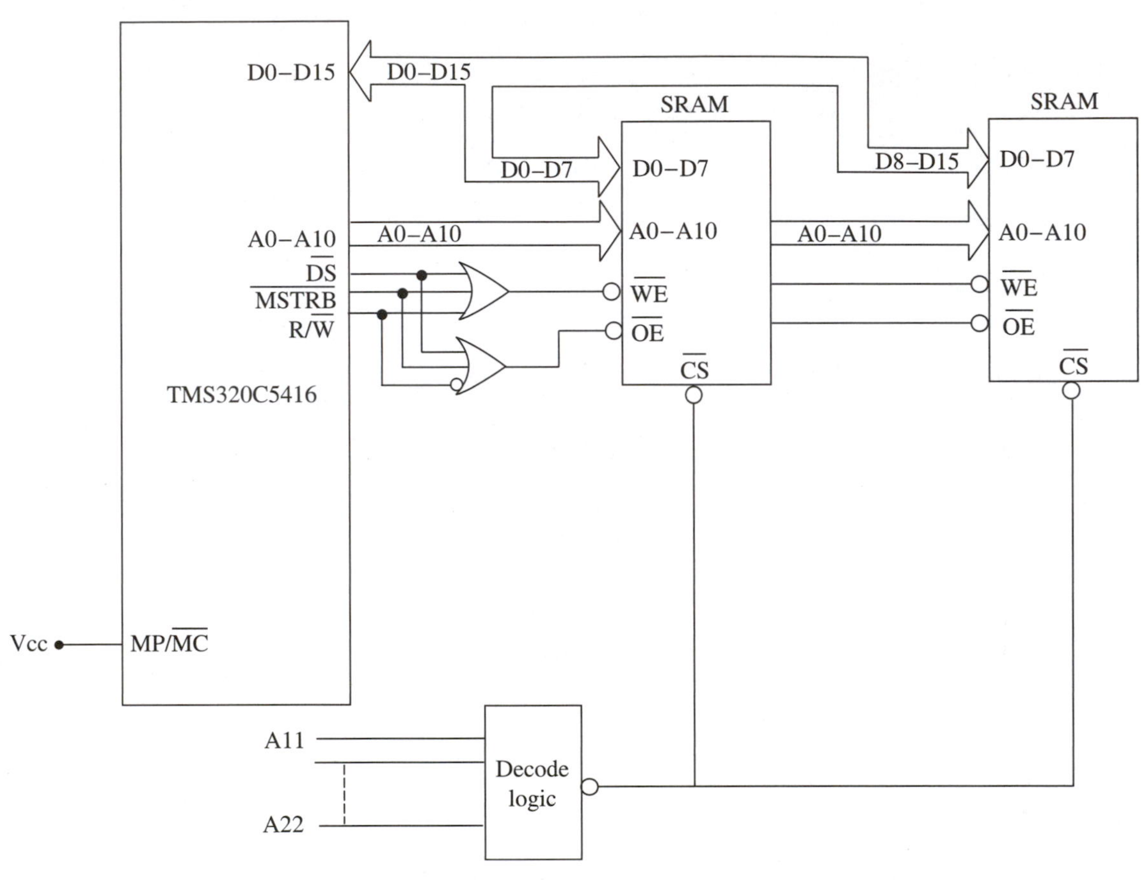


**Figure 9.7** Schematic of a 2K × 16 SRAM memory system for Example 9.3

Solution Figure 9.7 shows the memory interface. The width of the data bus for memory chips is 8 bits, but the width of the data bus for the processor is 16 bits. Hence, D0–D7 of the processor is connected to D0–D7 of the first memory chip and D8–D15 to D0–D7 of the second memory chip to create the 16-bit data bus. Output enable and write enable for the memory chips are generated by combining the $\overline{DS}$, $\overline{MSTRB}$, and R/$\overline{W}$ signals of the TMS320C5416 processor. Address lines A11–A22 are used in the decode logic to generate chip-select signals for the memory devices. These must all be logic 0 to generate the chip select for the two devices so that the memory responds to the desired address range.

▷ **Example 9.4** Interface an 8K × 16 program ROM to the C5416 DSP in the address range 7FE000h–7FFFFFh.

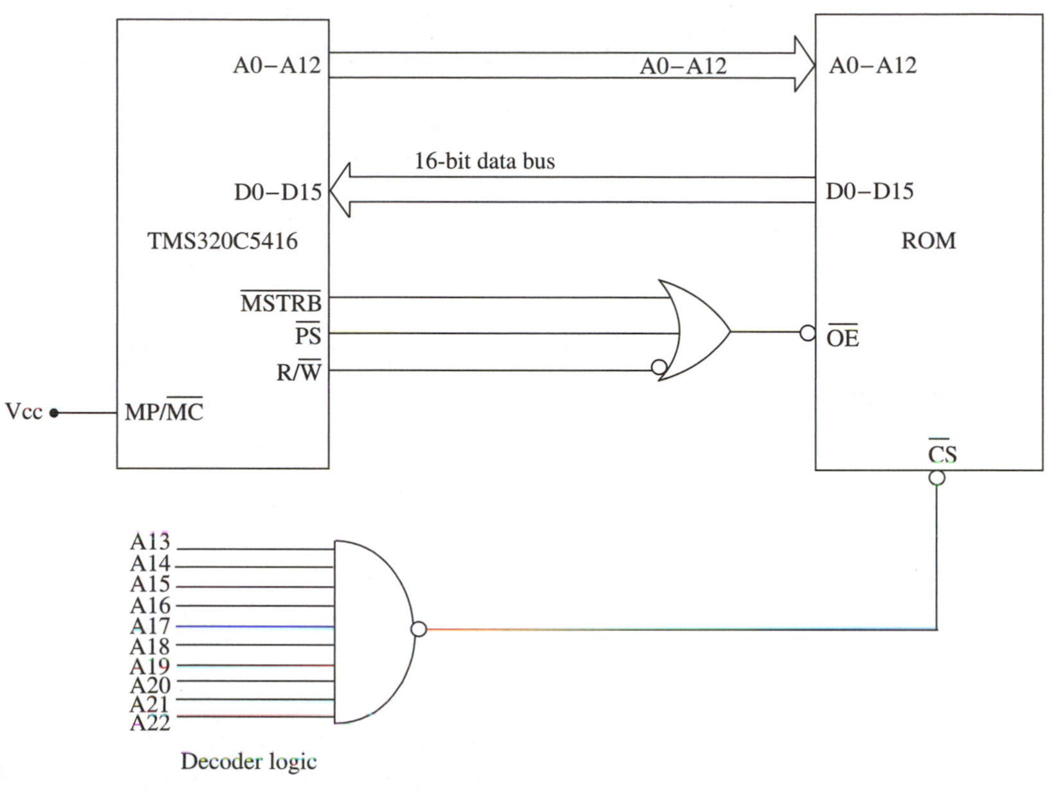


**Figure 9.8** Schematic of an 8K × 16 ROM memory interface circuit for Example 9.4

Solution Data flow takes place in only one direction while interfacing a ROM to a processor. Hence, generating only the output-enable signal is required for the memory device. Address lines A13–A22 are used to generate the chip-select control signal. Figure 9.8 shows the memory interface.

## 9.5 **Parallel I/O Interface**

Parallel I/O ports are used for interfacing external devices, such as A/D and D/A converters, to the DSP processor. Accessing I/O ports requires the use of PORTR and PORTW instructions. The PORTR (port read) instruction is used

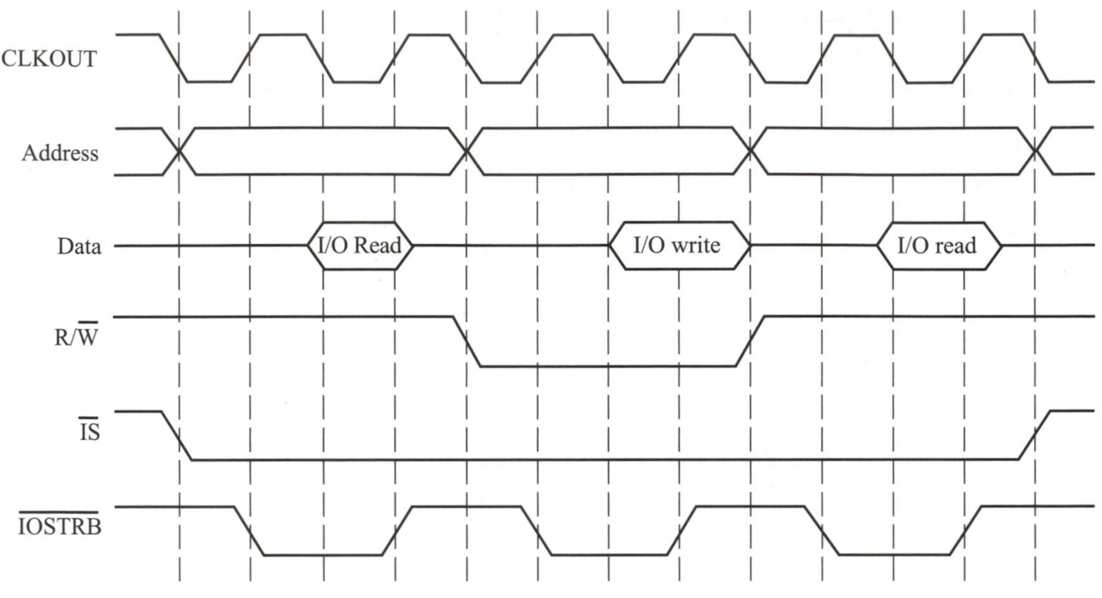


**Figure 9.9** I/O interface signals for a read–write–read sequence of operations

(Courtesy of Texas Instruments Inc.)

to read a peripheral connected to an input port. The data so read is placed in the specified data memory location. Similarly, the instruction PORTW (port write) is used to send the contents of the specified data memory location to an output port.

The timing diagram in Figure 9.9 shows the signals that are involved in an I/O transaction. This timing diagram is similar to a memory timing diagram except that a few different control signals are involved. The processor uses the $\overline{IS}$ signal to indicate an I/O operation. At least two clock cycles are required for performing the I/O read and write operations. During these operations, the $\overline{IOSTRB}$ signal remains low. This signal can be used to control the output enables of the external devices used to implement the I/O ports. Similar to a memory interface circuit, wait states can be inserted to interface slow peripherals.

We talk about three types of parallel I/O operations with a processor. These are unconditional I/O, programmed I/O, and interrupt I/O. Unconditional I/O is the simplest of the three types. This technique is used with devices that do not have any handshake signals. Programmed I/O and interrupt I/O are more sophisticated approaches, as these involve special signals and capabilities. In the next two sections, we discuss details of the programmed and interrupt I/O.

# 9.6 Programmed I/O

In this method, the CPU keeps polling the external device until it is ready for transmitting or receiving data. Software polling is used in programmed I/O to ascertain the readiness of the external device for a data transfer to or from the processor. C54xx devices have dedicated pins for this purpose. Control signals are sent and received via these pins by software. In addition, C54xx has two registers, named GPIOCR and GPIOSCR. GPIOCR is a general-purpose I/O register that is used to program the signals for I/O interfacing. GPIOSCR is a status register used to read the status of the handshake signals. Although these dedicated pins vary from one device to another, every version of the C54xx family has at least two dedicated pins for performing the I/O operations. These signals, as shown in Table 9.1, are $\overline{\text{BIO}}$ and XF. $\overline{\text{BIO}}$ is an input to the processor and XF is the output.

Using software, $\overline{\text{BIO}}$ can be used to monitor the status of an external peripheral. The XF signal is used to control the peripheral. This mode of communication using $\overline{\text{BIO}}$ and XF signals is asynchronous and is helpful in making the processor communicate with devices that are slower than the processor itself. Data length can be 8 bits or 16 bits. On detecting a low $\overline{\text{BIO}}$, the processor reads the peripheral data using the PORTR instruction. In turn, it informs the peripheral via XF about the completion of the transaction, allowing the processor to initiate the next transfer.

Figure 9.10 shows an example of an interface between an A/D converter

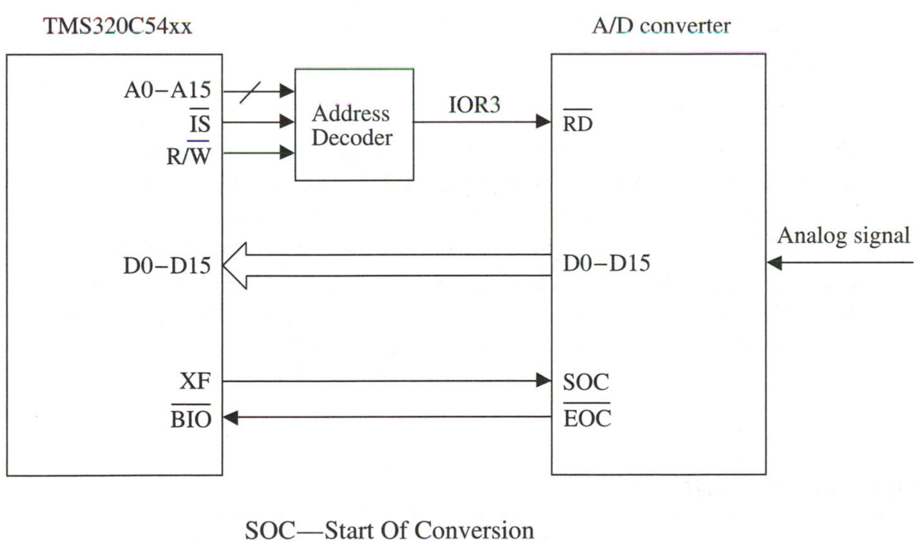


**Figure 9.10** An A/D converter interface in the programmed I/O mode

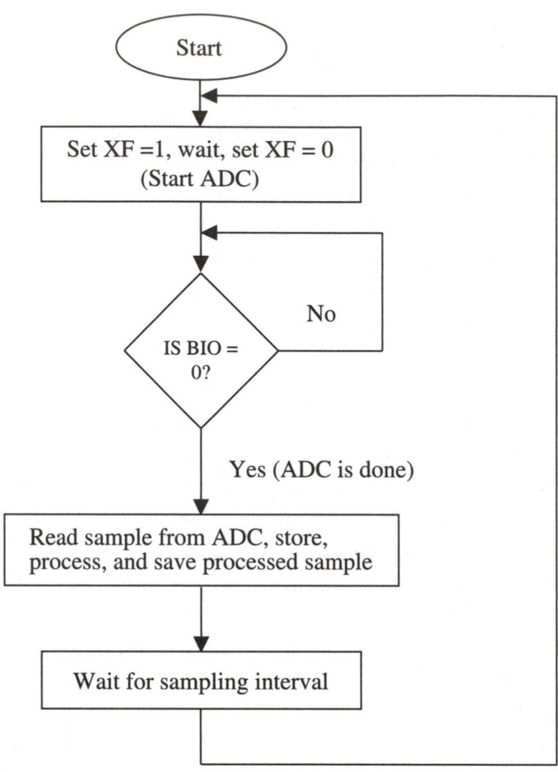


**Figure 9.11** Flow chart of the diagram for software polling for the programmed I/O interface of Figure 9.10

and the TMS320C54xx processor in the programmed I/O mode. Notice that XF is used to start the A/D conversion and $\overline{\text{BIO}}$ is used to determine its completion before the data is read.

The flow chart of the algorithm to implement the software polling used by the processor to communicate with the A/D converter is shown in Figure 9.11. The critical consideration in the implementation of this scheme is to control the time between any two consecutive XF or SOC pulses. This time is the sampling interval and must remain constant for all samples.

## 9.7 Interrupts and I/O

An interrupt is the signal that a DSP processor receives requesting it to execute a specific interrupt subroutine called a *service routine*. If certain conditions are satisfied, the processor suspends its current program and branches to execute the interrupt service routine. It resumes its previous activity after completing the service routine. Interrupt signals can be external or internal to the processor. Typically, these are requests for data exchange between the processor and a peripheral, such as a converter or another processor.

An interrupt request initiates a special processing by the processor. The request may be in the form of an electrical signal applied to the processor or may be by execution of an interrupt instruction. An interrupt instruction initiates what is called a *software interrupt*. The electrical interrupt signal initiates a hardware interrupt.

The table in Figure A.10 in Appendix A, called an *interrupt vector table*, lists all the interrupts that TMS320C5416 is capable of handling. As can be seen from the table, interrupt numbers are assigned to on-chip peripherals and to interrupt request signals. Each interrupt is assigned a priority and a memory location in the table. Priority is used to service the interrupt with higher priority when two requests are received simultaneously. The interrupt locations are used to branch to the service routines.

An example of a software interrupt is the instruction SINT18. In the TMS320C5416, this corresponds to software interrupt #18. The program counter branches to the software interrupt #18 at address location och. After executing the subroutine, it gets back to the suspended program. Hardware interrupt requests come from devices both external and internal to the processor. For instance, timer interrupt is an internal hardware interrupt, whereas INT2 is an external hardware interrupt.

Interrupts are also classified as maskable and nonmaskable. Maskable interrupts are the ones that can be masked by software, and as a result, the C54xx DSP ignores the requests for these interrupts and continues with its current task. However nonmaskable interrupts cannot be masked and the processor has to service these requests. In the case of the TMS320C54xx processors, the hardware interrupts $\overline{\text{RS}}$ and $\overline{\text{NMI}}$ are nonmaskable interrupts.

## 9.7.1 Handling of Interrupts

A flow chart of the interrupt handling by the C54xx processors is shown in Figure 9.12. Interrupt handling is done in three phases: receiving the interrupt request, acknowledging the interrupt request, and executing the interrupt service routine.

Servicing an interrupt depends on the pending interrupt status indicated by the bits of the memory-mapped register IFR (interrupt flag register), masked/unmasked status as indicated by the corresponding bit in the memory-mapped register IMR (interrupt mask register), and the global enable INTM bit in the status register ST1. The memory-mapped register IFR has bits corresponding to various interrupts. Whenever an interrupt request is made, the corresponding bit in IFR is set until the CPU recognizes the interrupt. IFR shows the pending external and internal interrupts. IMR is a register that is

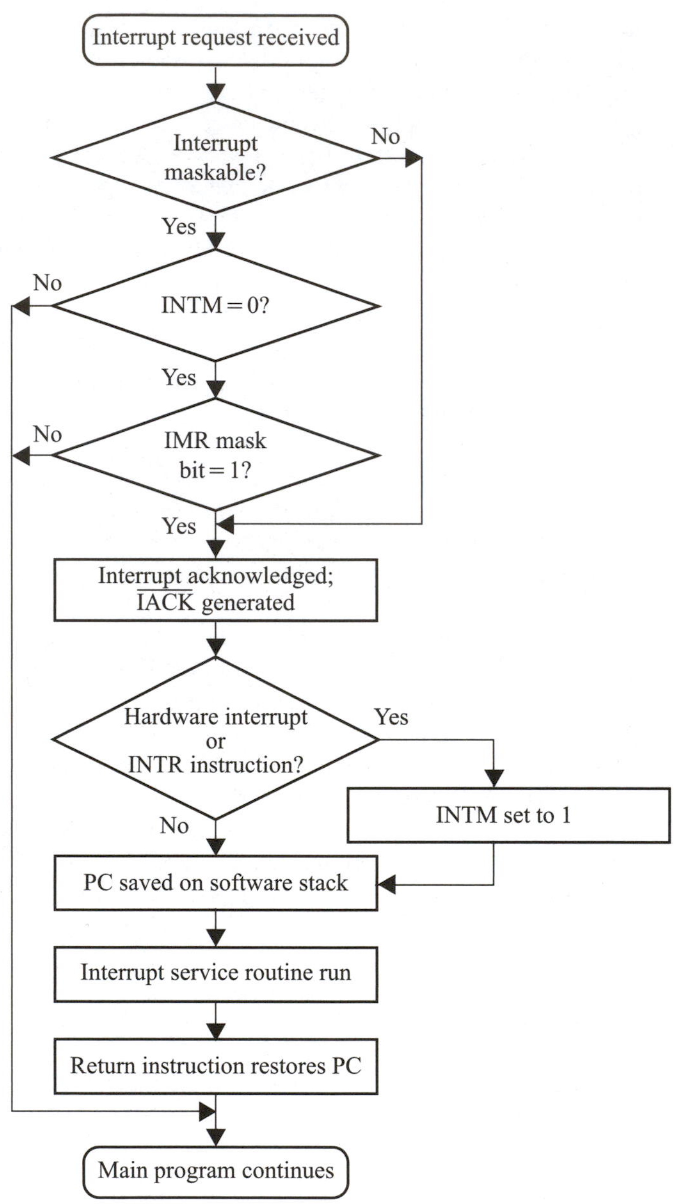


**Figure 9.12** A flow chart of interrupt handling by the processor

(Courtesy of Texas Instruments Inc.)

used for masking external and internal interrupts. An interrupt is unmasked by making 1 the corresponding bit in the IMR. The INTM bit in ST1 enables or disables all interrupts globally. If INTM is 0, the processor does not recognize any maskable interrupt.

As the processor receives the interrupt request, the corresponding bit in IFR is set high. An interrupt request is acknowledged depending upon certain conditions. First, if the interrupt is nonmaskable, it is acknowledged immediately. Maskable interrupts are first checked for priority, and then the INTM bit in the ST1 is checked to see if all the interrupts are globally enabled. The corresponding bit in IMR is then checked to see if it is masked or not. If the INTM is 0 and the IMR mask bit is 1, the processor sends acknowledgment by means of the $\overline{\text{IACK}}$ signal.

To service the interrupt, the program counter's current contents are pushed into the stack. This provides the mechanism for the execution to return to the interrupted program. The INTM bit is set to 1 to disable interrupts during the service routine. The instruction execution control transfers to the interrupt request location in the interrupt vector table. In the interrupt vector table, we write a branch instruction to transfer the execution control to the corresponding interrupt service routine (ISR). After completion of the execution of the ISR, the saved contents of the PC are popped from the stack and loaded back onto the PC. In this way, the CPU then starts executing the suspended program. Also, the return instruction in the service routine re-enables the interrupts by clearing the INTM bit.

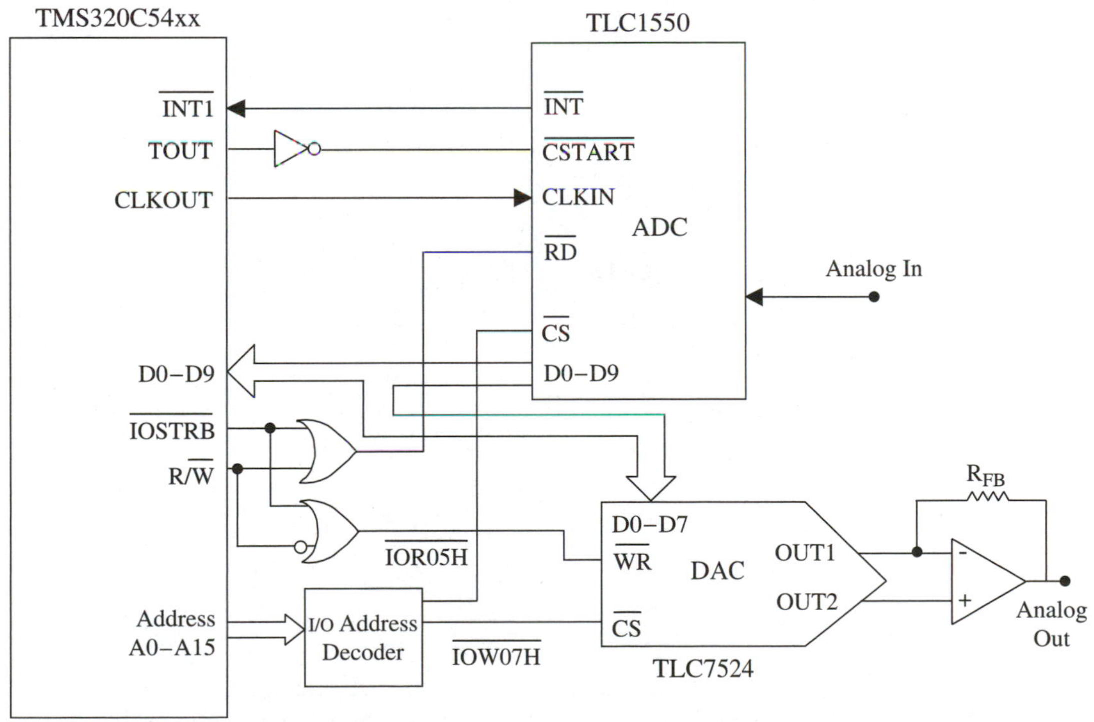


**Figure 9.13** Circuit for interfacing TLC1550 (ADC) and TLC7524 (DAC) to the TMS320C54xx

▷ **Example 9.5** Interface the TMS320C54xx to a 10-bit ADC (TLC1550) and an 8-bit DAC (TLC7524). The sampled signal read from the ADC is to be written to the DAC after adjusting its size. The start of the conversion is to be initiated by the TOUT signal of the timer.

Solution The ADC and the DAC can be connected to the DSP as shown in Figure 9.13. The rate of generation of TOUT is the sampling frequency for the ADC. Conversion is initiated by TOUT, and as soon as it is completed, $\overline{\text{INT}}$ goes low and the DSP receives the interrupt request on INT1. DSP suspends its current program and services the interrupt by initiating the execution of the ISR for INT1. The interrupt service routine involves the reading of the sampled data from the port for the ADC data and writing it to the port for the DAC. Before writing the data to the output port, it is shifted to the right by 2 bits, because the output from the ADC is a 10-bit word, whereas the DAC can receive only 8-bit words.

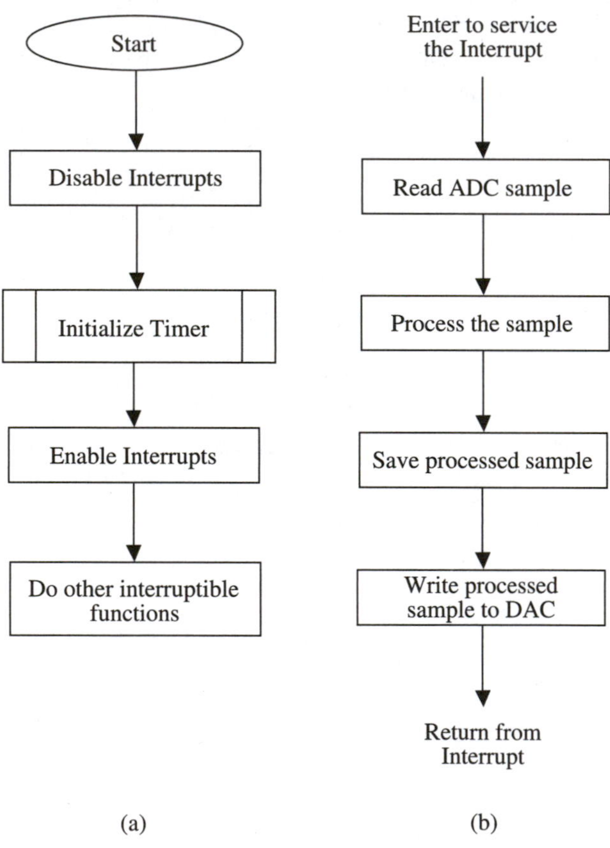


**Figure 9.14** Flow charts for the main program (a) and the interrupt service routine (b) for Example 9.5

```
**********************************************************************************
*
* PROGRAM FOR EXAMPLE 9.5 (File: ex9p5.asm)
*
* DESCRIPTION:  This C54xx program reads an input signal applied to the ADC and outputs
*               it to the DAC. The ADC is read and the data is written to the DAC in
*               the interrupt service routine for INT1.
*
* AUTHOR:       Avtar Singh, SJSU
*
**********************************************************************************
                .ref _c_int00
                .mmregs                         ; memory mapped reg definitions

buffer:         .bss sample, 1                  ; data buffer

                .text
_c_int00:
                stm #0x0500, SP                 ; init SP to 0x0500
                ssbx INTM                       ; disable all interrupts
                call init_DSP                   ; init DSP processor
                call init_timer                 ; init timer
                stm #0xFFFF, IFR                ; clear any pending interrupt
                orm #0002h, IMR                 ; unmask INT1 interrupt
                rsbx INTM                       ; enable all interrupts

wait_main:      ;You may insert code here to be executed during interrupt wait
                b wait_main                     ; wait for an INT1 interrupt
;---------------------------------------------------------------------------------
; Processor Intialization Routine
;---------------------------------------------------------------------------------
PMST_VAL        .set 00A0h                      ; Interrupt vect at 80h,
                                                ; MP/(MC*) = 0, OVLY = 1
BSCR_VAL        .set 0000h                      ; 64K mem bank, no extra cycles
                                                ; between consecutive reads
SWWSR_VAL       .set 2000h                      ; I/O wait states = 2 clocks

                .text
init_DSP:
                ld #0, DP                       ; Data page = 0
                stm #0, CLKMD
                stm #0, CLKMD
                stm #0x4007, CLKMD              ; Processor speed = 5xcrgst.Freq.
```

**Figure 9.15** Program for Example 9.5 *(continued)*

```
                stm #PMST_VAL, PMST          ; Init Processor Mode Status Reg
                stm #BSCR_VAL, BSCR          ; Set Bank Switching Wait States
                stm #SWWSR_VAL, SWWSR        ; Set S/W Wait State Reg
                ssbx OVM                     ; Saturate on overflow
                ssbx SXM                     ; Select sign extension mode
                ret                          ; Return
                nop
                nop
;-----------------------------------------------------------------------------------
; Timer Initialization Routine
; Timer out (TOUT) frequency = CPU Clock/(PRD+1) = sampling freq
;-----------------------------------------------------------------------------------
PRD_value       .set 9999                    ; PRD value for 8 KHz TOUT
TCR_value       .set 0000                    ; TCR value to start timer

                .text
init_timer:
                stm PRD_value, PRD           ; init PRD register
                stm TCR_value, TCR           ; start the timer
                ret                          ; return
                nop
                nop
;-----------------------------------------------------------------------------------
; Interrupt Service Routine
; This reads the 10-bit ADC sample, converts it to an 8-bit sample and
; writes it to the 8-bit DAC.
;-----------------------------------------------------------------------------------
ADC_Data_In     .set 05h                     ; ADC data-in I/O address
DAC_Data_Out    .set 07h                     ; DAC data-out I/O address

                .text
INT1_ISR:
                portr ADC_Data_In, sample    ; read the ADC data
                ld sample, -2, A             ; convert 10-bit data to-8 bit
                stl A, sample                ; save as 8-bit data

                                             ; Place for any DSP algorithm

                portw sample, DAC_Data_Out   ; write data to DAC
                ret                          ; return
                nop
                nop
```

**Figure 9.15** Continued

```
;-------------------------------------------------------------------------------
; Interrupt Vector Table
;-------------------------------------------------------------------------------
                .sect  ".vectors"
RESET:          B _c_int00                  ; Reset vector
                NOP
                NOP
NMI:            RET                         ; Nonmaskable Interrupt vector
                NOP
                NOP
                NOP
                .space 4*15*16              ; Space for unused vectors
INT1:           B INT1_ISR                  ; INT1 Vector
                NOP
                NOP
                .space 4*12*16              ; Space for unused vectors

                .end
```

**Figure 9.15** Continued

Figure 9.14(a) and 9.14(b) show the flow charts of the main program and the interrupt service routine, respectively. Figure 9.15 shows the program for the application. Notice in the program that we need to initialize the processor and the timer. The timer is initialized for generating the TOUT signal at the sampling frequency. We also must set up the interrupt vector table to service the INT1 request. As shown in the program, the service routine uses a memory location "sample" to save the sample value before sending it to the DAC.

## 9.8 Direct Memory Access (DMA)

*Direct memory access* (DMA) is the method of data transfer between regions in the memory space, or between memory and a peripheral, without any intervention by the CPU. Transfer of data can be to and from internal memory, internal peripherals, or external devices. DMA works in the background of the CPU operation. A DMA controller, which may be a part of the DSP device, manages the DMA operation. In this way, the DMA speeds up the overall processing as the two activities, signal transfer and the processing in the CPU, are carried out simultaneously.

TMS320C54xx devices have up to six independent programmable DMA channels for direct data transfer. Each channel connects a source location and a destination location. Therefore, six different source locations can be connected to the corresponding six destination locations. However, at a given time during the DMA operation, only one of the six channels can be used for signal transfer. Each channel has to be enabled before it can be used and each is assigned a priority. A high-priority DMA channel is serviced before a low-priority channel if they both request service at the same time. When multiple channels are enabled and have the same priority level, then the enabled channels are serviced in a circular pattern. As transfer of data involves read and write operations, it is necessary to specify the source and destination address locations for each channel separately. Transfer is in the form of blocks of data where each block consists of frames. Each frame consists of data elements, which can be 16 or 32 bits each. The sizes of the block, frame, and elements are programmed for each channel. DMA transfer for a channel can be programmed to be triggered by some specific event, such as the transmit interrupt.

The total number of CPU clock cycles required to complete a DMA transfer depends on the source and destination locations, external interface conditions such as wait states and bank-switching cycle etc., and the number of active DMA channels. A single data element transfer between two internal memory locations takes four CPU clock cycles, two cycles for read and two for write. In cases where external access is required, data transfer depends on the conditions of the external interface.

### 9.8.1 DMA Operation Configuration

Prior to transfer of data, the DMA registers have to be configured suitably. Configuration involves specifying details such as which channel is to be used for transfer, mode of transfer, source and destination addresses, assignment of priorities to different channels, and the sizes of the block, frame, and data element. A number of registers need to be programmed with configuration information. These registers along with their addresses are shown in the table of Figure A.9 in Appendix A.

The most important registers to be configured are the DMA channel priority and enable control register (DMPREC) and the channel context registers. The 16-bit DMPREC controls the enabling of the DMA channels and channel priorities. Six bits of this register are used to assign channel priorities and another six to enable each of the channels.

Each DMA channel has a set of five channel context registers to configure the operation of that channel. These are the channel source address register (DMSSEC), the channel destination address register (DMDST), the channel element count register (DMCTR), the channel sync select and frame count register (DMSFC), and the channel transfer code control register (DMMCR).

The DMSRC and DMDST of each channel hold the source and the destination addresses, respectively, for that channel. The DMCTR holds the number of data elements to be transferred in a frame. The DMSFC determines which synchronization events will be used to trigger the DMA transfers, the word size (16 bit or 32 bit) for the transfer, and the frame count. The DMMCR is a 16-bit register that controls the transfer mode and is used to specify the source and destination spaces, such as program memory, data memory, or I/O space. The user should consult the Reference Set to determine the contents to be programmed into the DMFC, DMMCR, and DMPREC registers [2].

## 9.8.2 Register Subaddressing

Register subaddressing is the technique used for configuring the DMA registers. As shown Figure 9.16, the stack of subaddressed registers is the set of DMA registers. To configure a DMA register, its code for configuration is loaded onto one of the two subbank access registers (DMSDI or DMSDN). Each DMA register has a unique subaddress. The subaddress of the DMA register to be configured is loaded into the subbank address register (DMSA). This directs the multiplexer to connect the subbank access registers (DMSDI or DMSDN) to the desired physical location, as shown in the figure. DMSDI is used when an automatic increment of the subaddress is required after each access. Therefore, DMSDI can be used to configure the entire set of registers. DMSDN is used if a single register access is desired. In this manner, just two memory-mapped registers, DMSDI and DMSDN, enable the user to have access to all DMA registers. However, addressing becomes a two-step process, one to set up the DMSA and the other to read or write to either DMSDN or DMSDI.

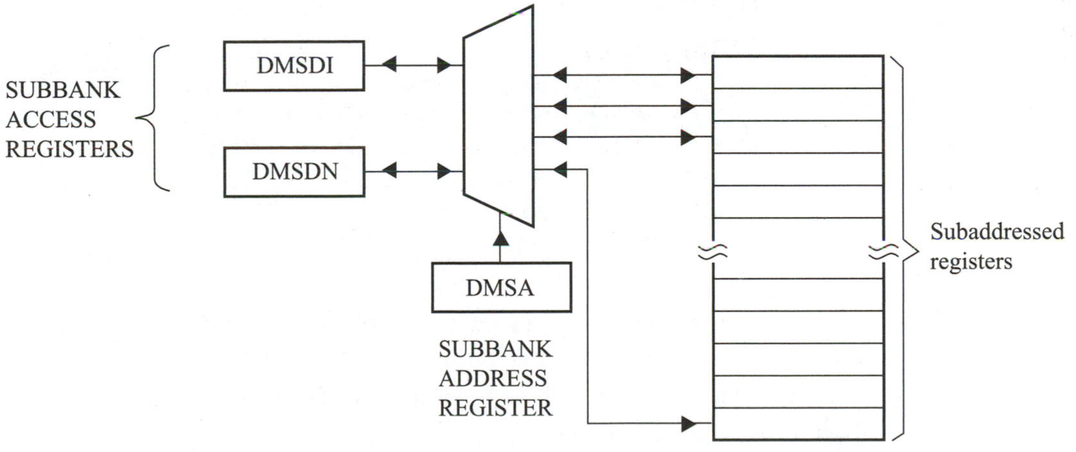


**Figure 9.16** Register subaddressing technique for configuring DMA operation

(Courtesy of Texas Instruments Inc.)

▷ **Example 9.6** Write code to show how the DMA channel 2 source address register can be initialized with the value 1111h.

**Solution** Since a single register is to be modified, subbank register DMSDN can be used. The TMS320C54xx code to achieve this is as follows:

```
DMSA    .set 55h              ; subbank address register address
DMSDN   .set 57h              ; subbank access register address
DMSRC2  .set 0Ah              ; subaddress
        STM  DMSRC2, DMSA     ; DMSA = address of DMSRC2
        STM  #1111h, DMSDN    ; write 1111h to DMSRC2
```

---

▷ **Example 9.7** Write TMS320C54xx code to show how the DMA channel 5 context registers can be initialized. Choose arbitrary values to be written to the registers.

**Solution** Since this is a case of configuring a set of registers, a subbank access register with auto increment (DMSDI) is used in this example. The code to achieve this is as follows. Note that only first subaddress in the sequential addresses of the context registers is needed.

```
DMSA    .set  55h  ; subbank address register address
DMSDI   .set  56h  ; subbank access register address
DMSRC5  .set  19h  ; subaddress of DMSRC5
DMDST5  .set  1Ah
DMCTR5  .set  1Bh
DMSFC5  .set  1Ch
DMMCR5  .set  1Dh

    STM  DMSRC5, DMSA    ; DMSA = first sub address
    STM  #2000h, DMSDI   ; write 2000h to DMSRC5
    STM  #3000h, DMSDI   ; write 3000h to DMDST5
    STM  #0010h, DMSDI   ; write 10h to DMCTR5
    STM  #0002h, DMSDI   ; write 2h to DMSFC5
    STM  #0000h, DMSDI   ; write 0h to DMMCR5
```

---

▷ **Example 9.8** Write a TMS320C54xx code to transfer a block of data from the program memory to the data memory. Following are the specifications:

| | |
|---|---|
| Source address: | 26000h in program space (extended memory page 2) |
| Destination address: | 07000h in data space |
| Transfer size: | 1000h single (16-bit) words |
| Channel use: | DMA channel #0 |

Solution The following code assumes that DMA registers have been defined with appropriate directives.

```
STM  DMSRCP, DMSA        ; set source program page
STM  #2h, DMSDN
STM  DMSRC0, DMSA        ; set source program address to 6000h
STM  #6000h, DMSDI       ; DMSA points to DMDST0
STM  #7000h, DMSDI       ; set destination address to 7000h
                         ; DMSA points to DMCTR0
STM  #(1000h–1), DMSDI   ; set for 1000h transfers
                         ; DMSA points to DMSFC0
STM  #00000h, DMSDI      ; configure DMSFC0
                         ; DMSA points to DMMCR0
STM  #00105h, DMSDI      ; configure DMMCR0
                         ; DMSA points to DMSRC0
STM  #00101h, DMPREC     ; configure DMPREC
```

## 9.9 Summary

In this chapter, we looked at the signals for parallel interfacing of memory and peripherals and studied various interfacing circuits for memory and data converters. Under memory interfacing, we considered various memory options such as SRAM, ROM, and flash. We also studied various types of I/O interfacing methods, including programmed I/O, interrupt I/O, and direct memory access.

## References


1. *TMS320C54xx DSP Reference Set, Volume 1*, Texas Instruments Inc., March 2001.
2. *TMS320C54xx DSP Reference Set, Volume 5*, Texas Instruments Inc., June 1999.
3. Texas Instruments Inc., *Application Report: Understanding C54x Memory Maps and Examining an Optimum C5000 Memory Interface*, SPRA607, November 1999.
4. Texas Instruments Inc., *Application Report: Connecting TMS320C54x DSP with Flash Memory*, SPRA585, August 1999.

## Assignments

1. What is the range of addresses that can be decoded if A19 is pulled low in a processor with 20 address lines?
2. Up to what limit can the program memory be extended in a processor with 20 address lines? How must the extended-memory be organized for addressing by a C54xx processor?
3. How many address lines are required to access all locations of an 16K × 16 SRAM?
4. If TMS320C54xx is reading a memory word operand from address FFF00h in an SRAM, specify the logic levels of the following signals while the read operation is being performed: A0–A19, R/$\overline{\text{W}}$, $\overline{\text{DS}}$, $\overline{\text{PS}}$, $\overline{\text{IS}}$, $\overline{\text{MSTRB}}$, and $\overline{\text{IOSTRB}}$.
5. Design a circuit to interface a 4K × 16 and a 2K × 16 memory chip to realize program memory space for the TMS320C54xx processor in the address ranges: 03FFFFh–03F000h and 05F800h–05FFFFh, respectively.
6. Design a circuit to interface 64K words of the program memory space from 0FFFFFh to 0F0000h for the TMS320C5416 processor using 16K × 16 memory chips.
7. Write an assembly language program for the system in Figure 9.10 using the programmed I/O approach as shown in the Figure 9.11.
8. Describe methods to implement the signal-processing subroutine block in Figure 9.11 so that a uniform sampling interval can be realized.
9. What are the various classifications of interrupts for the TMS320C5416 processor?
10. How does the interrupt handling in the TMS320C54xx DSP differ for a software and a hardware interrupt?
11. Redraw the circuit of Figure 9.13 for a 16-bit ADC and a 16-bit DAC. Use INT2 for the signal sample transfer.
12. Write a program for the circuit of Problem 9.11. Let the sampling rate be 1/4096th of the processor clock. The DAC output (at the same sampling rate) is to be generated by averaging the immediate four input samples as received from the ADC.
13. How does DMA help in increasing the processing speed of a DSP processor?
14. For TMS320C54xx DSP operating at a clock frequency of 100 MHz, how many 16-bit data elements can be transferred between two internal memory locations per second in the DMA mode?
15. Write a TMS320C54xx code to initialize the DMA channel 5 destination register to #5555h without using autoincrement. Rewrite the code using autoincrement for the same operation.

**16.** Write a TMS320C54xx code to transfer a block of data from the program memory to the data memory. Following are the specifications:

| | |
|---|---|
| Source address: | 6000h in program space |
| Destination address: | 8000h in data space |
| Transfer size: | 800h single (16-bit) words |
| Channel use: | DMA channel #1 |

# Chapter 10

## Interfacing Serial Converters to a Programmable DSP Device

## 10.1 Introduction

In the previous chapter, we studied the parallel peripheral interface of programmable DSP devices. In a DSP system, in addition to the parallel interface, there is provision to interface serial peripherals. In the serial interfacing mode, data transfer takes place bit by bit. The serial data transfer may be synchronous or asynchronous. Synchronous serial transfer allows faster data communication but requires a clock signal as the timing reference.

In this chapter, we study the synchronous serial interface as provided in the TMS320C5416 DSP. This device provides three multichannel buffered serial ports (McBSP). We also study how to interface the DSP to an audio CODEC PCM3002 that provides a serial analog-to-digital converter (ADC) and a serial digital-to-analog converter (DAC). This is the device that is used on the C5416 DSK board. Specifically, the following topics are considered:

Synchronous serial interface

A multichannel buffered serial port (McBSP)

McBSP programming

A CODEC interface circuit

CODEC programming

A CODEC–DSP interface example

## 10.2 Synchronous Serial Interface

The synchronous serial interface of the C54xx DSP [1] allows it to communicate with the serial peripherals. Such an interface is shown in Figure 10.1 for a device called an analog input/output CODEC. The CODEC consists of A/D

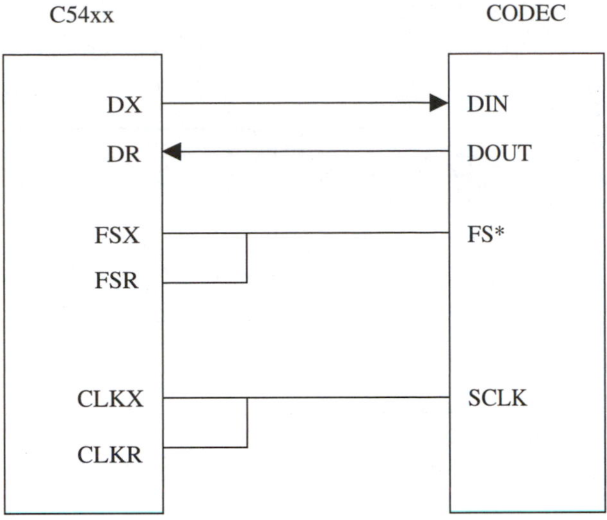


**Figure 10.1** Synchronous serial interface (SSI) between the C54xx and a CODEC device

and D/A converters. The signals used in the interface are shown in Figure 10.1. On the DSP device the DX data line transmits the serial data to the CODEC, and the DR receives it from the CODEC. The receive data is timed with reference to the clock signal CLKR, and the transmit data with respect to the clock signal CLKX. The start of the respective data (the first bit) is synchronized to the frame sync signals FSR and FSX. Similar to the DSP device, the corresponding signal pins are provided on the CODEC device.

Figure 10.2(a) is the timing diagram for the receive operation for the interface. Data reception starts with the FSR pulse. A bit is received for each clock pulse of the CLKR. After receiving all bits, 8 in this case, the processor generates a RRDY signal to indicate that the word of data is ready in the data receive register of the serial port. The status signal RRDY can be read by the processor to determine if a word of data has been received.

Similar to receive timing is the transmit timing shown in Figure 10.2(b). Here the transmission starts with FSX and the completion is indicated by XRDY changing from logic 0 to logic 1. The XRDY indicates that the previously placed data word has been transmitted and the port is ready to transmit the next word, if so desired.

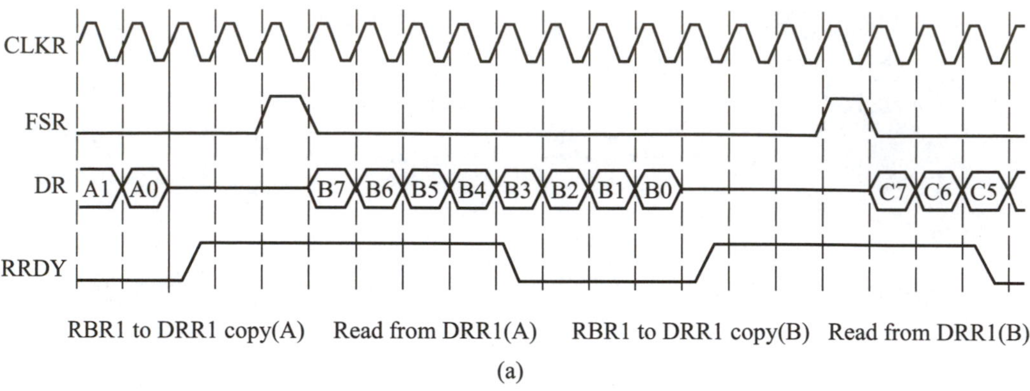


**Figure 10.2(a)** Receive operation timing for the SSI

(Courtesy of Texas Instruments Inc.)

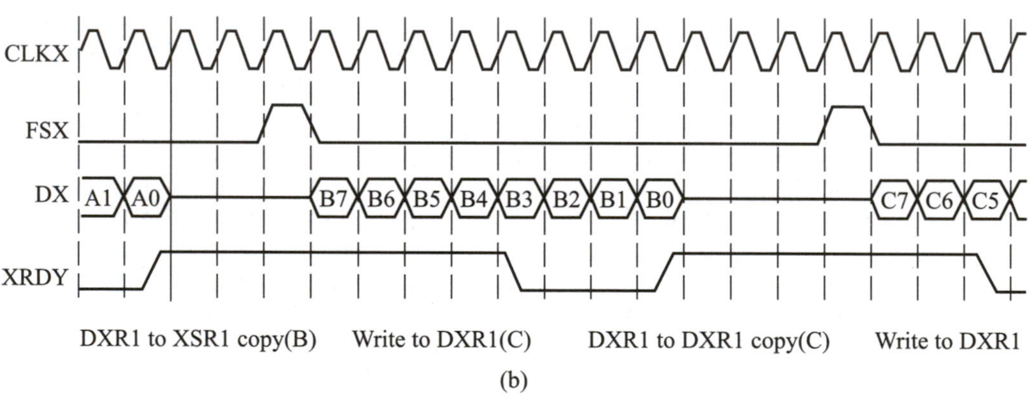


**Figure 10.2(b)** Transmit operation timing for the SSI

(Courtesy of Texas Instruments Inc.)

## 10.3 A Multichannel Buffered Serial Port (McBSP)

McBSP is a full-duplex synchronous serial port. Three such ports are provided on the TMS320C5416 DSP. McBSP can be used to interface synchronous serial peripherals such as a CODEC. The block diagram of Figure 10.3 shows the structure of this port. We will briefly discuss the McBSP here. For details, the reader is advised to read the manual referenced at the end of this chapter [2].

The incoming data enters the port through the DR line into the receive shift register, RSR, where it is assembled into a word that is transferred to re-

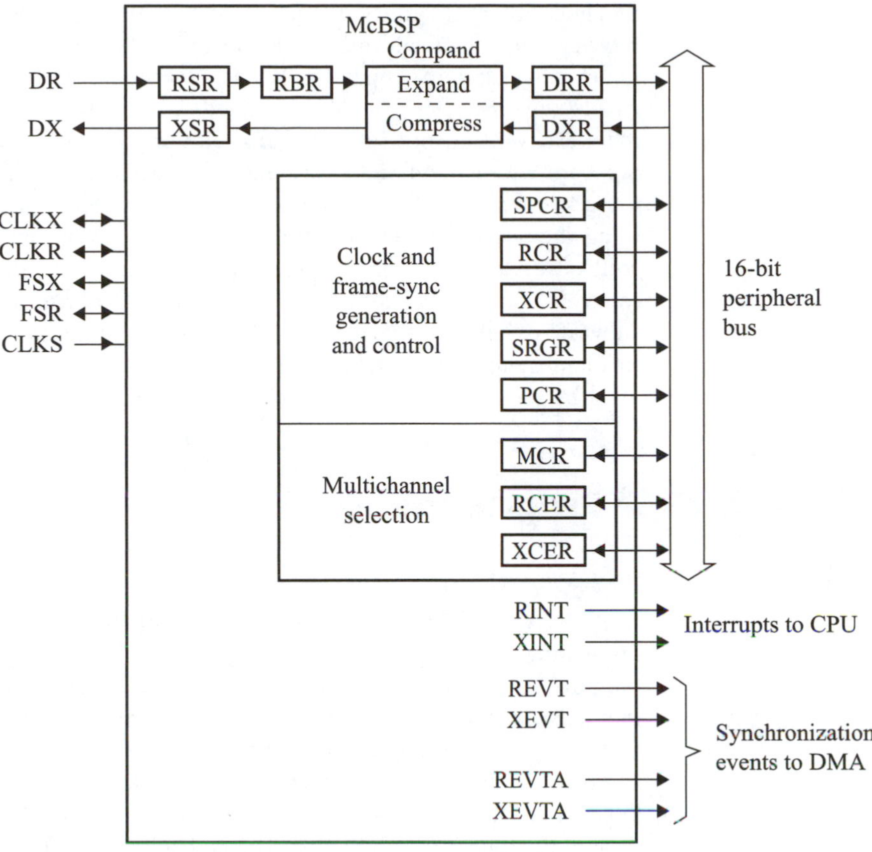


**Figure 10.3** Block diagram of the McBSP of C54xx

(Courtesy of Texas Instruments Inc.)

ceive buffer register, RBR. From the buffer register it is transferred to the data receive register, DRR. The DSP processor reads the data from the memory-mapped register DRR using an internal peripheral data bus. The port informs the processor about the data in DRR using receive interrupt request, RINT, or using the DMA signals. The DRR status is recorded in the serial port control register 1, as the RRDY bit, so that the processor can determine when the data is ready for transfer. The DSP can send the data to the outside world using the memory-mapped data transmit register, DXR. The data written to DXR is transferred to the transmit shift register, XSR, for shifting out 1 bit at a time. The port informs the processor about the data in DXR using transmit interrupt request, XINT, or using the DMA signals. The DXR status is recorded in the serial port control register 2, as the XRDY bit, so that the processor can determine when the data has been transmitted.

There are six memory-mapped registers associated with each port. These registers with their addresses are shown in the table of peripheral memory-mapped registers in Appendix A. Each register is of 16-bit length. There are two receive registers to enable received data lengths up to 32 bits. Similarly, there are two transmit registers for each port. There are two more registers—SPSA for address and SPSD for data—associated with each port. It is by using these two registers that we can access subbank control registers for programming the serial port. The control registers are shown in the table for McBSP control registers and subaddresses in Appendix A. For instance, to write data to receive control register 2 (RCR22) of McBSP2 whose subaddress is 0x0003h, we write 0x0003h to register SPSA2 at memory address 0x0034h and the data to memory-mapped register SPSD2 at address 0x0035h. A similar sequence must be used while reading a subbank register.

## 10.4 McBSP Programming

In order to configure the McBSP, one needs to write appropriate data to the control registers. The functions of the bits of these registers are described in the manual (see chapter reference [2]), which should be consulted to program the port. A sample program is shown in Figure 10.4. This program configures the McBSP2 to work with serial 20-bit input data and serial 20-bit output data and will be used in the example at the end of this chapter.

From the manual and Appendix A, we can see that the control register SPCR12 enables or disables the receiver. Similarly, SPCR22 serves to enable or disable the transmitter function. The control register RCR12 selects the 20-bit data mode for the receiver, and RCR22 specifies that FSR will be used to start receiving the data bits. Similarly, the control registers XCR12 and XCR22 select the corresponding functions for the transmitter. Finally, the PCR2 defines clocks and frame sync pulses to be external and active high. This register also specifies other functions of the pins of the serial port, as indicated in the program of Figure 10.4.

## 10.5 A CODEC Interface Circuit

The PCM3002 [4] is a device that can be directly connected to the synchronous serial port of the DSP. It provides 16/20-bit oversampling sigma–delta A/D and D/A converters. The maximum sampling rate that can be implemented with this device is 48 KHz. Figure 10.5(a) shows the building blocks of the CODEC device. The detailed block diagram of Figure 10.5(b) shows the internal architecture of the PCM3002. As you can see from the block diagram, the device provides stereo ADC and DAC with single-ended voltage input and output for the left and right channels. The CODEC can be programmed for

```
*******************************************************************
*
* initMcBSP2.asm
*
* This module initializes the serial port McBSP2 on the C5416 DSK.
*
* Author:  Avtar Singh, SJSU
*
*******************************************************************
           .include  "regs.asm"
           .def      initMcBSP2

* Define the default values for the registers of McBSP2.

           ; Serial Port Control Register 1 (0010 0000 0010 0000)

           ; Bit15 = 0: Digital loopback disabled
           ; Bit14-13 = 01: Right-justify, sign extend
           ; Bit12-11 = 00: Clock stop disabled
           ; Bit10-8 = 00: Reserved
           ; Bit7 = 0: DX enabler off
           ; Bit6 = 0: A-bis mode disabled
           ; Bit5-4 = 10: RINT driven by frame sync
           ; Bit3 = 0: No sync error
           ; Bit2 = 0: RBRs not in overrun condition
           ; Bit1 = 0: Receiver not ready
           ; Bit0 = 0: Receiver in disabled and in reset state

VAL_SPCR1  .set      2020h

           ; Serial Port Control Register 2 (0000 0000 0000 0000)

           ; Bit15-10: = 00h: Reserved
           ; Bit9 = 0: Free running mode disabled
           ; Bit8 = 0: Soft mode disabled
           ; Bit7 = 0: Frame sync not generated
           ; Bit6 = 0: Disable sample rate generator
           ; Bit5-4 = 00: XINT driven by XRDY
           ; Bit3 = 0: No sync error
           ; Bit2 = 0: XSRs empty
           ; Bit1 = 0: Transmitter not ready
           ; Bit0 = 0: Transmitter in disabled and in reset state
```

**Figure 10.4** A program to initialize the McBSP2 *(continued)*

```
VAL_SPCR2  .set      0000h

           ; Receive Control Register 1 (0000 0000 0110 0000)

           ; Bit15 = 0: Reserved
           ; Bit14-8 = 0000000: 1 word per frame
           ; Bit7-5 = 011: 20 bit receive word
           ; Bit4-0 = 00000: Reserved

VAL_RCR1   .set      0060h

           ; Receive Control Register 2 (0000 0000 0110 0001)

           ; Bit15 = 0: Single phase frame
           ; Bit14-8 = 00h: 1 word per frame
           ; Bit7-5 = 011: 20 bit receive word
           ; Bit4-3 = 00: No companding
           ; Bit2 = 0: Receive frame sync pulses not ignored
           ; Bit1-0 = 01: 1-bit data delay

VAL_RCR2   .set      0061h

           ; Transmit Control Register 1 (0000 0000 0110 0000)

           ; Bit15 = 0: Reserved
           ; Bit14-8 = 00h: 1 word per frame
           ; Bit7-5 = 011: 20 bit transmit word
           ; Bit4-0 = 0h: Reserved

VAL_XCR1   .set      0060h

           ; Transmit Control Register 2 (0000 0000 0110 0000)

           ; Bit15 = 0: Single phase frame
           ; Bit14-8 = 00h: 1 word per frame
           ; Bit7-5 = 011: 20 bit transmit word
           ; Bit4-3 = 00: No companding
           ; Bit2 = 0: Transmit frame sync pulses not ignored
           ; Bit1-0 = 00: 0-bit data delay

VAL_XCR2   .set      0060h

           ; Pin Control Register (0000 0000 0000 1100)
```

**Figure 10.4** Continued

```
            ; Bit15-14 = 00: Reserved
            ; Bit13 = 0: DX, FSX, and CLKX are serial port pins
            ; Bit12 = 0: DR, FSR, CLKR, and CLKS are serial port pins
            ; Bit11 = 0: External transmit frame sync
            ; Bit10 = 0: External receive frame sync
            ; Bit9 = 0: External transmit clock
            ; Bit8 = 0: External receive clock
            ; Bit7 = 0: Reserved
            ; Bit6 = 0: CLKS status
            ; Bit5 = 0: DX status
            ; Bit4 = 0: DR status
            ; Bit3 = 1: FSX active high
            ; Bit2 = 1: FSR active high
            ; Bit1 = 0: Transmit data sampled on rising edge of CLKX
            ; Bit0 = 0: Receive data sampled on rising edge of CLKR

VAL_PCR     .set      000Ch

* This procedure initializes the McBSP2 for use with the PCM3002 codec
* on the C5416 DSK.

            .text

initMcBSP2:
            stm       #SPCR1, MCBSP2_SPSA       ; Disable McBSP2 RX
            ldm       MCBSP2_SPSD, A
            and       #0FFFEh, A
            stlm      A, MCBSP2_SPSD

            stm       #SPCR2, MCBSP2_SPSA       ; Disable McBSP2 TX
            ldm       MCBSP2_SPSD, A
            and       #0FFFEh, A
            stlm      A, MCBSP2_SPSD

            stm       #SPCR1, MCBSP2_SPSA       ; Set SPCR1
            stm       #VAL_SPCR1, MCBSP2_SPSD

            stm       #SPCR2, MCBSP2_SPSA       ; Set SPCR2
            stm       #VAL_SPCR2, MCBSP2_SPSD

            stm       #RCR1, MCBSP2_SPSA        ; Set RCR1
            stm       #VAL_RCR1, MCBSP2_SPSD
```

**Figure 10.4** Continued

```
        stm     #RCR2, MCBSP2_SPSA       ; Set RCR2
        stm     #VAL_RCR2, MCBSP2_SPSD

        stm     #XCR1, MCBSP2_SPSA       ; Set XCR1
        stm     #VAL_XCR1, MCBSP2_SPSD

        stm     #XCR2, MCBSP2_SPSA       ; Set XCR2
        stm     #VAL_XCR2, MCBSP2_SPSD

        stm     #PCR, MCBSP2_SPSA        ; Set PCR
        stm     #VAL_PCR, MCBSP2_SPSD

        ret
```

**Figure 10.4** Continued

digital de-emphasis, digital attenuation, soft mute, digital loop-back, and the power-down mode for the ADC and the DAC.

An analog signal is applied to the combination of a delta–sigma modulator and a decimation filter to convert it to a corresponding digital signal. The input signal is sampled at a 64X oversampling rate, eliminating the need for a sample-and-hold circuit and also simplifying the need for an antialiasing filter. A decimation filter is used to reduce the digital data rate to the sampling rate before generating the output bitstream. A highpass filter removes the dc components of the signal.

The delta–sigma modulator in conjunction with an interpolation filter forms the DAC, which converts the serial digital signal to the corresponding analog signal. The interpolation filter is used to increase the sampling rate to

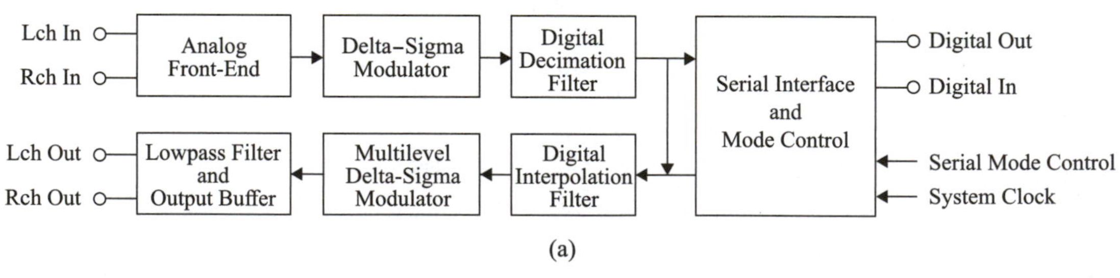


**Figure 10.5(a)** Block diagram for the PCM3002 CODEC

(Courtesy of Burr–Brown Corporation)

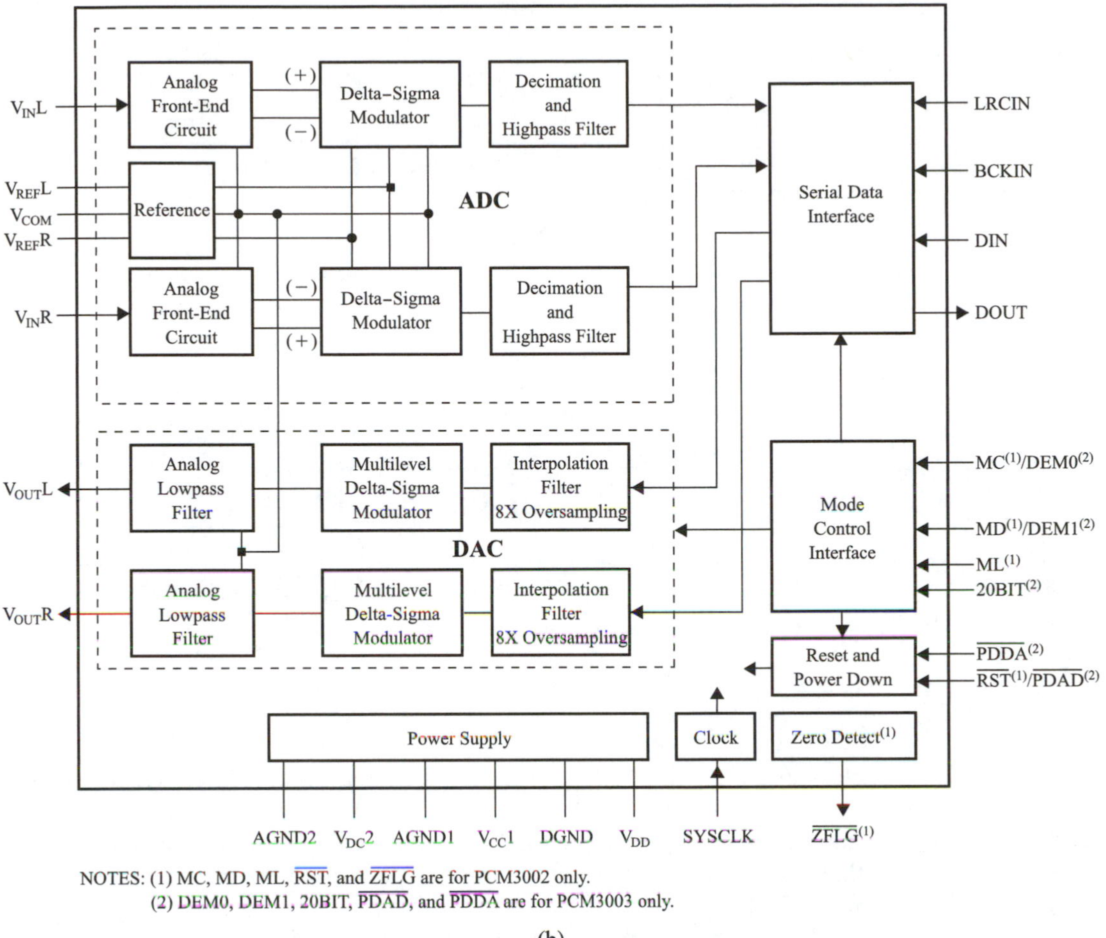


(b)

**Figure 10.5(b)** Details of the PCM3002 CODEC

(Courtesy of Burr–Brown Corporation)

the one needed by the modulator. The converted signal is filtered with an analog lowpass filter to generate the analog output.

As shown in the Figure 10.5(b), there are two distinct parts of the CODEC device: one to handle the serial data transfers, and the other for its initialization and to set it to work in the desired mode. The two blocks, the serial data interface and the mode control interface, handle these two functions.

A block diagram of how the PCM3002 CODEC device is used in the C5416 DSK board is shown in Figure 10.6. The CPLD on the DSK provides the system

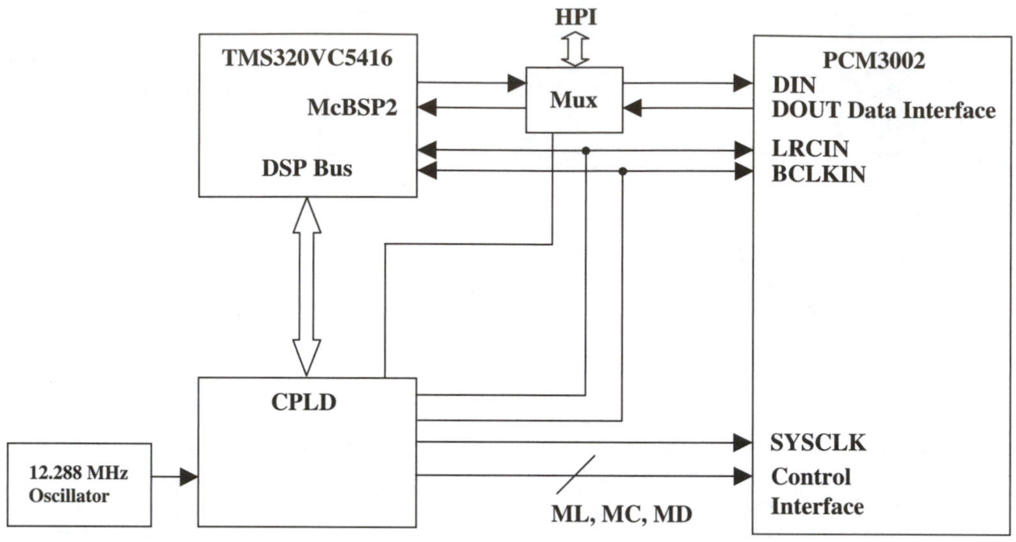


**Figure 10.6** Block diagram showing the PCM3002 interface to the TMS320VC5416 in the DSK (Courtesy of Spectrum Digital Inc.)

clock and the other timing signals for the mode control interface. It also controls the choice of using the McBSP2 port on the DSP for connection either to the host PC (HPI) or to the PCM3002. The CPLD has user-accessible registers that can be loaded to define the various parameters of the CODEC data and control interfaces [3].

The system clock for generating various timing signals for the CODEC is its SYSCLK. This clock must be $256f_s$, or $384f_s$, or $512f_s$, where $f_s$ is the sampling frequency. The CODEC detects the system clock and uses it to generate the internal clock at $256f_s$ for the digital filters and delta-sigma modulators. In the C5416 DSK board, the SYSCLK is supplied by the CPLD-generated clock CODEC_SYSCLK, which is generated from the 12.288 MHz CODEC_CLK.

The data interface of the CODEC and the DSP is by way of DIN for data input, DOUT for data output, BCKIN for data bit clock, and LRCIN for frame sync signal for the left and right channels. The data bit clock and the frame sync signals are generated by the CPLD from the CODEC_CLK and applied to the CODEC and the DSP. The timing for the data input and output is shown in Figure 10.7 for the four possible data formats. The frequency of the LRCIN signal is the ADC/DAC sampling frequency. The bits are transferred using the bit clock BCLKIN. In the CPLD, the BCKIN and LRCIN are generated from the 12.228-MHz oscillator clock called the CODEC_CLK, which is also the default CODEC_SYSCLK, applied to the CODEC device. The corresponding

**FORMAT 0: PCM2002/3003**

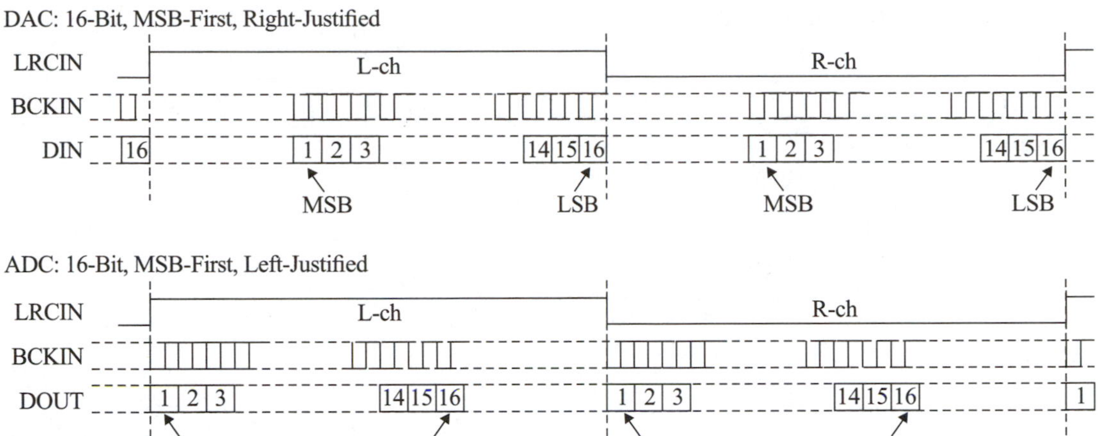


**FORMAT 1: PCM2002/3003**

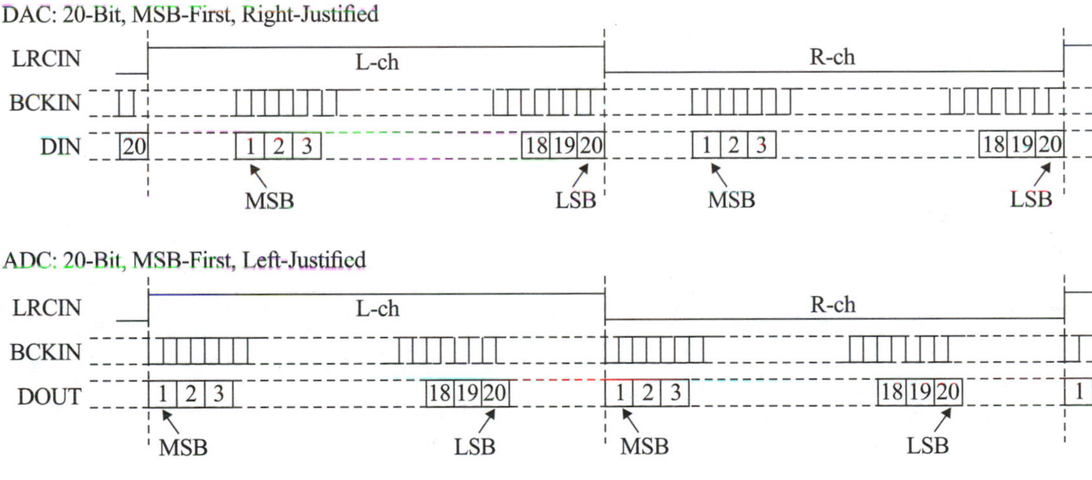


**Figure 10.7** Data transmission formats for the PCM3002 CODEC (*continued*)
(Courtesy of Burr–Brown Corporation)

default bit clock BCLKIN frequency is 3.0122 MHz (or one-fourth of the CODEC_SYSCLK), and the sampling frequency is 48 KHz. The default frequencies can be changed by dividing the CODEC_CLK by 2, 4, 6, or 8. This provides the capability to change the sampling rate to one of five rates, the smallest being 6 KHz and the largest 48 KHz.

**FORMAT 2: PCM3002 Only**

DAC: 20-Bit, MSB-First, Left-Justified

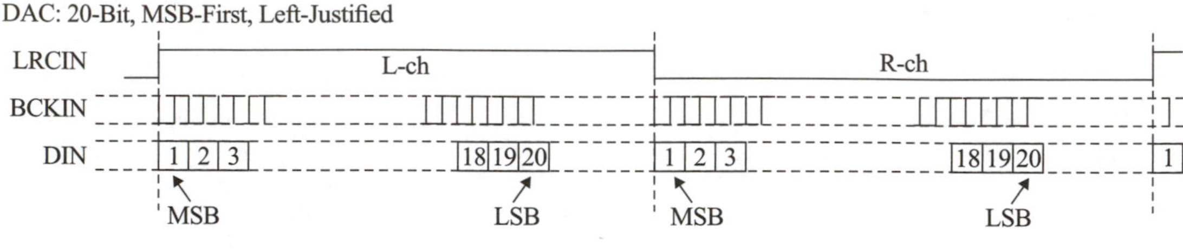


ADC: 20-Bit, MSB-First, Left-Justified

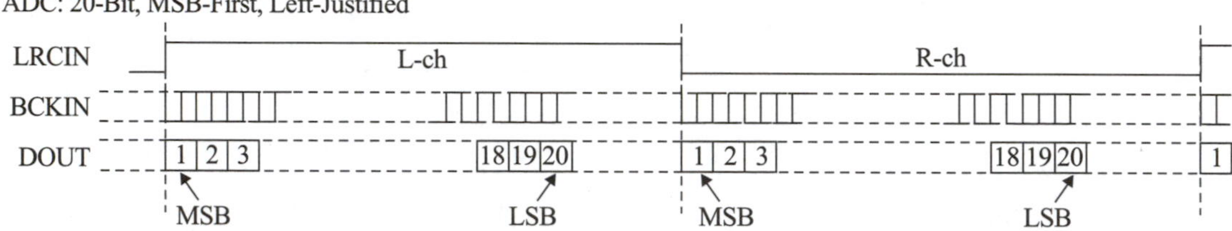


**FORMAT 3: PCM3002 Only**

DAC: 20-Bit, MSB-First, $I^2S$

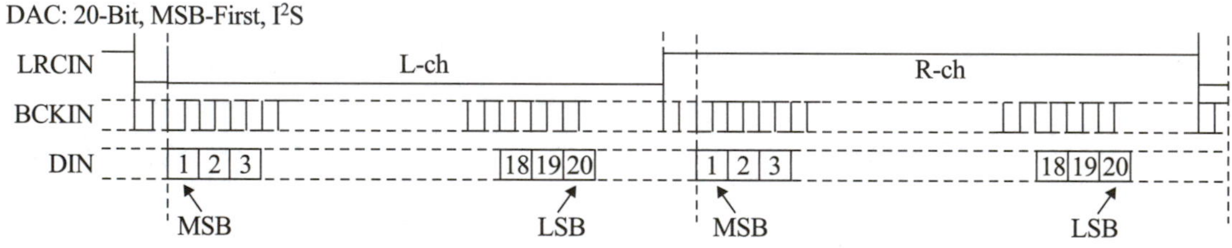


ADC: 20-Bit, MSB-First, $I^2S$

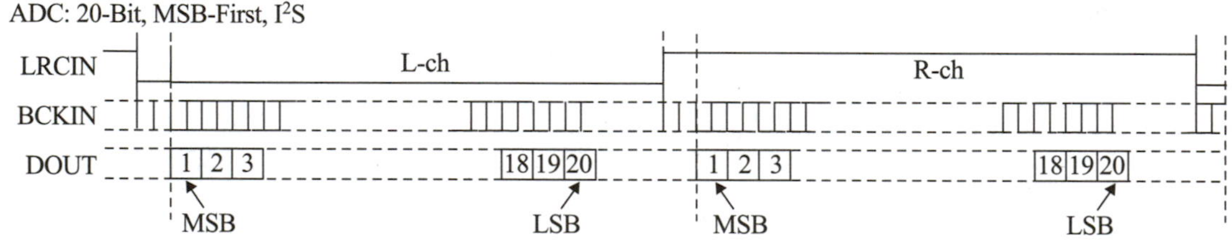


**Figure 10.7** Continued

▷ **Example 10.1** Determine the timing parameters for a 16-bit data communication in a DSK configured for a clock divisor of 6. The oscillator clock (CODEC_CLK) is at 12.288 MHz.

Solution CPLD input clock (CODEC_CLK) = 12.288 MHz

CPLD output clock for the PCM3002 (CODEC_SYSCLK) = 12.288 MHz/6 = 2.044 MHz

Sampling frequency $f_s = 2.044$ MHz/256 = 8 KHz.

Sampling interval = 1/8K = 125 msec

Bit clock frequency (BCLKIN) = 2.044 MHz/4 = 511 KHz

Bit clock period = 1/511K = 1.96 μsec

Time to communicate 16 bits of data = 16 × 1.96μ = 31.31 μsec

Thus, in each 125 msec of time, the data is communicated just for 2 × 31.31 μsec for both channels.

## 10.6 CODEC Programming

To configure the CODEC we send control data using the mode control interface signals as shown in the timing diagram of Figure 10.8. The mode bits represented by the signal MD are sent using the mode clock signal MC. The mode load signal ML defines the start and end of latching the bits into the CODEC device. In the DSK these signals are generated in the CPLD from the oscillator clock. The 16-bit mode control data that is transferred comes from the CPLD and is placed into one of the four registers of the CODEC device to program it.

The four program registers of the PCM3002 are shown in Figure 10.9(a). The description of the various bits of these registers is shown in Figure 10.9(b). For a detailed description the reader is advised to consult the data sheet for the CODEC device (see reference 4 at the end of this chapter). In the program registers the two bits indicated as A1A0 specify the register to which the data in other bits refer. For instance, for register 0 these bits are 00. Register 0 can be loaded to control the attenuation to be applied to the DAC for the left channel. Similarly, register 1 can be loaded with the attenuation data for the DAC of the right channel. The number loaded in the 8 bits of either of these two registers applies the attenuation to the two channels according to the equation

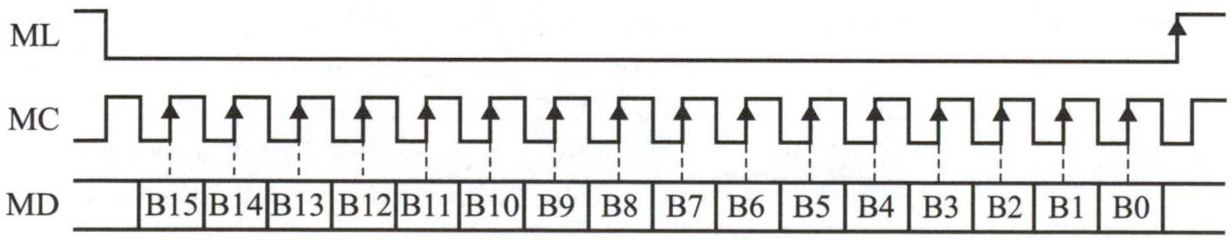


**Figure 10.8** Mode control interface signal timing for the PCM3002 CODEC
(Courtesy of Burr–Brown Corporation)

| | B15 | B14 | B13 | B12 | B11 | B10 | B9 | B8 | B7 | B6 | B5 | B4 | B3 | B2 | B1 | B0 |
|---|---|---|---|---|---|---|---|---|---|---|---|---|---|---|---|---|
| REGISTER 0 | res | res | res | res | res | A1 | A0 | LDL | AL7 | AL6 | AL5 | AL4 | AL3 | AL2 | AL1 | AL0 |
| REGISTER 1 | res | res | res | res | res | A1 | A0 | LDR | AR7 | AR6 | AR5 | AR4 | AR3 | AR2 | AR1 | AR0 |
| REGISTER 2 | res | res | res | res | res | A1 | A0 | PDAD | BYPS | PDDA | ATC | IZD | OUT | DEM1 | DEM0 | MUT |
| REGISTER 3 | res | res | res | res | res | A1 | A0 | res | res | res | LOP | res | FMT1 | FMT0 | LRP | res |

**Figure 10.9(a)** Program registers for the PCM3002 CODEC

(Courtesy of Burr–Brown Corporation)

| REGISTER NAME | BIT NAME | DESCRIPTION |
|---|---|---|
| Register 0 | A (1:0) | Register Address "00" |
| | res | Reserved, should be set to "0" |
| | LDL | DAC Attenuation Data Load Control for Lch |
| | AL (7:0) | Attenuation Data for Lch |
| Register 1 | A (1:0) | Register Address "01" |
| | res | Reserved, should be set to "0" |
| | LDR | DAC Attenuation Data Load Control for Rch |
| | AR (7:0) | DAC Attenuation for Rch |
| Register 2 | A (1:0) | Register Address "10" |
| | res | Reserved, should be set to "0" |
| | PDAD | ADC Power-Down Control |
| | PDDA | DAC Power-Down Control |
| | BYPS | ADC High-Pass Filter Operation Control |
| | ATC | DAC Attenuation Data Mode Control |
| | IZD | DAC Infinite Zero Detection Circuit Control |
| | OUT | DAC Output Enable Control |
| | DEM (1:0) | DAC De-emphasis Control |
| | MUT | Lch and Rch Soft Mute Control |
| Register 3 | A (1:0) | Register Address "11" |
| | res | Reserved, should be set to "0" |
| | LOP | ADC/DAC Analog Loop-Back Control |
| | FMT (1:0) | ADC/DAC Audio Data Format Selection |
| | LRP | ADC/DAC Polarity of LR-dock Selection |

**Figure 10.9(b)** Definition of the bits of the program registers of the PCM3002 CODEC

(Courtesy of Burr–Brown Corporation)

$$\text{Attenuation} = 20 \log(ATT/255)$$

where $ATT$ is the value represented by the 8 attenuation bits in register 0 or register 1.

Either the LDL bit in register 0 or the LDR in register 1 can use the attenuation data to control the two channels.

The bits in register 2 are meant to select the power down mode for the ADC and DAC, the ADC highpass filter bypass control, DAC attenuation channel control, DAC infinite zero detection circuit control, DAC output enable control, DAC deemphasis control, and the DAC soft mute control. To enable or select a mode, the corresponding bit or bits are made 1. For the deemphasis control, the two bits used are as follows: 00 selects deemphasis 44.1 KHz, 01 deselects deemphasis, 10 selects 48 KHz deemphasis, and 11 selects 32 KHz deemphasis.

Register 3 provides ADC/DAC loopback control, audio data format selection, and polarity selection for the LRCIN signal. A 1 in the LOP bit enables the loopback. A 1 in the LRP bit selects the left channel when LRCIN is low and the right channel when it is high. The data format is selected by the two bits FMT1 and FMT0. The 00 on these two bits selects the format 0 for the data as received from the ADC or applied to the DAC. These data formats are shown in Figure 10.7 and provide four different ways to communicate data.

The CPLD that provides data for the four program registers and other controls on the DSK board has eight registers accessible from the DSP. These registers are shown in Figure 10.10. These registers are each 8 bits wide and are located in the I/O space of the C5416. For instance, the registers at I/O addresses 2 and 3 hold the CODEC programming data. For details of the bits of these registers, the reader should consult the DSK manual [3], which is also available in the DEBUG environment of the CCS. The most significant bit in the miscellaneous register at the I/O address 6 must be checked each time any new data is written to the CPLD registers for programming the CODEC.

The sampling frequency can be changed by loading the divisor, for the CODEC clock, to the CODEC-CLK register at the I/O address 7. The sequence of steps that need to be followed is: stop the clock, load the divisor, start the clock, and select the divisor. The bits of the CODEC-CLK register need to be loaded appropriately to accomplish these steps. The other CPLD registers are there for configuring the memories and for communicating with the user switches and the LEDs of the DSK.

## 10.7 A CODEC–DSP Interface Example

In this section, we write a simple application that involves configuring McBSP2 and the PCM3002 on the DSK board. The configured system is used

| I/O Add | Name | Bit 7 | Bit 6 | Bit 5 | Bit 4 | Bit 3 | Bit 2 | Bit 1 | Bit 0 |
|---|---|---|---|---|---|---|---|---|---|
| 0 | USER_ REG | USR_ SW3<br>R | USR_ SW2<br>R | USR_ SW1<br>R | USR_ SW0<br>R | USR_ LED3<br>R/W<br>0 (Off) | USR_ LED2<br>R/W<br>0 (Off) | USR_ LED1<br>R/W<br>0 (Off) | USR_ LED0<br>R/W<br>0 (Off) |
| 1 | DC_REG | DC_ DET<br>R | DC_ ID_CTL<br>R/W<br>0 | DC_ STAT1<br>R | DC_ STAT0<br>R | DC_ RST<br>R/W<br>0 (No Reset) | 0 | DC_ CNTL1<br>R/W<br>0 (Low) | DC_ CNTL0<br>R/W<br>0 (Low) |
| 2 | CODEC_L | CODEC_L_CMD [7..0]<br>R/W<br>0 | | | | | | | |
| 3 | CODEC_H | CODEC_H_CMD [15..8]<br>R/W<br>0 | | | | | | | |
| 4 | VERSION | CPLD_VER [3..0]<br>R | | | | 0 | BOARD_VER [2..0]<br>R | | |
| 5 | DM_ CNTL | DM_ SEL<br>R/W<br>0 (internal) | MEM TYPE_DS<br>R/W<br>0 (FLASH) | MEM TYPE_PS<br>R/W<br>0 (FLASH) | DM_ PG4<br>R/W<br>0 (Page 0) | DM_ PG3<br>R/W<br>0 (Page 0) | DM_ PG2<br>R/W<br>0 (Page 0) | DM_ PG1<br>R/W<br>0 (Page 0) | DM_ PG0<br>R/W<br>0 (Page 0) |
| 6 | MISC | CODEC_ RDY<br>R<br>0 (Ready) | 0 | 0 | 0 | 0 | DC_ WIDE<br>R/W<br>0 (18 bits) | DC32_ DDD<br>R/W<br>0 (Even) | BSP2_ SEL<br>R/W<br>0 (CODEC) |
| 7 | CODEC_ CLK | 0 | 0 | 0 | 0 | DIV_ SEL<br>R/W | CLK_ STOP<br>R/W | CLK_ DIV1<br>R/W | CLK_ DIV0<br>R/W |

**Figure 10.10** CPLD register definitions in the DSK5416

(Courtesy of Spectrum Digital Inc.)

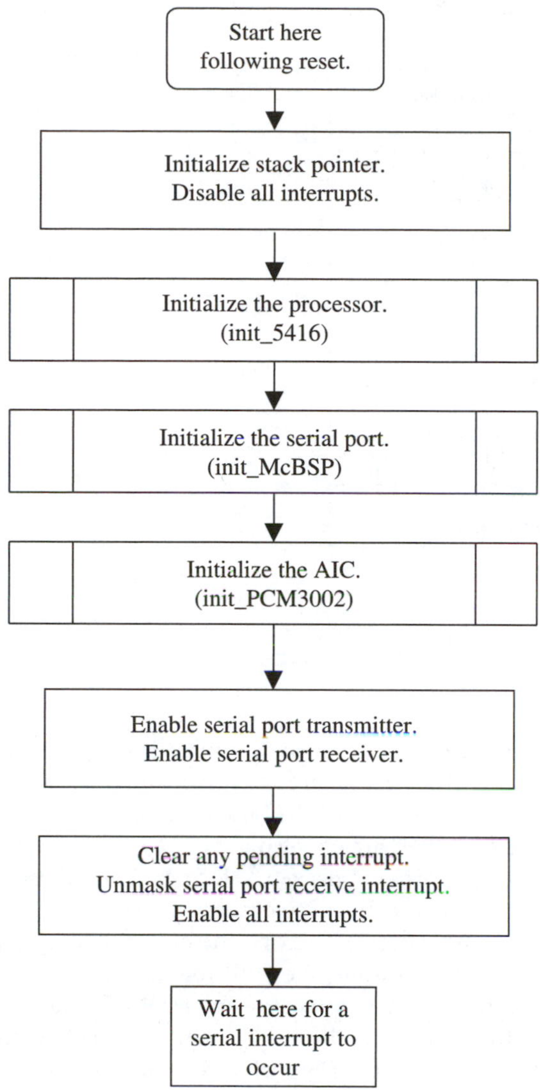


**Figure 10.11(a)** Main program flow chart for the signal loopback program

to implement a signal loopback by reading a signal applied to the ADC and writing it to the DAC. The application can be easily extended to include any kind of processing on the signal read from the ADC before sending it to the DAC. The flow chart of the main program is shown in Figure 10.11(a).

The main program starts by initializing the stack pointer and disabling the interrupts. Establishing the stack allows using subroutines. The disabled

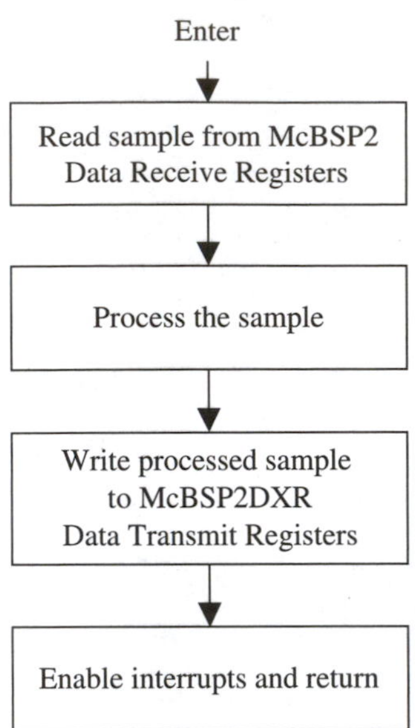


**Figure 10.11(b)** Receive interrupt service routine flow chart

interrupt system ensures that during initialization interrupts will be ignored. This is followed by three subroutines that initialize the processor, the serial port, and the PCM3002. After the initializations are done, the serial port transmitter and the receiver are enabled. Next, any pending interrupt is cleared and the receive interrupt is unmasked before enabling the interrupts. At the end, the processor waits for the receive interrupt to occur.

When a receive interrupt occurs, the corresponding service routine is executed. In the service routine, the DRR registers are read into the accumulator. The word so read is written back to the DXR after formatting it for the DAC. The return from the service routine, with interrupts enabled, makes the program wait for the next interrupt, which occurs after the ADC provides the next sample to the port. It is in this routine that any signal processing on the received signal can be implemented. The receive interrupt service routine flow chart is shown in Figure 10.11(b).

The entire program, shown in Figure 10.11(c), consists of the main module signalLBmain.asm; interrupt vector module C5416vec.asm; initialization modules initC5416.asm, initMcBSP2.asm, and initPCM3002.asm; and the module regs.asm that defines various constants used in the program. Notice that some of the definitions in the regs.asm module are specific to the DSK-implemented registers, such as DSP_CPLD_CODEC_L. These registers are described in the DSK manual [3] and in the Help facility of the DSK software Code Composer Studio (CCS).

```
*************************************************************************************
*
* signalLBmain.asm
*
* This program reads an input signal from the ADC and writes it to the DAC on the
* DSK5416 board. This main module includes the entry-point for the program.
*
* Author:           Avtar Singh, SJSU
*
*************************************************************************************
                    .include  "regs.asm"
                    .ref      initC5416
                    .ref      initMcBSP2
                    .ref      initPCM3002
                    .def      _c_int00
                    .def      brint2_isr
                    .ref      sample_receive
                    .ref      sample_transmit

VAL_SP              .set      0x0500                 ; initial stack address

                    .data

sample_upper_word   .word     0                      ; received sample
sample_lower_word   .word     0

                    .text

* The entry-point for the program

_c_int00:
                    stm       #VAL_SP, SP            ; Define the stack
                    ssbx      INTM                   ; Disable all interrupts

                    call      initC5416              ; Init the DSP processor
                    call      initMcBSP2             ; Init the McBSP2 port
                    call      initPCM3002            ; Init the DSK CODEC

                    stm       #SPCR1, MCBSP2_SPSA    ; Enable McBSP2 receiver
                    orm       #0001h, MCBSP2_SPSD

                    stm       #SPCR2, MCBSP2_SPSA    ; Enable McBSP2 transmitter
                    orm       #0001h, MCBSP2_SPSD
```

**Figure 10.11(c)** A signal loopback implementation program for the DSK5416

```
                    stm       #0FFFFh, IFR             ; Clear pending interrupts
                    orm       #040h, IMR               ; Unmask McBSP2 RX int

                    rsbx      INTM                     ; Enable all interrupts

wait_main:          idle      1                        ; Wait for an RX interrupt
                    b         wait_main
                    nop
                    nop
                    nop

* Interrupt service routine for McBSP2 Receiver

brint2_isr:
                    call      sample_receive           ; Receive the sample
                    nop                                ; Process the sample
                    call      sample_transmit          ; Transmit the sample
                    rete

* This procedure receives a 20-bit value from the ADC
* Return with A (LSBs) = 20 bit received sample

sample_receive:
                    pshm      AR5

                    ldm       MCBSP2_DRR2, B           ; Retrieve upper 16 bits
                    stm       #sample_upper_word, AR5
                    stl       B, *AR5+                 ; Save upper bits locally

                    ldm       MCBSP2_DRR1, A           ; Retrieve lower 16 bits
                    and       #0FFFFh, A
                    stl       A, *AR5                  ; Save lower bits locally

                    sftl      B, 15
                    or        B, 1, A                  ; Construct the sample in A

                    popm      AR5

                    ret

* This procedure sends a 20-bit value in A (LSBs) to the DAC.
```

**Figure 10.11(c)** Continued

```
sample_transmit:
                    stlm      A, MCBSP2_DXR1          ; Transmit lower 16 bits
                    sfta      A, -16
                    stlm      A, MCBSP2_DXR2          ; Transmit upper bits

                    ret

                    .end
*****************************************************************************************
*
* initC5416.asm
*
* This module initializes the processor.
*
* Author:           Avtar Singh, SJSU
*
*****************************************************************************************
                    .include  "regs.asm"

                    .def      initC5416

* Define values for the DSP registers.

                    ; Processor Mode Status Register (0000 0000 1110 1000)

                    ; IPTR = 000000001: Vector table resides at address 0080h
                    ; MP/MC* = 1: Enable microprocessor mode
                    ; OVLY = 1: On-chip RAM addressable in data space, but not in
                              program space
                    ; AVIS = 0: Address visibility mode
                    ; DROM = 1: On-chip ROM not mapped into data space
                    ; CLKOUT = 0: CLOCKOUT off
                    ; SMUL = 0: Saturation on multiplication
                    ; SST = 0: Saturation on store

VAL_PMST            .set      00E8h

                    ; Software Wait State Register (0111 1111 1111 1111)

                    ; XPA = 0: Extended program address control bit
                    ; I/O = 111: Base wait states for I/O accesses
                    ; Data = 111: Base wait states for upper external data access
                    ; Data = 111: Base wait states for lower external data access
                    ; Program = 111: Base wait states for upper extern prog access
                    ; Program = 111: Base wait states for lower extern prog access
```

**Figure 10.11(c)** Continued

```
VAL_SWWSR           .set      7FFFh

                    .text

initC5416:
                    ld        #0, DP                ; Data page = 0
                    stm       #4007, CLKMD          ; DSP clock = 5xPLL

                    stm       #VAL_PMST, PMST       ; Init PMST
                    stm       #VAL_SWWSR, SWWSR     ; Init SWWSR

                    ssbx      SXM                   ; Enable sign extension

                    ret
*****************************************************************************************
*
* initMcBSP2.asm
*
* This module initializes the serial port McBSP2 on the C5416 DSK.
*
* Author:           Avtar Singh, SJSU
*
*****************************************************************************************
                    .include  "regs.asm"
                    .def      initMcBSP2

* Define the default values for the registers of McBSP2.

                    ; Serial Port Control Register 1 (0010 0000 0010 0000)

                    ; Bit15 = 0: Digital loopback disabled
                    ; Bit14-13 = 01: Right-justify, sign extend
                    ; Bit12-11 = 00: Clock stop disabled
                    ; Bit10-8 = 00: Reserved
                    ; Bit7 = 0: DX enabler off
                    ; Bit6 = 0: A-bis mode disabled
                    ; Bit5-4 = 10: RINT driven by frame sync
                    ; Bit3 = 0: No sync error
                    ; Bit2 = 0: RBRs not in overrun condition
                    ; Bit1 = 0: Receiver not ready
                    ; Bit0 = 0: Receiver in disabled and in reset state
```

**Figure 10.11(c)** Continued

```
VAL_SPCR1           .set      2020h

                    ; Serial Port Control Register 2 (0000 0000 0000 0000)

                    ; Bit15-10: = 00h: Reserved
                    ; Bit9 = 0: Free running mode disabled
                    ; Bit8 = 0: Soft mode disabled
                    ; Bit7 = 0: Frame sync not generated
                    ; Bit6 = 0: Disable sample rate generator
                    ; Bit5-4 = 00: XINT driven by XRDY
                    ; Bit3 = 0: No sync error
                    ; Bit2 = 0: XSRs empty
                    ; Bit1 = 0: Transmitter not ready
                    ; Bit0 = 0: Transmitter in disabled and in reset state

VAL_SPCR2           .set      0000h

                    ; Receive Control Register 1 (0000 0000 0110 0000)

                    ; Bit15 = 0: Reserved
                    ; Bit14-8 = 0000000: 1 word per frame
                    ; Bit7-5 = 011: 20 bit receive word
                    ; Bit4-0 = 00000: Reserved

VAL_RCR1            .set      0060h

                    ; Receive Control Register 2 (0000 0000 0110 0001)

                    ; Bit15 = 0: Single phase frame
                    ; Bit14-8 = 00h: 1 word per frame
                    ; Bit7-5 = 011: 20 bit receive word
                    ; Bit4-3 = 00: No companding
                    ; Bit2 = 0: Receive frame sync pulses not ignored
                    ; Bit1-0 = 01: 1-bit data delay

VAL_RCR2            .set      0061h

                    ; Transmit Control Register 1 (0000 0000 0110 0000)

                    ; Bit15 = 0: Reserved
                    ; Bit14-8 = 00h: 1 word per frame
                    ; Bit7-5 = 011: 20 bit transmit word
                    ; Bit4-0 = 0h: Reserved
```

**Figure 10.11(c)** Continued

```
VAL_XCR1            .set      0060h

                    ; Transmit Control Register 2 (0000 0000 0110 0000)

                    ; Bit15 = 0: Single phase frame
                    ; Bit14-8 = 00h: 1 word per frame
                    ; Bit7-5 = 011: 20 bit transmit word
                    ; Bit4-3 = 00: No companding
                    ; Bit2 = 0: Transmit frame sync pulses not ignored
                    ; Bit1-0 = 00: 0-bit data delay

VAL_XCR2            .set      0060h

                    ; Pin Control Register (0000 0000 0000 1100)

                    ; Bit15-14 = 00: Reserved
                    ; Bit13 = 0: DX, FSX, and CLKX are seial port pins
                    ; Bit12 = 0: DR, FSR, CLKR, and CLKS are serial port pins
                    ; Bit11 = 0: External transmit frame sync
                    ; Bit10 = 0: External receive frame sync
                    ; Bit9 = 0: External transmit clock
                    ; Bit8 = 0: External receive clock
                    ; Bit7 = 0: Reserved
                    ; Bit6 = 0: CLKS status
                    ; Bit5 = 0: DX status
                    ; Bit4 = 0: DR status
                    ; Bit3 = 1: FSX active high
                    ; Bit2 = 1: FSR active high
                    ; Bit1 = 0: Transmit data sampled on rising edge of CLKX
                    ; Bit0 = 0: Receive data sampled on rising edge of CLKR

VAL_PCR             .set      000Ch

* This procedure initializes the McBSP2 for use with the PCM3002 codec on the C5416 DSK.

                    .text

initMcBSP2:
                    stm       #SPCR1, MCBSP2_SPSA       ; Disable McBSP2 RX
                    ldm       MCBSP2_SPSD, A
                    and       #0FFFEh, A
                    stlm      A, MCBSP2_SPSD
```

**Figure 10.11(c)** Continued

```
                stm     #SPCR2, MCBSP2_SPSA     ; Disable McBSP2 TX
                ldm     MCBSP2_SPSD, A
                and     #0FFFEh, A
                stlm    A, MCBSP2_SPSD

                stm     #SPCR1, MCBSP2_SPSA     ; Set SPCR1
                stm     #VAL_SPCR1, MCBSP2_SPSD

                stm     #SPCR2, MCBSP2_SPSA     ; Set SPCR2
                stm     #VAL_SPCR2, MCBSP2_SPSD

                stm     #RCR1, MCBSP2_SPSA      ; Set RCR1
                stm     #VAL_RCR1, MCBSP2_SPSD

                stm     #RCR2, MCBSP2_SPSA      ; Set RCR2
                stm     #VAL_RCR2, MCBSP2_SPSD

                stm     #XCR1, MCBSP2_SPSA      ; Set XCR1
                stm     #VAL_XCR1, MCBSP2_SPSD

                stm     #XCR2, MCBSP2_SPSA      ; Set XCR2
                stm     #VAL_XCR2, MCBSP2_SPSD

                stm     #PCR, MCBSP2_SPSA       ; Set PCR
                stm     #VAL_PCR, MCBSP2_SPSD

                ret
******************************************************************************************
*
* initPCM3002.asm
*
* This module initializes the PCM3002 codec on the C5416 DSK.
*
* Author:         Avtar Singh, SJSU
*
******************************************************************************************
                .include  "regs.asm"
                .def      initPCM3002
                .def      sampling_rate_set

* Define values for the codec clock register (in the CPLD) and the control registers of
* the PCM3002 codec.
```

**Figure 10.11(c)** Continued

```
                    ; Codec clock Register (0000 1010)

                    ; Bit7-4 = 0000: Reserved
                    ; Bit3 = 1: Clock divisor selected
                    ;        0: No divisor, 48 KHz sampling rate
                    ; Bit2 = 0: Clock enabled
                    ; Bit1-0 = 00: Clock divisor for 24 KHz sampling rate
                    ;          01: Clock divisor for 12 KHz sampling rate
                    ;          10: Clock divisor for 8 KHz sampling rate
                    ;          11: Clock divisor for 6 KHz sampling rate

VAL_CLK_REG         .set      12h

                    ; Register 0 (0000 0001 1111 1111)
                    ; Bit15-11 = 00000: Reserved
                    ; Bit10-9 = 00: Register address 0
                    ; Bit8 = 1: Enable DAC attenuation data LDL
                    ; Bit7-0 = 11111111: 0 dB left channel attenuation

VAL_REG0            .set      01ffh

                    ; Register 1 (0000 0011 1111 1111)

                    ; Bit15-11 = 00000: Reserved
                    ; Bit10-9 = 01: Register address 1
                    ; Bit8 = 1: Enable DAC attenuation data LDR
                    ; Bit7-0 = 11111111: 0 dB right channel attenuation

VAL_REG1            .set      03ffh

                    ; Register 2 (0000 0100 1000 0010)

                    ; Bit15-11 = 00000: Reserved
                    ; Bit10-9 = 10: Register address 2
                    ; Bit8 = 0: Disable ADC power-down control (PDAD)
                    ; Bit7 = 1: Bypass high-pass filter
                    ; Bit5 = 0: Individual channel attenuation control
                    ; Bit4 = 0: Infinite zero detection disabled
                    ; Bit3 = 0: DAC outputs enabled
                    ; Bit2-1 = 01: De-emphasis off
                    ; Bit0 = 0: Mute disabled
```

**Figure 10.11(c)** Continued

```
VAL_REG2            .set      0482h

                    ; Register 3 (0000 0110 0000 1000)

                    ; Bit15-11 = 00000: Reserved
                    ; Bit10-9 = 11: Register address 3
                    ; Bit8-6 = 000: Reserved
                    ; Bit5 = 0: Loop-back disabled
                    ; Bit4 = 0: Reserved
                    ; Bit3-2 = 10: Format 2
                    ; Bit1 = 0: Left is H, Right is L
                    ; Bit0 = 0: Reserved

VAL_REG3            .set      0608h

                    .text

* This procedure initializes the PCM3002 codec on the C5416 DSK via the CPLD.
* The procedure uses location 60h as scratch pad

initPCM3002:
                    portr     DSK_CPLD_MISC, 60h       ; Select codec
                    andm      #0FFFEh, 60h
                    portw     60h, DSK_CPLD_MISC

                    call      sampling_rate_set        ; Set Sampling rate

                    ld        #VAL_REG0, A             ; Program codec reg0
                    call      CPLD_write

                    ld        #VAL_REG1, A             ; Program codec reg1
                    call      CPLD_write

                    ld        #VAL_REG2, A             ; Program codec reg2
                    call      CPLD_write

                    ld        #VAL_REG3, A             ; Program codec reg3
                    call      CPLD_write

                    ret
```

**Figure 10.11(c)** Continued

```
* This procedure sets the clock for the PCM3002 codec.
* The following sequence is specified:
*
* 1. Set the CLK_STOP bit to 1.
* 2. Set the CLK_DIV1 and CLK_DIV0 bits to the sampling rate value, keeping the CLK_STOP
*    bit as 1.
* 3. Reset the CLK_STOP bit to 0.
* 4. Set the DIV_SEL bit to 1.
*
* Enter with A = #VAL_CLK_REG to specify the sampling rate.
*
* The procedure uses location 60h as scratch pad

sampling_rate_set:
                    portr    DSK_CPLD_CODEC_CLK, 60h  ; Stop the clock
                    orm      #04h, 60h
                    portw    60h, DSK_CPLD_CODEC_CLK

                    ld       #VAL_CLK_REG, A          ; Get Sample rate value
                    bc       NoDivisor, AEQ           ; Check if highest rate

                    and      #03h, A                  ; Select the divisor bits
                    or       #04h, A                  ; Keep the clock stopped
                    stl      A, 60h                   ; Set the clock divisor
                    portw    60h, DSK_CPLD_CODEC_CLK

                    andm     #0FBh, 60h               ; Resume the clock
                    portw    60h, DSK_CPLD_CODEC_CLK

                    orm      #08h, 60h                ; Select the divisor
                    portw    60h, DSK_CPLD_CODEC_CLK

                    b        sampling_rate_done

NoDivisor:
                    st       #00h, 60h                ; Resume the clock
                    portw    60h, DSK_CPLD_CODEC_CLK

sampling_rate_done:
                    ret
```

**Figure 10.11(c)** Continued

```
* This procedure transmits a 16-bit control word to the PCM3002 via the CPLD.
* The procedure uses location 60h as scratch pad
*
* Argument A: 16-bit control word

CPLD_write:
                stl     A, 60h              ; Write low control byte
                portw   60h, DSK_CPLD_CODEC_L

                stl     A, -8, 60h          ; write high control byte
                portw   60h, DSK_CPLD_CODEC_H

CODEC_WAIT:
                portr   DSK_CPLD_MISC, 60h
                andm    #80h, 60h           ; Get the CODEC_RDY bit
                ld      60h, A
                bc      CODEC_WAIT, ANEQ    ; wait till all bits sent

                ret
******************************************************************************************
*
* C5416vec.asm
*
* This module contains the interrupt vector table for the signal loopback program.
*
* Author:       Avtar Singh, SJSU
*
******************************************************************************************
                .ref    _c_int00
                .ref    brint2_isr

                .sect   ".vectors"

RESET:          b       _c_int00            ; Reset vector
                nop
                nop

NMI:            rete                        ; Nonmaskable Interrupt vector
                nop
                nop
                nop
```

**Figure 10.11(c)** Continued

```
                    .space    14*4*16                  ; Space for unused s/w interrupts

                    .space    6*4*16                   ; Space for unused h/w interrupts

BRINT2:             b         brint2_isr               ; Receive Interrupt Vector
                    nop
                    nop

BXINT2:             rete                               ; Transmit Interrupt Vector
                    nop
                    nop
                    nop

                    .space    16*4*2
*****************************************************************************************
*
* regs.asm
*
* This module defines constants for the TMS320C54xx DSP and the C5416 DSK Board.
*
* Adapted from regs1.h available in TI literature
*
* Author:           Avtar Singh, SJSU
*
*****************************************************************************************
                    .mmregs

*
* McBSP0 Registers
*
MCBSP0_DRR2         .set      0020h                    ; McBSP0 Data Rx Reg2
MCBSP0_DRR1         .set      0021h                    ; McBSP0 Data Rx Reg1
MCBSP0_DXR2         .set      0022h                    ; McBSP0 Data Tx Reg2
MCBSP0_DXR1         .set      0023h                    ; McBSP0 Data Tx Reg1
MCBSP0_SPSA         .set      0038h                    ; McBSP0 Sub Bank Addr Reg
MCBSP0_SPSD         .set      0039h                    ; McBSP0 Sub Bank Data Reg

*
* McBSP1 Registers
*
MCBSP1_DRR2         .set      0040h                    ; McBSP1 Data Rx Reg2
MCBSP1_DRR1         .set      0041h                    ; McBSP1 Data Rx Reg1
MCBSP1_DXR2         .set      0042h                    ; McBSP1 Data Tx Reg2
```

**Figure 10.11(c)** Continued

```
MCBSP1_DXR1         .set      0043h                      ; McBSP1 Data Tx Reg1
MCBSP1_SPSA         .set      0048h                      ; McBSP1 Sub Bank Addr Reg
MCBSP1_SPSD         .set      0049h                      ; McBSP1 Sub Bank Data Reg

*
* McBSP2 Registers
*
MCBSP2_DRR2         .set      0030h                      ; McBSP2 Data Rx Reg2
MCBSP2_DRR1         .set      0031h                      ; McBSP2 Data Rx Reg1
MCBSP2_DXR2         .set      0032h                      ; McBSP2 Data Tx Reg2
MCBSP2_DXR1         .set      0033h                      ; McBSP2 Data Tx Reg1
MCBSP2_SPSA         .set      0034h                      ; McBSP2 Sub Bank Addr Reg
MCBSP2_SPSD         .set      0035h                      ; McBSP2 Sub Bank Data Reg

*
* McBSP0, McBSP1 and McBSP2 Subbank Addressed Registers
*
SPCR1               .set      0000h                      ; Ser Port Ctrl Reg1
SPCR2               .set      0001h                      ; Ser Port Ctrl Reg2
RCR1                .set      0002h                      ; Rx Ctrl Reg1
RCR2                .set      0003h                      ; Rx Ctrl Reg2
XCR1                .set      0004h                      ; Tx Ctrl Reg1
XCR2                .set      0005h                      ; Tx Ctrl Reg2
SRGR1               .set      0006h                      ; Sample Rate Gen Reg1
SRGR2               .set      0007h                      ; Sample Rate Gen Reg2
MCR1                .set      0008h                      ; Multichan Reg1
MCR2                .set      0009h                      ; Multichan Reg2
RCERA               .set      000Ah                      ; Rx Chan Enable Reg Part A
RCERB               .set      000Bh                      ; Rx Chan Enable Reg Part B
XCERA               .set      000Ch                      ; Tx Chan Enable Reg Part A
XCERB               .set      000Dh                      ; Tx Chan Enable Reg Part B
PCR                 .set      000Eh                      ; Pin Ctrl Reg

*
* CPLD Registers (DSK5416)
*
DSK_CPLD_USER_REG   .set      0000h                      ; User LEDs and Switches Reg
DSK_CPLD_DC_REG     .set      0001h                      ; Daughter Card Register
DSK_CPLD_CODEC_L    .set      0002h                      ; CODEC_L_CMD Register
DSK_CPLD_CODEC_H    .set      0003h                      ; CODEC_H_CMD Register
DSK_CPLD_VERSION    .set      0004h                      ; Version Register
DSK_CPLD_DM_CNTL    .set      0005h                      ; Memory Control Register
DSK_CPLD_MISC       .set      0006h                      ; Miscellaneous Register
DSK_CPLD_CODEC_CLK  .set      0007h                      ; CODEC Clock Register
```

**Figure 10.11(c)** Continued

```
/****************************************************
*
* Signal loopback program command file (signalLB.cmd)
*
****************************************************/
MEMORY
{
   PAGE 0:  DARAMV:  origin = 0080h,  length = 0080h
   PAGE 0:  DARAMP:  origin = 1000h,  length = 1000h
   PAGE 1:  DARAMD:  origin = 4000h,  length = 0B000h
}

SECTIONS
{
   .text     > DARAMP PAGE 0
   .vectors  > DARAMV PAGE 0
   .data     > DARAMD PAGE 1
}
```

**Figure 10.11(d)** The command file for the loopback program

To build the program for the DSK, the command file shown in Figure 10.11(d) can be used.

To test the program functionality, a signal can be applied to the microphone input on the DSK. A speaker connected to the analog output should receive the signal when the program is loaded to the board and run. A PC can provide this test setup if its speaker output is applied to the microphone input of the DSK (using an appropriate cable) and the speaker output of the DSK is connected to another speaker or the one disconnected from the PC. Any audio file played on the PC with the DSK program running can be heard on the speakers. The program can also be tested with an input signal from a signal generator. There are also programs available that can be run to generate a test signal on the PC. One such program can be downloaded from the site in the reference at end of this chapter [5].

## 10.8 Summary

In this chapter, we looked at the serial peripheral interfacing using the multichannel buffered serial port (McBSP). We also considered a specific serial peripheral, PCM3002, that provides 16-bit synchronous serial ADC and DAC. The chapter ends with an example of the DSK to illustrate the interface and the associated program.

## References


1. *TMS320C54xx DSP Reference Set, Volume 1*, Texas Instruments Inc., March 2001.
2. *TMS320C54xx DSP Enhanced Peripheral Reference Set, Volume 5*, Texas Instruments Inc., SPRU302, June 1999.
3. *TMS320VC5416 DSK Technical Reference*, Spectrum Digital Inc., 506005-0001 Rev. A, March 2002.
4. Burr–Brown Corporation, PCM3002/3003 Data Sheet, January 2000.
5. NCH Tone Generator Software, www.nch.com


## Assignments

**10.1** Frame sync is generated by dividing the 8.192-MHz clock by 256 for the serial communication. Determine the sampling rate and the time a 16-bit sample takes when transmitted on the data line.

**10.2** What is the address for the PCR register of McBSP2? Write an instruction sequence to write to it data defined by PCR_VAL.

**10.3** Write an instruction sequence to reset and disable the transmitter and receiver for the McBSP2.

**10.4** Which registers and which bits need to be changed to implement an 8-bit transmission and reception for the McBSP2?

**10.5** A PCM3002 is programmed for the 12-KHz sampling rate. Determine the divisor *N* that should be written to the CPLD of the DSK and the various clock frequencies for the setup.

**10.6** Determine the timing parameters for a 20-bit data communication at 8 KHz.

**10.7** Which bits and register are used to program the analog input gain? Determine the bit setting to obtain a 0-dB gain.

**10.8** Which bits and register of the PCM3002 are used to program the application of a 48-KHz deemphasis to the DAC output of the PCM3002? Determine the bit setting.

**10.9** What are the maximum and the minimum sampling rates that can be implemented for the PCM3002 on the 5416 DSK? Determine the bits, their value, and the register that needs to be programmed to achieve the maximum and minimum sample rate settings.

**10.10** Modify the program in Figure 10.11(c) to change the sampling rate to 12 KHz.

**10.11** Modify the program in Figure 10.11(c) to output the absolute value of the signal sampled at the input.

**10.12** Modify the program in Figure 10.11(c) to incorporate the FIR filter implemented in Chapter 7, Section 7.3.

**10.13** Determine, using CCS debug capability, the processing time per sample for the filter implemented in Problem 12. Assume that the DSP is running at 80 MHz. Based on this measurement and the consideration for the CODEC device, what is the maximum sampling frequency that can be implemented? Also determine the highest signal frequency that can be handled for processing.

**10.14** Implement the FFT program of Chapter 8 so as to process a real-time signal to compute its spectrum and display it on an oscilloscope. Compute the spectrum each time a new sample is received. Determine the maximum sampling rate that can be used in the implementation on the DSK.

**10.15** Repeat Problem 14 for computing the spectrum, each time, after receiving the block of samples used in FFT calculations.

# Chapter 11

# Applications of Programmable DSP Devices

## 11.1 Introduction

As commercial programmable DSPs are becoming more and more powerful in terms of their speed and functionality and are available at lower and lower costs, there is an explosion of applications in which these devices are increasingly used. These applications span a wide spectrum of areas, such as automotive, control, communication, entertainment, instrumentation, and medicine. Typical applications include toys, medical instruments, speech synthesis and recognition systems, audio equalizers, echo cancellers, and robotic controllers. These applications exploit such capabilities of the programmable DSP devices as high speed and throughput, facility to carry out complex computations with precision, ease of programming, and ability to interface with host processors and external peripherals. In this chapter, we look into a few representative applications and study their requirements to see how these are met by systems implemented using DSPs. Following are the representative applications considered in this chapter:

An ECG processing system
A speech processing system
An image processing system
A position control system
A power measurement system

## 11.2 A DSP System

Digital signal processors are computational devices that process digital representation of input signals and produce digital representation of signals as

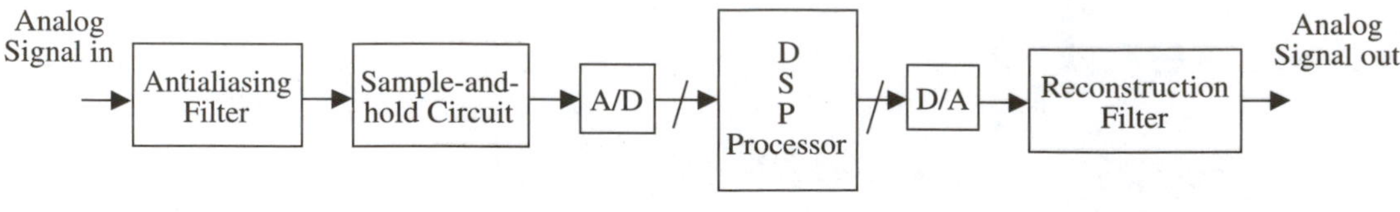


**Figure 11.1** The block diagram of a DSP system

outputs. The difference between these devices and the general-purpose processors lies in the fact that DSPs process data representing real-world signals, whereas the general-purpose processors deal with applications requiring large volumes of stored data. Since real-world signals are mostly analog, they have to be converted into digital signals before being processed by the DSP and, likewise, DSP output needs to be converted back to analog for use in the real world. Figure 11.1 shows the block diagram depicting the processing blocks of a typical DSP system. We have discussed this system in previous chapters. It consists of the DSP processor between the analog front end and the analog back end. The analog front end consists of an antialiasing filter, a sample-and-hold circuit, and an analog-to-digital converter feeding into the DSP. The back end consists of a digital-to-analog converter to convert the digital output to its analog value, followed by a reconstruction filter.

The block diagram of Figure 11.1 applies to almost all DSP systems. All or just some of the blocks shown in the figure may realize a particular system. Implementations may differ in details such as the signal frequency spectrum, the sampling rate, memory requirements, and the computational complexity. In the application examples that follow, we look at the nature and computational complexity of the algorithm to be implemented with a view to understanding how the processing power and other features of the programmable DSPs are utilized in each case. Description and design of the analog front and back ends as well as the analog-to-digital and the digital-to-analog converters are beyond the scope of this book.

## 11.3 DSP-Based Biotelemetry Receiver

Biotelemetry is a process by which physiological information or signals are transferred from one remote location to another, typically using radio frequency links. The importance of biotelemetry becomes obvious when we consider monitoring life in remote or inaccessible locations such as an astronaut in space or a baby in mother's womb. The biomedical signals at the source are encoded, modulated, and then transmitted. At the receiver end, the signals are demodulated, decoded, displayed, and analyzed to extract diagnostic information for evaluation.

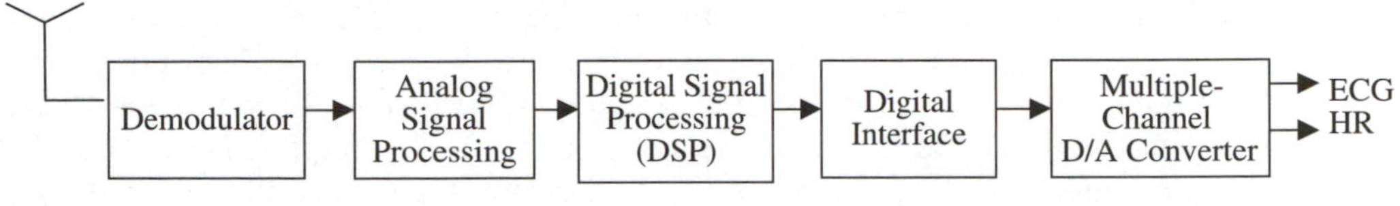


**Figure 11.2** A DSP-based biotelemetry receiver system

The block diagram shown in Figure 11.2 shows a scheme that can be used to implement a biotelemetry receiver [1]. The DSP device receives the demodulated signal as obtained from the demodulator and analog processing circuits. The device can be programmed to decode the received signal by inverting the process of encoding used in the transmitter and thus generate the corresponding biomedical signals. The decoded signals can be presented to a D/A converter to generate analog signals.

### 11.3.1 Pulse Position Modulation (PPM)

PPM is a scheme that can be used to encode a single signal or multiple signals. The position of a pulse encodes the sample value of a signal. A PPM signal that encodes two signals in addition to providing a fixed sampling rate is shown in Figure 11.3. The PPM signal requires a sync signal (two pulses) to

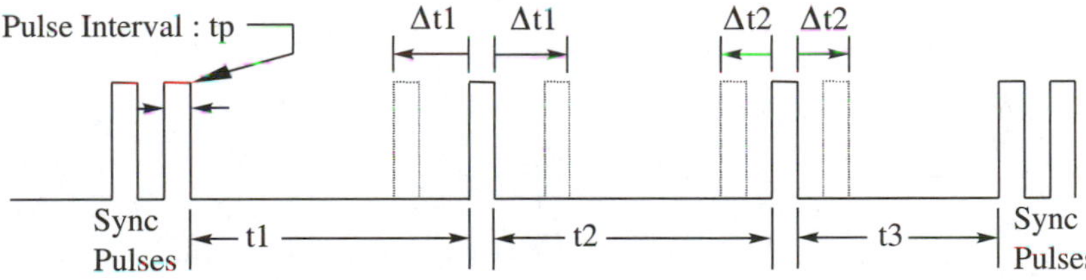


| Parameter | Function | Duration (μsec)—an Example |
|---|---|---|
| t1 | Encodes signal 1 | 1000 |
| t2 | Encodes signal 2 | 800 |
| t3 | Compensation interval | 1700 |
| Each pulse interval (tp) | | 100 |
| Sync interval | | 3 × 100 |
| Total: t1 + t2 + t3 + 5tp | Sampling interval | 4000 |

**Figure 11.3** A PPM signal for encoding two biomedical signals

mark the beginning of a cycle for encoding two or more signals. As shown in the figure, t1 encodes one signal, and t2 encodes the other. The time interval t3 is simply needed to keep the sampling interval constant to provide a fixed sampling rate. In the example shown, the fixed sampling rate is 2.5 KHz. The example encoding can be modified to encode three signals by incorporating another time interval for the third signal or by superimposing the third signal in either of the intervals t1 or t2. The superimposed signals should be distinguishable in the frequency domain so that it can be separated in the receiver. For instance, the system can be used to encode ECG, temperature, and pressure signals. Temperature being the lowest frequency signal, it is combined with the highest frequency ECG signal and encoded as interval t2.

### 11.3.2 Decoding Scheme for the PPM Receiver

The schematic diagram in Figure 11.4 shows how a DSP device can be used to decode a PPM signal to recover the encoded biomedical signals. The decoding requires measurements of time intervals in a PPM signal. The DSP device timer can be used for time measurement. To initiate the measurement process, the pulses in the PPM signal can be used to generate interrupt signals for the DSP device, which then are used to start or terminate the timer. This approach avoids using an A/D converter to handle the PPM signal, but it requires that the DSP device be fast enough so as not to miss a pulse or introduce time measurement error.

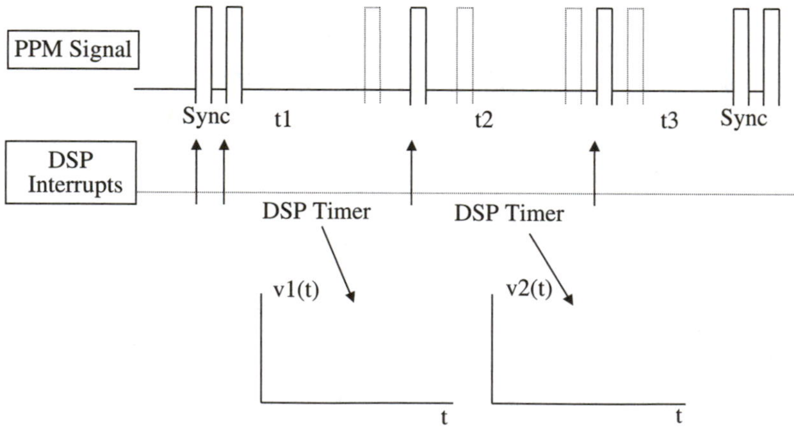


**Figure 11.4** A DSP-based decoding scheme for a PPM signal

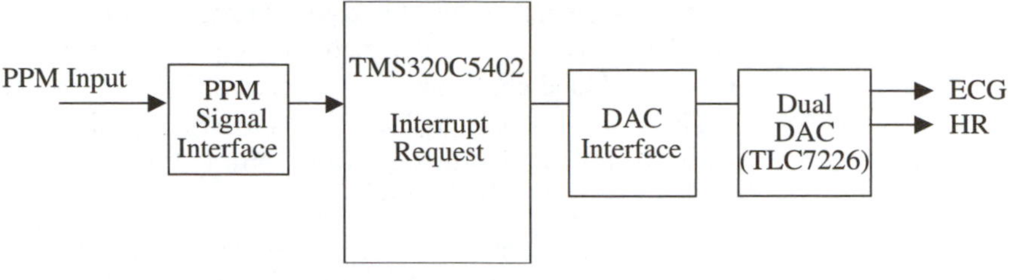


**Figure 11.5** A DSP-based biotelemetry receiver implementation

### 11.3.3 Biotelemetry Receiver Implementation

The block diagram in Figure 11.5 shows the system used for implementation. The PPM signal is processed in the analog domain before it is applied to the interrupt system of the signal processor. The DSP device is interfaced to appropriate digital-to-analog converters so that signals can be generated for analog display monitoring devices. The signal processor in the system is the TMS320C5402. An EPROM device can provide storage for the operating system as well as the decoding software. In order for the DSP to generate the two recovered biomedical signals, a dual-channel parallel digital-to-analog converter can be used.

Two types of software programs are stored in the EPROM. One is the software for decoding PPM signals to generate the encoded biomedical signals. The other software allows providing debugging capability using a PC connected to a parallel port similar to a DSK. In fact, a DSK can be used to debug the software before building the receiver.

### 11.3.4 ECG Signal Processing for Heart Rate Determination

The most important information contained in an ECG signal is the associated heart rate. Determining the heart rate involves determining the time interval between QRS complexes. Therefore, we need a reliable algorithm to detect the QRS complexes so that the QRS interval can be determined to compute the heart rate.

A nonlinear transformation is used to enhance the QRS complex so that it can be detected reliably with a threshold detector. The transformation in our implementation uses absolute values of the first and second derivatives of the signal as follows:

$$y1(n) = |x(n) - x(n-1)|$$

$$y2(n) = |x(n-2) - 2x(n-1) + x(n)|$$

$$y3(n) = y1(n) + y2(n)$$

where $x(n)$ refers to the ECG signal sample, $y1(n)$ is the absolute value of the first derivative, $y2(n)$ is the absolute value of the second derivative, and $y3(n)$ is the combined absolute first and second derivatives.

The transformed signal is filtered to remove high-frequency noise components. To accomplish this, we use a simple IIR filter as follows

$$y4(n) = \alpha(y3(n) - y4(n-1)) + y4(n-1)$$

where $\alpha$, a number less than 1, is the IIR filter coefficient. Its value is chosen based on the smoothing effect that should be used to discard high frequencies. The $y4(n)$ in the difference equation denotes the filtered transformed signal.

A QRS complex is detected using a threshold detector. Processing typical ECG signals by the above algorithm and determining the mean of half of the peak amplitudes of the filtered signals determines the threshold for the detector. This estimated threshold value is then used to detect the QRS complexes in a given ECG waveform.

The time interval between two complexes is the QRS interval. Finally, the heart rate (HR) in beats per minute (BPM) is computed using the formula

$$\text{HR} = (\text{Sampling rate} \times 60)/\text{QRS interval}$$

The sampling rate is determined from the time duration of a PPM cycle or depends upon the modulation technique. To produce a heart rate value accurate on an average, the computed heart rate can also be filtered using an appropriate filter. Figure 11.6 shows the ECG and HR waveforms generated by the system.

## 11.4 A Speech Processing System

Depending on the objective of speech processing, the techniques of processing differ. For instance, if the objective is to understand speech characteristics, analysis-type algorithms are used. To improve the speech quality, filtering algorithms are employed. Here, we consider a technique called *pitch period estimation*. Pitch period estimation (or, equivalently, fundamental frequency estimation) is one of the most important problems in speech processing. Pitch detectors are used in vocoders, speech identification and verification systems, and in aids to the handicapped. Because of its importance, many solutions have been proposed to this problem. Here, we present pitch estimation using the autocorrelation technique implemented on the DSP. Before describing the algorithm for pitch detection, we introduce the concept of how speech is generated and classified.

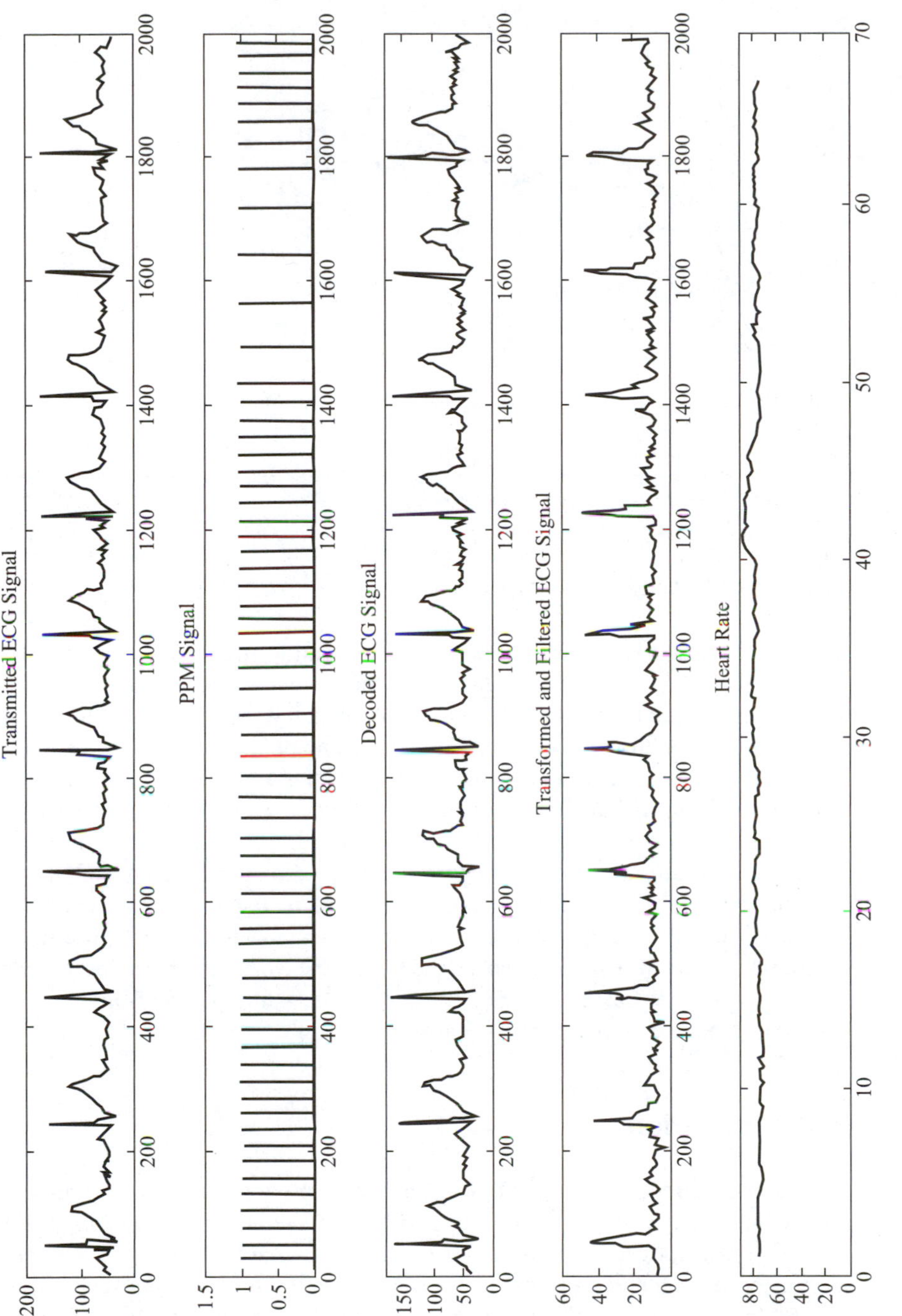


**Figure 11.6** ECG signal and heart rate generated by the DSP telemetry receiver from the PPM signal

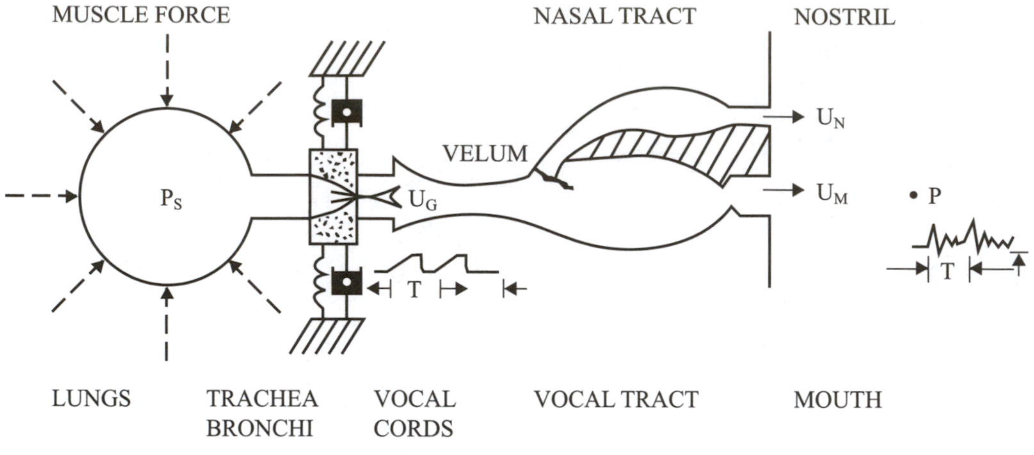


**Figure 11.7** A schematic diagram of the human vocal apparatus

### 11.4.1 A Digital Model for Production of Speech Signal

A schematic diagram of the human vocal apparatus is shown in Figure 11.7. The vocal tract is an acoustic tube that is terminated at one end by the vocal chords and at the other end by the lips. An ancillary tube, the nasal tract, can be connected or disconnected by the movement of the velum. The shape of the vocal tract is determined by the position of the lips, jaws, the tongue, and the velum. Sounds can be generated in different ways. Voiced sounds are produced by exciting the vocal tract with quasi-periodic pulses of air pressure caused by vibration of the vocal chords. Unvoiced or the fricative sounds are produced by forming a constriction somewhere in the vocal tract and forcing air through the constriction, thereby creating turbulence that produces a source of noise to excite the vocal tract [2]. The vocal tract can be characterized by its natural frequencies (or formants), which correspond to resonance in the sound transmission characteristics of the vocal tract.

### 11.4.2 Autocorrelation

In the voiced intervals, the speech signal is characterized by a sequence of peaks that occur periodically at the fundamental frequency of the speech signal. In contrast, during unvoiced intervals the peaks are relatively smaller and do not occur in any discernible pattern. Autocorrelation is a common method of obtaining the pitch of the speech signal. Periodicity in the autocorrelation function indicates the periodicity of the speech signal.

Speech is not a stationary signal but the properties of the speech signal remain fixed over relatively long time intervals. However, the major limitation of the autocorrelation representation is that it retains too much of the information in the speech signal. Techniques known as spectrum flattening techniques are applied to the speech signal before performing the autocorrelation so as to filter out extraneous details. The block diagram of a clipping autocorrelation pitch detector is shown in Figure 11.8.

## Autocorrelation Computation

The computation of the autocorrelation function for a three-level center-clipped signal is particularly simple [3]. If we denote the output of the three-level center clipper as $y(n)$, then the product terms $y(n+m)y(n+m+k)$ in the autocorrelation function [4]

$$R_n(K) = \sum_{m=0}^{N-k-1} y(n+m)y(n+m+k)$$

can have only three different values; that is,

$$\begin{aligned} y(n+m)y(n+m+k) &= 0 \quad \text{if } y(n+m) = 0 \text{ or } y(n+m+k) = 0, \\ &= +1 \quad \text{if } y(n+m) = y(n+m+k), \text{ and} \\ &= -1 \quad \text{if } y(n+m) \neq y(n+m+k) \end{aligned}$$

The three-level clipping scheme is shown in Figure 11.9. The algorithm for pitch period estimation is summarized below:

The speech signal is filtered with a 900 Hz lowpass analog filter and sampled at the rate of 10 KHz.

Filtered signal segments, each of length 30 msec (300 samples), are selected at 10-msec intervals. Thus, the segments overlap by 20 msec.

The average of absolute magnitudes is computed with a 100-sample rectangular window. The peak signal level in each frame is compared to a threshold determined by measuring the peak signal level for 50 msec of background noise, as shown in the block "compute silence level threshold" in the block diagram. If the peak signal level is above the threshold, signifying that the segment is speech, not noise, then the algorithm proceeds as follows; otherwise the segment is classified as silence and no further action is taken.

The clipping level is determined as a fixed percentage (e.g., 68%) of the minimum of the maximum absolute values in the first and last 100 samples of the speech segment.

Using this clipping level, the speech signal is processed by a three-level center clipper, and the correlation function is computed over a range spanning the expected range of pitch periods.

The largest peak of the autocorrelation function is located and the peak value is compared to a fixed threshold (e.g., 30% of $R_n(0)$). If the peak falls

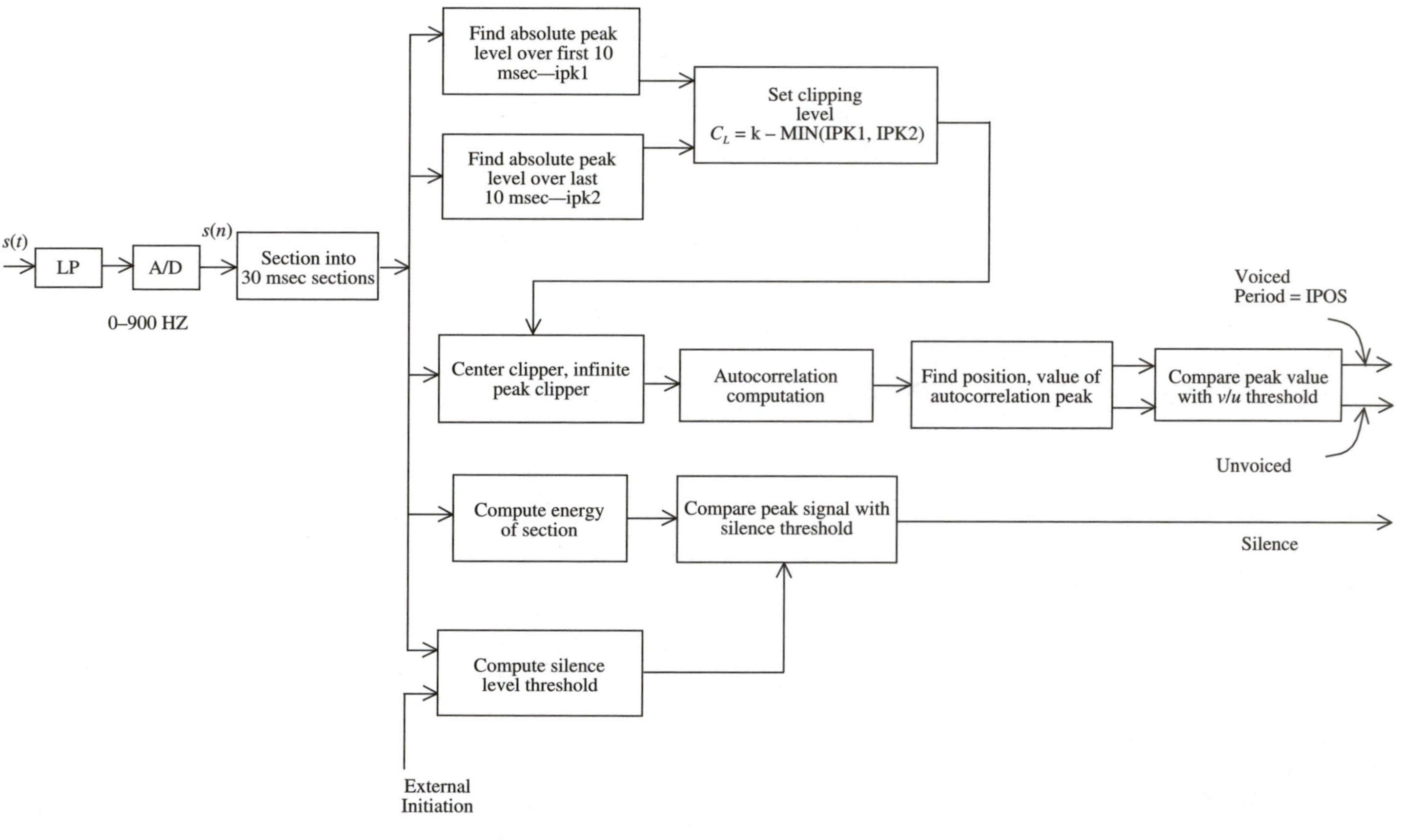


**Figure 11.8** The block diagram of a clipping autocorrelation pitch detector

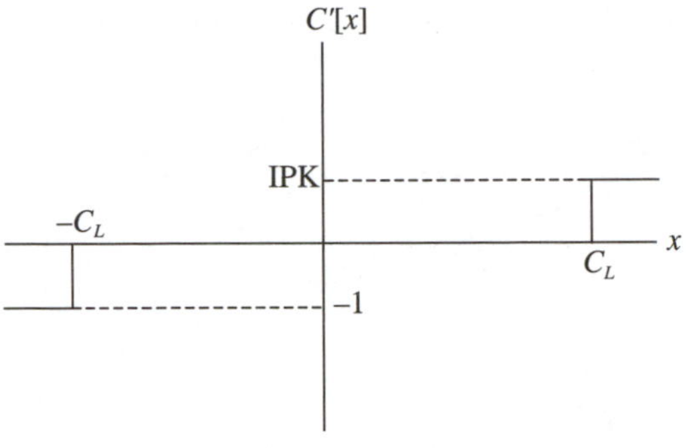


**Figure 11.9** Three-level center-clipped signal

below threshold, the segment is classified as unvoiced, and if it is above the threshold, the pitch period is defined as the location of the largest peak.

### 11.4.3 Implementation on the TMS320C54xx Processor

Speech samples were recorded using voice recorder software in Windows 98. The signal was sampled at 16 KHz in 16-bit mono format. The autocorrelation module is the most computation-intensive section for pitch detection. For this reason DSP was used to compute a 400-point autocorrelation for a 480-sample segment. For the sampling frequency of 16 KHz, 30 msec of speech corresponds to 480 samples, and it takes about 17200 clock cycles or 0.17 msec for the TMS320C5402 running at 100 MHz. Timing can be improved by using a lower sampling rate and thereby reducing the section size. Reduction of window size for computation of autocorrelation or using adaptive methods for determining the frame size will further reduce the computations involved. Figure 11.10 shows the autocorrelation output of a voiced speech signal and Figure 11.11 that of an unvoiced speech signal. The complete implementation of the block diagram shown in Figure 11.8 is left as an exercise for the implementor.

## 11.5 An Image Processing System

Images represent huge amounts of data. Image processing applications such as high-definition television, video conferencing, computer communication, and so forth require large storage and high-speed channels for handling the huge volumes of data. In order to reduce the storage and communication channel bandwidth requirements to manageable levels, data compression

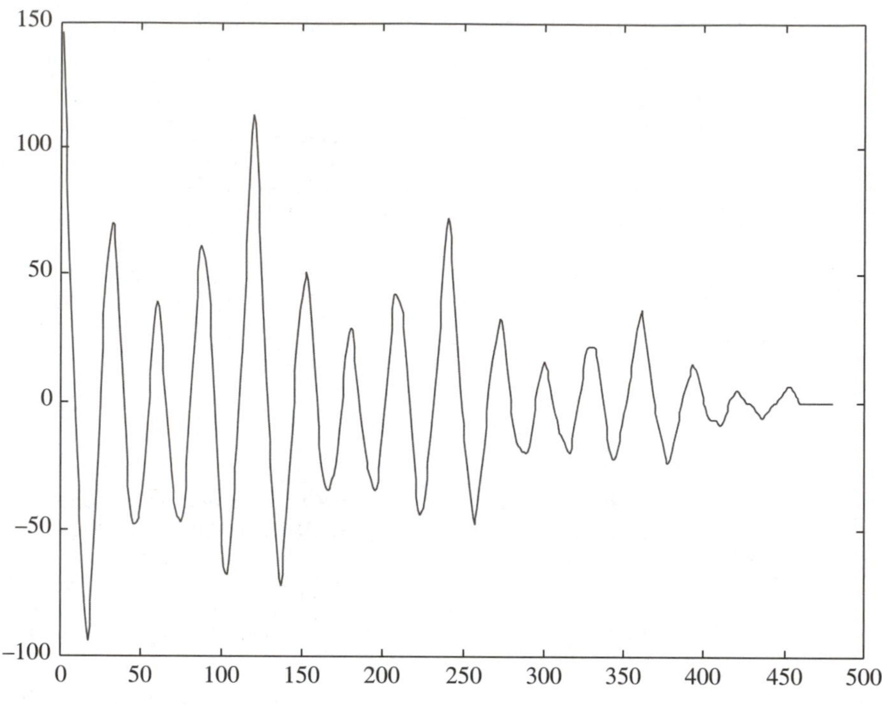


**Figure 11.10** Typical autocorrelation output for a voiced speech segment

techniques are imperative. Data compression on the order of 20 to 50 is feasible depending on the actual picture contents and techniques adopted for compression.

JPEG, which stands for Joint Photographic Experts Group, the name of the committee that wrote the standard, is a still-image compression standard. JPEG is used to compress either full-color or gray-scale images of natural or real-world scenes. It works well on pictures such as photographs and naturalistic artwork, not so well on lettering, simple cartoons, and line drawings. JPEG is "lossy," meaning that the decompressed image is not exactly the same as the original. JPEG is designed to exploit known limitations of the human eye, notably the fact that small color changes are perceived less accurately than small changes in brightness. Thus, JPEG is intended for compressing images that will be looked at by humans. The usefulness of JPEG is that the degree of lossiness can be adjusted by varying the compression parameters. This means that the image maker can trade off file size against image quality. JPEG achieves image compression by methodically throwing away visually insignificant image information. This information includes the high-frequency components of the image, which are less important to image content than the

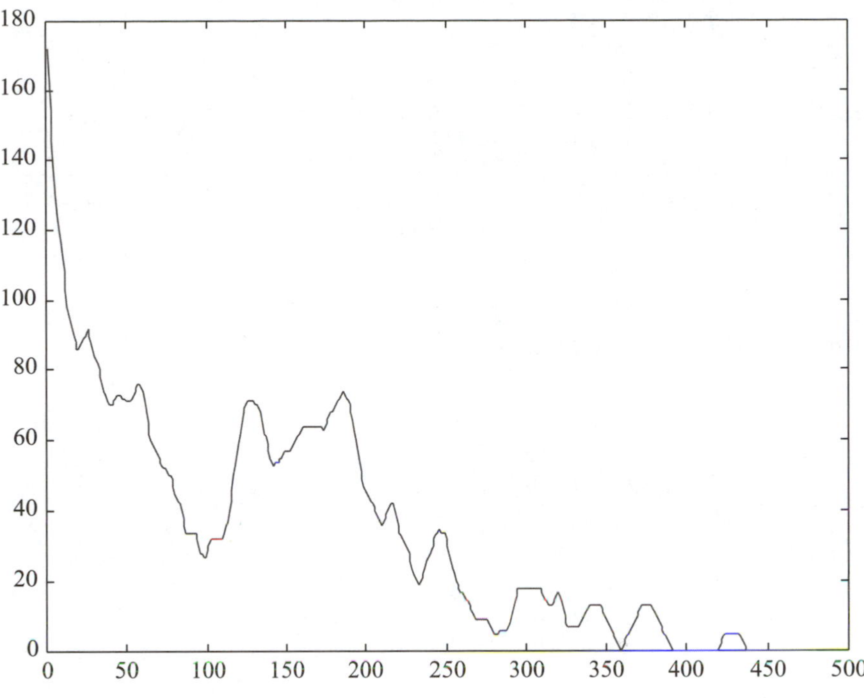


**Figure 11.11** Typical autocorrelation output for an unvoiced speech segment

low-frequency components. When an image is compressed using JPEG, the discarded high-frequency component cannot be retrieved. Another important aspect of JPEG is that decoders can trade off decoding speed against image quality, by using approximations to the required calculations.

### 11.5.1 JPEG Algorithm Overview

The original image is divided into $8 \times 8$ blocks. Each $8 \times 8$ block is transformed by the forward discrete cosine transform (DCT), which extracts the various frequency components and their relative amplitudes of the two-dimensional image signal represented by the $8 \times 8$ block into a set of 64 values, referred to as DCT coefficients [5]. Each of the 64 coefficients is then quantized using a quantizing table, which allocates more bits for coefficients corresponding to more dominant frequency components and fewer or zero bits for insignificant frequency components. The resulting 64 values (including zero values) are further coded by a process known as entropy encoding, wherein based on the statistical probability of occurrence of these long sequences, shorter codes are allotted to long-running sequences of 0s and 1s.

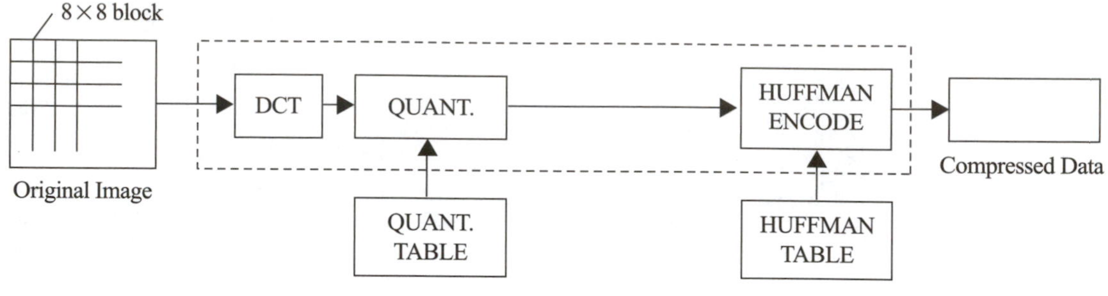


**Figure 11.12** The block diagram of the JPEG encoder

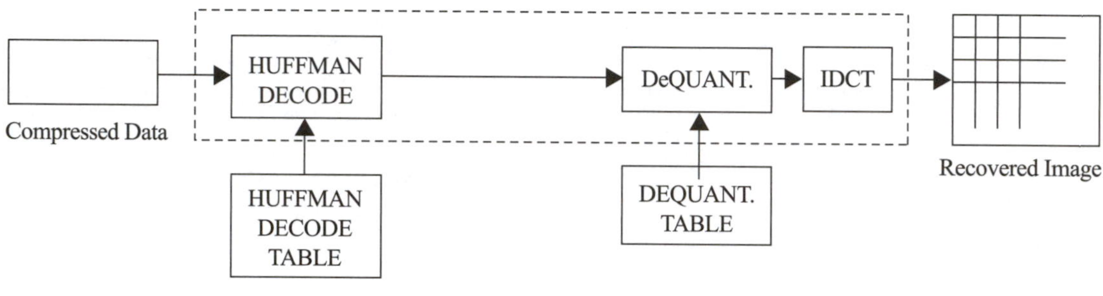


**Figure 11.13** The block diagram of the JPEG decoder

This way, the two-dimensional image data is converted to a bitstream of much smaller size compared to the original image data retaining most of the image features while discarding the insignificant information not easily discernible by the human eye. Figure 11.12 shows the block diagram of a JPEG encoder. The JPEG decoding process is the reverse of encoding and it is shown in Figure 11.13 [6].

### 11.5.2 JPEG Encoding

As mentioned above, the first step in JPEG encoding is computing the forward DCT of the $8 \times 8$ image block. We obtain the 64 DCT coefficients after applying the forward DCT on the two-dimensional image matrix. One of these values is referred as the dc coefficient and the other 63 as the ac coefficients. The forward DCT is computed from the equation

$$f_{v,u} = \frac{1}{4} c_u c_v \sum_{x=0}^{7} \sum_{y=0}^{7} f_{y,x} \cos \frac{(2x+1)u\pi}{16} \cos \frac{(2y+1)v\pi}{16}$$

The second step is quantization. Each of the 64 coefficients is quantized using one of 64 corresponding values from a quantization table. After quantization, the dc coefficient and the ac coefficients are prepared for entropy coding, which is also known as Huffman coding. The previous dc coefficient is subtracted from the current dc coefficient, and the difference is encoded. The 63 quantized ac coefficients undergo no such differential encoding, but are converted into a one-dimensional zig-zag sequence before being coded. Since many coefficients are zero, runs of zeros are identified and coded efficiently.

### 11.5.3 JPEG Decoding

In the reverse processes of Huffman decoding, dequantization and the inverse DCT are used to recover the original image data. The Huffman decoding table is used to recover the compressed data from the bitstream format to 64 16-bit data. The values in the dequantization table are the inverse of the values in the quantization table. The inverse DCT equation is

$$f_{y,x} = \frac{1}{4}\sum_{u=0}^{7}\sum_{y=0}^{7} c_u c_v f_{v,u} \cos\frac{(2x+1)u\pi}{16} \cos\frac{(2y+1)v\pi}{16}$$

After IDCT, decoding of the $8 \times 8$ image block is completed. The last procedure is to combine the $8 \times 8$ blocks to create the image.

### 11.5.4 Encoding and Decoding of JPEG Using the TMS320C54xx

For implementing the DCT of an $8 \times 8$ block, the FDCT algorithm by Lee [7] is used. The signal flow graph for computing the 8-point DCT using Lee's DCT algorithm is shown in Figure 11.14. The IDCT is obtained using the same flow graph by reversing the direction of the arrows and inputs given from the opposite side.

The matrix used for quantization and dequantization is shown in Figure 11.15. Notice the large quantization steps at the high-frequency end of the matrix compared to the smaller values at the low-frequency end.

For the implementation described here, the Huffman-coding and -decoding algorithms were programmed in C and interfaced to the DSP codes for DCT/quantization and IDCT/dequantization, respectively. After merging, the entire program was run in the TMS320C5402 processor. Encoding an image of a $256 \times 256$ size requires approximately 150,000,000 instruction cycles, or 150 msec in the TMS320VC5402, with an instruction cycle of 10 ns. The time taken for decoding is about the same. Figure 11.16 shows a sample image before and after being processed by the JPEG encoder and decoder. The two images look very much alike.

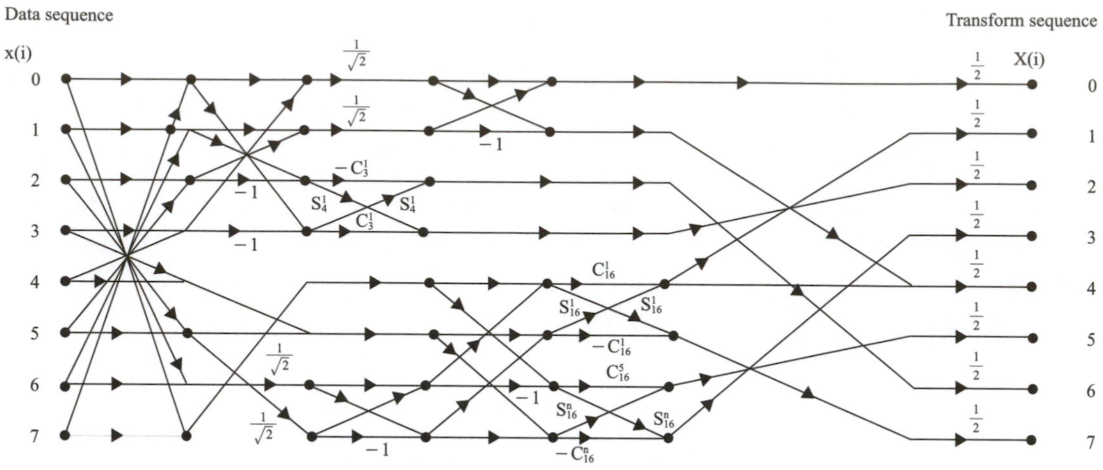


**Figure 11.14** Signal flow graph for an 8-point DCT algorithm

$$
\begin{bmatrix}
16 & 11 & 10 & 16 & 24 & 40 & 51 & 61 \\
12 & 12 & 14 & 19 & 26 & 58 & 60 & 55 \\
14 & 13 & 16 & 24 & 40 & 57 & 69 & 56 \\
14 & 17 & 22 & 29 & 51 & 87 & 80 & 62 \\
18 & 22 & 37 & 56 & 68 & 109 & 103 & 77 \\
24 & 35 & 55 & 64 & 81 & 104 & 113 & 92 \\
49 & 64 & 78 & 87 & 103 & 121 & 120 & 101 \\
72 & 92 & 95 & 98 & 112 & 100 & 103 & 99
\end{bmatrix}
$$

**Figure 11.15** Matrix used for quantization and dequantization

## 11.6 A Position Control System for a Hard Disk Drive

One important application for digital signal processors is the positioning of a read/write head on a hard disk. The DSP provides the computational capability, while a microcontroller handles the driver's functions for positioning of the head. Today, the single-chip solution using a DSP offers low cost, improved reliability, and low power consumption for hard disk controllers.

Details of the control system are shown in Figure 11.17. The parameter to be controlled is the drive input of a servomotor that determines the position of the read/write head on the disk. The controller issues the appropriate commands to the servomotor via the DAC. The servomotor, in turn, moves

**Figure 11.16** A sample image before and after JPEG processing: (a) raw image and (b) the image after JPEG compression and decompression

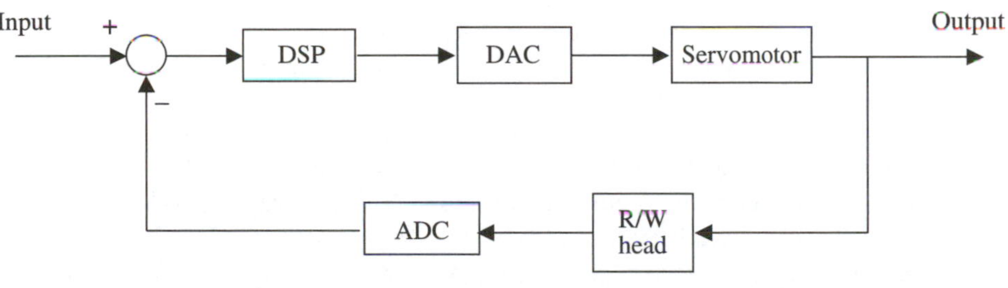


**Figure 11.17** The block diagram of a hard disk drive servo control system

the read/write head from the current position to the desired track on the disk. The design objective is to keep the position error minimized at all times. The DSP controller incorporates the algorithm to minimize the position error and use the position error to control the motor.

With the constant increase in disk storage capacity, there is a steady increase in the number of tracks and a decrease in their widths. The demand for accurate head position and tracking requires a more frequent sampling of the head position than would have been otherwise needed. Another reason for increasing the sampling rate is the decrease in the time constant of the process to be controlled. Therefore, the disk controller must be capable of high

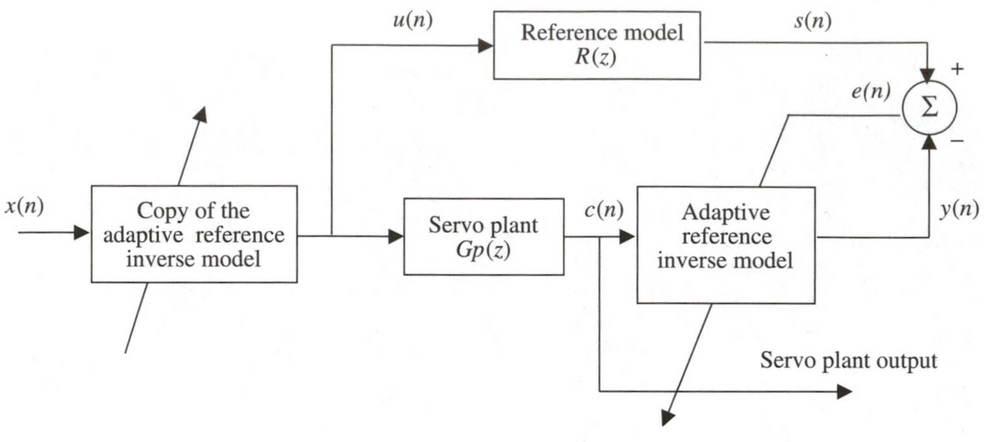


**Figure 11.18** An adaptive scheme for head positioning in a servo control system in the presence of noise

sampling rates in addition to a math-intensive algorithm for the digital control of the servomotor.

Figure 11.18 shows an adaptive scheme for head positioning in the presence of environmental variations [8]. According to this scheme, the servo-plant output, $c(n)$, must follow a reference (desired) model output, $s(n)$. A digital controller, $D(z)$ and the servo plant, $G(z)$, comprise the reference model, $R(z)$, while a servomotor, $G(s)$, and the DAC comprise the servo plant, $Gp(z)$. The adaptive reference inverse model is an inverse model of the servo plant, which, when combined with the servo plant and the reference model, gives an output, $y(n)$, that follows the reference model output $s(n)$. The adaptive reference inverse model is computed offline. Once an adaptive reference inverse model is obtained, it is incorporated into the control system. The servo plant is driven by the output obtained from a copy of the adaptive reference inverse model, which is updated after each seek operation. This ensures that the servo plant follows the same profile as the reference model at all times.

The reference model transfer function as given in reference [8] is

$$R(z) = C(z)/E(z) = \frac{0.01524z + 0.0147}{z^2 - 1.6847z + 0.7147}$$

This reference model may now be used to derive the adaptive reference inverse model of the servo plant. Figure 11.19 illustrates the adaptive reference inverse modeling technique. This particular model incorporates a 40-tap transversal filter whose coefficients are updated according to the least-mean-square algorithm. The compromise between the accuracy and computational complexity dictates the choice of the number of taps in the transversal filter. The following equations describe this model:

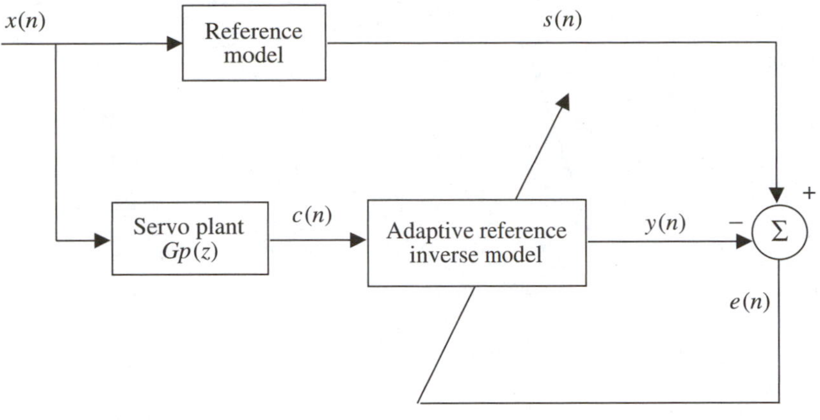


**Figure 11.19** Adaptive inverse modeling scheme

*Servo plant output:*

$$c(n) = 0.0048x(n-1) + 0.0046x(n) + 1.9c(n-1) - 0.9094c(n-2)$$

*Reference model output:*

$$s(n) = 0.01524x(n-1) + 0.0147x(n-2) + 1.68476s(n-1) - 0.7147s(n-2)$$

*Adaptive reference inverse model output:*

$$y(n) = \sum w(i)c(n-i+1); \quad i = 1 \text{ to } 40$$

*Error:* $e(n) = s(n) - y(n)$

*Weight vector update:*

$$w_i = w_i + \mu e c(n-i+1), \quad i = 1, 40$$

The weight vectors, $w_i$, which represent the adaptive reference inverse model, are obtained by performing 500 iterations of the adaptive loop. The parameter $\mu$, which determines the rate of convergence for obtaining the weight vector, is chosen empirically, in this case, to be 0.05. The input to the system is assumed to be a step. Once the adaptive reference of the inverse model is obtained, it can be applied to the control system of Figure 11.18. Due to the adaptive reference inverse model, any variations in the internal variables of the servo plant result in corresponding changes in the coefficients of the adaptive reference inverse model. Hence, the servo output follows the reference at all times.

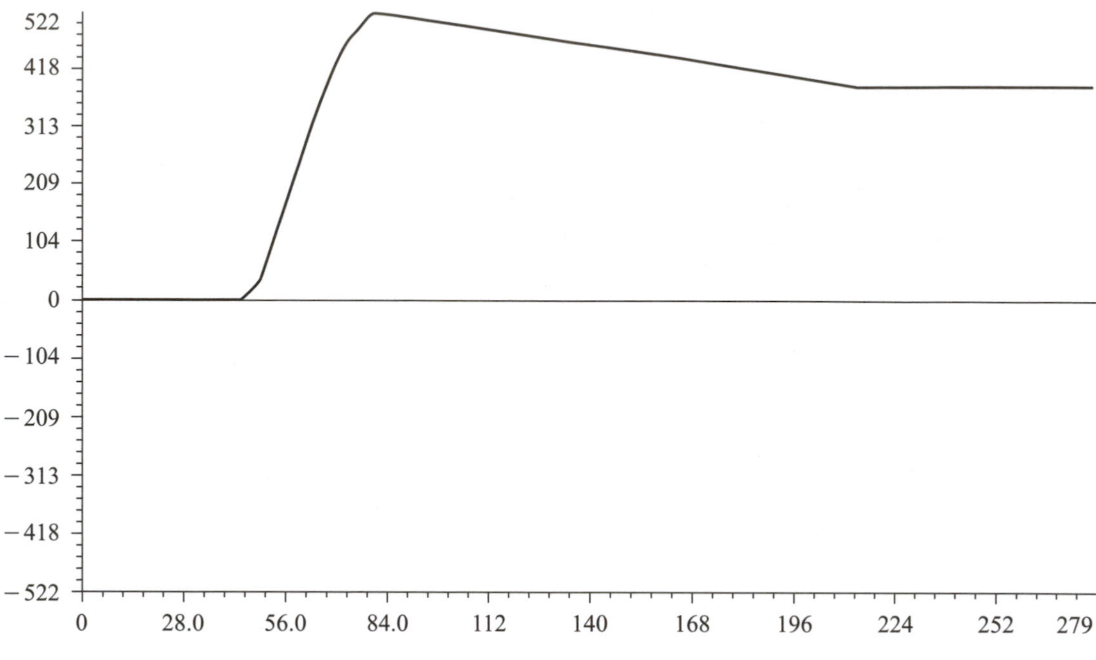


**Figure 11.20** The control signal $c(n)$ as obtained from the implementation of the control system on the TMS320C5402 processor

The initial 40 coefficients of the transversal filter were obtained from a MATLAB program and were then used in the final assembly language code of the DSP. Once an input impulse is given to a DSP, the output settles down in about 201300 instruction cycles, which is 0.2 msec for the TMS320C5402, for which each instruction cycle takes 10 nsec. Figure 11.20 shows the graph for $c(n)$ obtained from the actual implementation of the control system on the TMS320C5402 processor.

## 11.7 DSP-Based Power Meter

Measurement of power is an important task in evaluating performance of a system or a household appliance. Power has been conventionally measured using older electromagnetic-mechanical systems. This project is about designing a power measuring system using modern DSP technology. The result of this approach can be a device that provides better performance at lower cost. The project details are available elsewhere in a report [9].

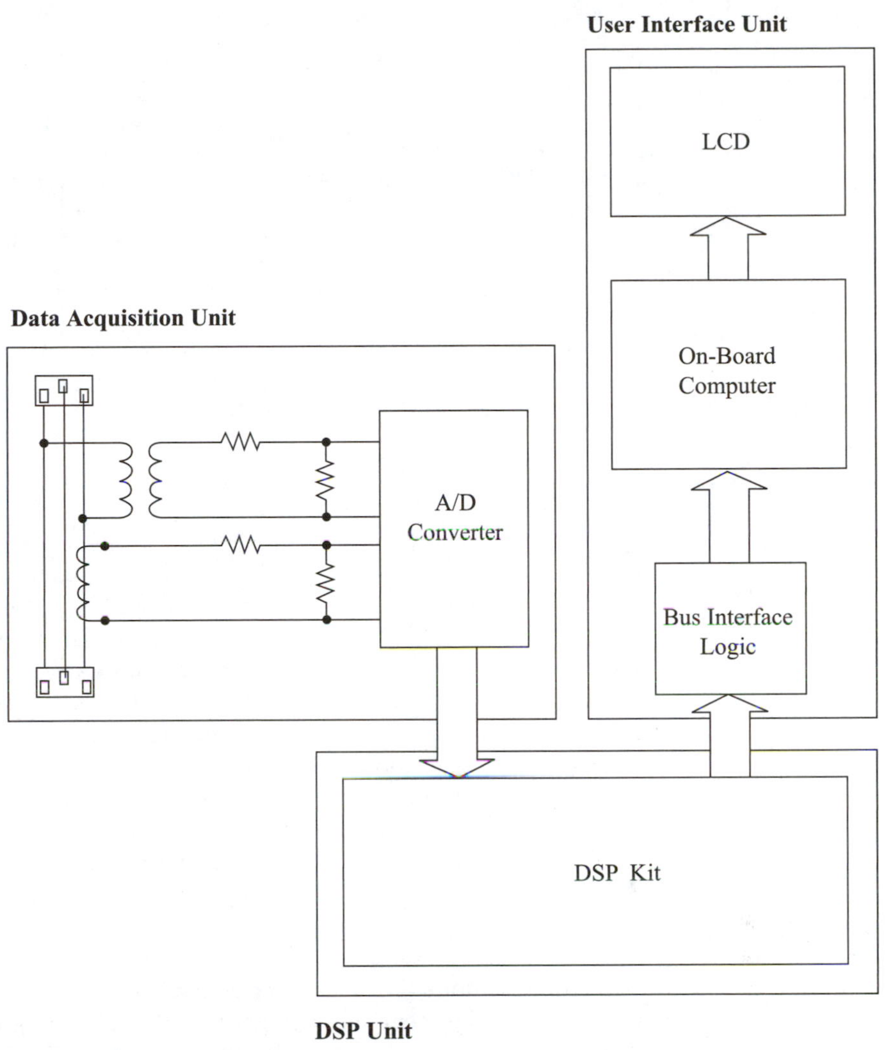


**Figure 11.21** Block diagram of a DSP-based power meter

### 11.7.1 Power Measurement System

Figure 11.21 shows a block diagram that can be implemented to measure power. The block diagram shows three functional units: the data acquisition unit, the DSP unit, and the user interface unit. The data acquisition unit gets the electrical signals representing power, the DSP unit processes the signals to compute power, and the user interface presents the results to the user for viewing graphically.

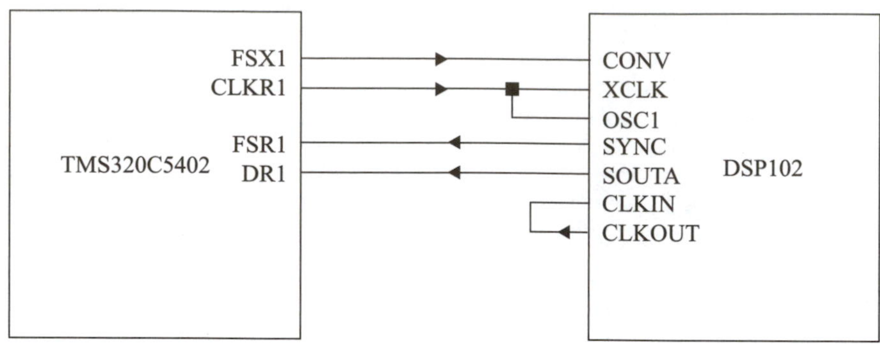


**Figure 11.22** ADC interface to the DSP using McBSPs

### Data Acquisition Unit

Data acquisition consists of acquiring the voltage signal using a voltage transformer and the current signal using a current transformer. The voltage transformer is used to transform the voltage signal to a value that can be handled by an A/D converter. Similarly, the current transformer produces voltage proportional to the current in the circuit. This voltage is fed to a second A/D converter. The A/D converters produce digital data at the selected sampling rate. The number of A/D bits specifies the resolution for the digital signal. The dual-channel ADC device (DSP102 from TI), with the maximum sampling rate of 200 KHz, 16-bit resolution, and serial interface, was used in this design.

### DSP Unit

For power computations, the Texas Instruments DSK5402 DSP board was used. The development software package, CCS, running on PC, was used to develop and download software for the DSP. The DSP's on-chip multichannel buffered serial ports (McBSP0 and McBSP1) provide the mechanism to collect data from the two A/D converters on the data acquisition unit. Figure 11.22 shows the interface between the A/D converters and the DSP.

Programming the McBSP registers can configure the clock and sampling frequencies. The sampling frequency was programmed for 12.2 KHz. The DSP is programmed to generate the A/D convert pulse. The A/D supplies two data samples as a 32-bit number after asserting the sync signal on the FSR of the serial port. The receiver has two 16-bit registers, DRR11 and DRR21, that receive the data every conversion cycle. From here, it is the DMA that transfers the signal data to the DSP memory. Two DMA channels are used for the two signals. The DMA is also used to transfer the data from the DSP to the user interface unit using another DMA channel and the transmitter register DXR10

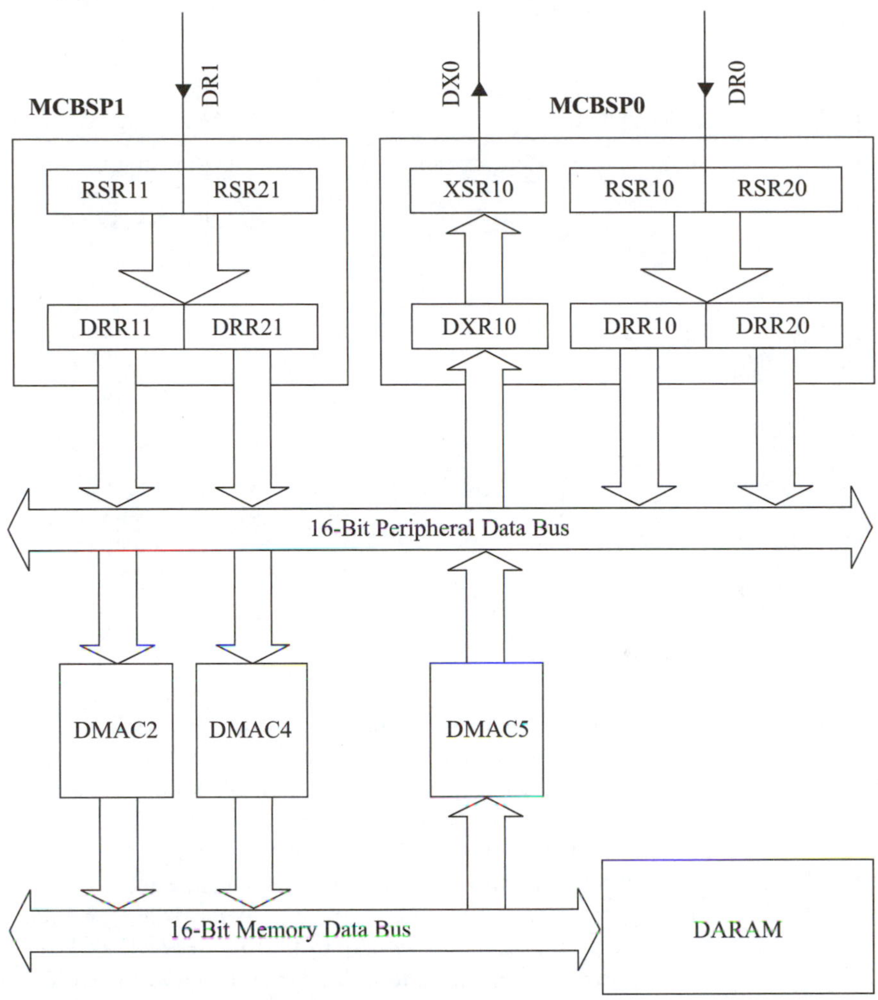


**Figure 11.23** DSP's DMA system for interfacing to ADCs and to the user interface unit

on the McBSP0 serial port. Figure 11.23 shows the details of the DMA interface for receiving the A/D data using McBSP1 and transmitting the computed signal data using the McBSP0.

### User Interface Unit

The user interface displays the signals received from the DSP. For this purpose, in this project a complete embedded computer system was used. In this way, the DSP can dedicate itself to analyzing the data, and the computer

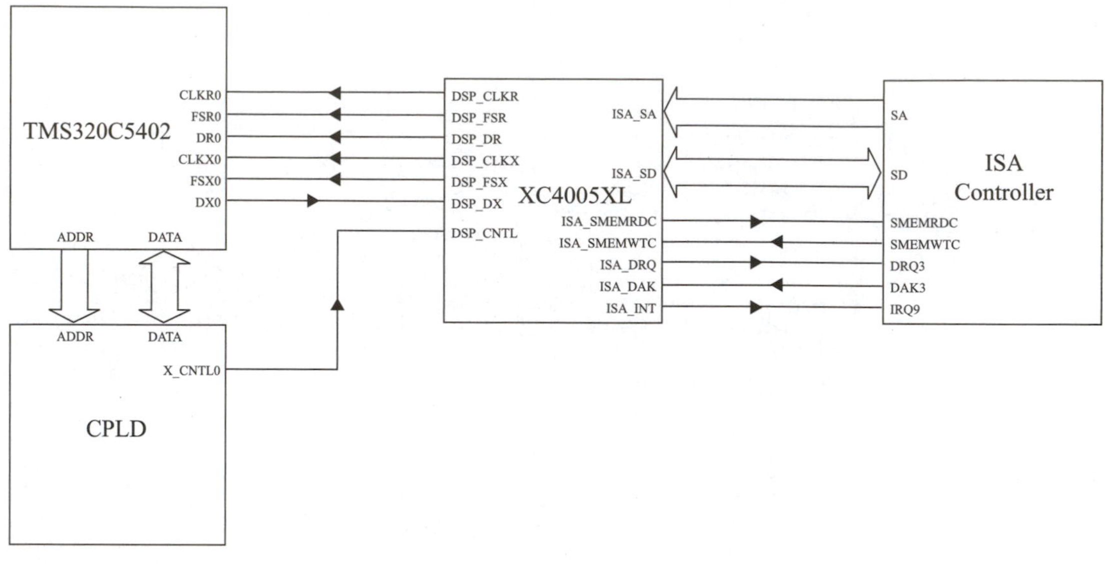


**Figure 11.24** DSP-to-computer interface logic

provides capability to display signals on a LCD display. This requires provision of two-way communication between the DSP and the computer. The interface was designed using Xilinx XC4005XL-PC84 FPGA. An EEPROM is used to configure the FPGA upon power-up.

The bus interface logic shown in Figure 11.24 has two main responsibilities. First, it controls the DSP and second, it supplies the DSP data to the computer. For the computer to control the DSP, the interface converts parallel data and delivers it serially to the DSP. For the DSP to send data to the computer, the opposite mechanism, that is, serial-to-parallel conversion, is used in addition to coordinating the DMA transfer to the computer.

The embedded computer uses a graphic controller to drive the LCD. The LCD provides a resolution of $640 \times 480$. To implement the interface, a few computer resources such as interrupt and memory locations are dedicated to the interface.

### 11.7.2 Software for the Power Meter

The software for the device consists of the system software and the application software. The system software consists of routines that manage the hardware, both for the DSP and the computer.

The DSP system software consists of modules, written in C, that manage the ADC and operate the computer interface logic. This software is stored in

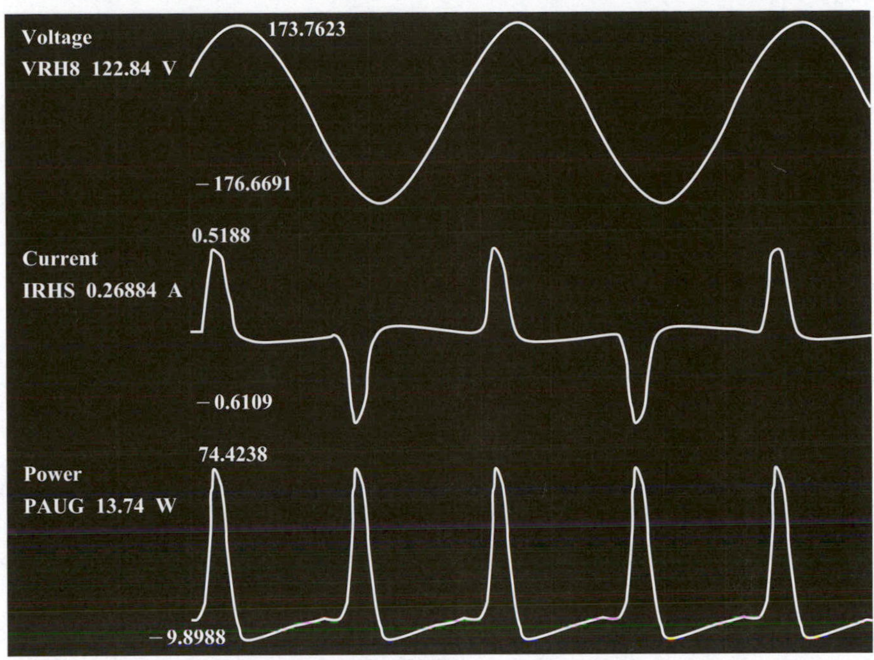


**Figure 11.25** Waveforms and computed quantities as displayed on the LCD screen

the flash memory of the DSK. A user interface device driver that runs on the computer provides read/write operations to the DSP and starts the DSP.

The application software running on the DSP uses the current and voltage signal data to compute the quantities in the following equations:

$$p(k) = v(k)i(k), \quad k = k, k+1, \ldots, k+N$$

$$V_{\text{rms}} = \sqrt{1/N \sum_{N} v^2(k)}$$

$$I_{\text{rms}} = \sqrt{1/N \sum_{N} i^2(k)}$$

$$P_{\text{avg}} = 1/N \sum_{N} p(k)$$

An example of these computed signals and quantities as displayed on the LCD screen is shown in Figure 11.25. The waveforms and the text are displayed using the user interface software running on the computer.

## 11.8 Summary

In this chapter, we have seen that the programmable DSP can be used for a variety of applications. Although these applications vary in the nature of the signals to be processed and their computational complexities, the architecture and other features of the DSP are suitable for implementing these and many other applications. As examples, we studied the use of the DSP for five representative applications.

## References


1. Singh, A., Hines, J., and Somps, C. "A digital Signal Processor Based Hand Held Multichannel, Multiple-Subject Biotelemetry System," *NASA Ames University Consortium Report*, NCC2-5112, 1996.
2. Rabiner, L. R., and Schafer, R. W. "Digital Representation of Speech Signals," *Proc. IEEE*, Vol. 63, pp. 662–677, April 1975.
3. Rabiner, L. R. "On the Use of Autocorrelation for Pitch Detection," *IEEE Trans. Acoust., Speech, and Signal Processing*, Vol. ASSP-25, No. 1, February 1977.
4. Rabiner, L. R., and Schafer, R. W. *Digital Processing of Speech Signals*, Prentice Hall Inc., 1978.
5. Rao, K. R., and Yip, P. *Discrete Cosine Transform: Algorithms, Advantages, Applications*, Academic Press, 1969.
6. Shi, Y. Q., and Sun, H. *Image and Video Compression for Multimedia Engineering: Fundamentals, Algorithms, and Standards*, CRC Press, 2000.
7. Lee, B. G. "A new algorithm to calculate the Discrete Cosine Transform," *IEEE Trans. Acoust. Speech, and Signal Processing*, Vol. ASSP-32, pp. 1243–1245, December 1984.
8. Nekoogar, F., and Moriarty, G. *Digital Control using Digital Signal Processing*, Prentice Hall Information and System Sciences Series, 1999.
9. Muico, U., and Larios, H. *DSP-Based Power Measurement Device*, EE198 Senior Project Report, San Jose State University, Fall 2001.

# Appendix A

## Architectural Details of TMS320VC5416 Digital Signal Processor

| 15–7 | 6 | 5 | 4 | 3 | 2 | 1 | 0 |
|---|---|---|---|---|---|---|---|
| IPTR | $MP/\overline{MC}$ | OVLY | AVIS | DROM | CLK OFF | SMUL | SST |
| **R/W-1FF** | **$MP/\overline{MC}$ Pin** | **R/W-0** | **R/W-0** | **R/W-0** | **R/W-0** | **R/W-0** | **R/W-0** |

**LEGEND: R = Read, W = Write**

| BIT NO. | BIT NAME | RESET VALUE | FUNCTION |
|---|---|---|---|
| 15–7 | IPTR | 1FFh | 15–7 IPTR 1FFh Interrupt vector pointer. The 9-bit IPTR field points to the 128-word program page where the interrupt vectors reside. The interrupt vectors can be remapped to RAM for boot-loaded operations. At reset, these bits are all set to 1; the reset vector always resides at address FF80h in program memory space. The RESET instruction does not affect this field. |
| 6 | $MP/\overline{MC}$ | $MP/\overline{MC}$ pin | Microprocessor/microcomputer mode. $MP/\overline{MC}$ enables/disables the on-chip ROM to be addressable in program memory space.<br>_ $MP/\overline{MC}$ = 0: The on-chip ROM is enabled and addressable.<br>_ $MP/\overline{MC}$ = 1: The on-chip ROM is not available. $MP/\overline{MC}$ is set to the value corresponding to the logic level on the $MP/\overline{MC}$ pin when sampled at reset. This pin is not sampled again until the next reset. The RESET instruction does not affect this bit. This bit can also be set or cleared by software. |

**Figure A.1** Processor Mode Status (PMST) Register (*continued*)

(Courtesy of Texas Instruments Inc.)

| | | | |
|---|---|---|---|
| 5 | OVLY | 0 | RAM overlay. OVLY enables on-chip dual-access data RAM blocks to be mapped into program space. The values for the OVLY bit are: _ OVLY = 0: The on-chip RAM is addressable in data space but not in program space. _ OVLY = 1: The on-chip RAM is mapped into program space and data space. Data page 0 (addresses 0h to 7Fh), however, is not mapped into program space. |
| 4 | AVIS | 0 | Address visibility mode. AVIS enables/disables the internal program address to be visible at the address pins. _ AVIS = 0: The external address lines do not change with the internal program address. Control and data lines are not affected and the address bus is driven with the last address on the bus. _ AVIS = 1: This mode allows the internal program address to appear at the pins of the 5416 so that the internal program address can be traced. Also, it allows the interrupt vector to be decoded in conjunction with IACK when the interrupt vectors reside on on-chip memory. |
| 3 | DROM | 0 | DROM enables on-chip DARAM4–7 to be mapped into data space. The DROM bit values are: _ DROM = 0: The on-chip DARAM4–7 is not mapped into data space. _ DROM = 1: The on-chip DARAM4–7 is mapped into data space. |
| 2 | CLKOFF | 0 | CLOCKOUT off. When the CLKOFF bit is 1, the output of CLKOUT is disabled and remains at a high level. |
| 1 | SMUL | N/A | Saturation on multiplication. When SMUL = 1, saturation of a multiplication result occurs before performing the accumulation in a MAC of MAS instruction. The SMUL bit applies only when OVM = 1 and FRCT = 1. |
| 0 | SST | N/A | Saturation on store. When SST = 1, saturation of the data from the accumulator is enabled before storing in memory. The saturation is performed after the shift operation. |

**Figure A.1** Continued

| 15 | 14 12 | 11 9 | 8 6 | 5 3 | 2 0 |
|---|---|---|---|---|---|
| XPA | I/O | Data | Data | Program | Program |
| R/W-0 | R/W-111 | R/W-111 | R/W-111 | R/W-111 | R/W-111 |

**LEGEND: R = Read, W = Write, 0/111 = Value after reset**

| BIT NO. | BIT NAME | RESET VALUE | FUNCTION |
|---|---|---|---|
| 15 | XPA | 0 | Extended program address control bit. XPA is used in conjunction with the program space fields (bits 0 through 5) to select the address range for program space wait states. |
| 14–12 | I/O | 111 | I/O space. The field value (0–7) corresponds to the base number of wait states for I/O space accesses within addresses 0000–FFFFh. The SWSM bit of the SWCR defines a multiplication factor of 1 or 2 for the base number of wait states. |
| 11–9 | Data | 111 | Upper data space. The field value (0–7) corresponds to the base number of wait states for external data space accesses within addresses 8000–FFFFh. The SWSM bit of the SWCR defines a multiplication factor of 1 or 2 for the base number of wait states. |
| 8–6 | Data | 111 | Lower data space. The field value (0–7) corresponds to the base number of wait states for external data space accesses within addresses 0000–7FFFh. The SWSM bit of the SWCR defines a multiplication factor of 1 or 2 for the base number of wait states. |
| 5–3 | Program | 111 | Upper program space. The field value (0–7) corresponds to the base number of wait states for external program space accesses within the following addresses:<br>• XPA = 0: xx8000 – xxFFFFh<br>• XPA = 1: 400000h – 7FFFFFh. The SWSM bit of the SWCR defines a multiplication factor of 1 or 2 for the base number of wait states. |
| 2–0 | Program | 111 | Program space. The field value (0–7) corresponds to the base number of wait states for external program space accesses within the following addresses:<br>• XPA = 0: xx0000 – xx7FFFh<br>• XPA = 1: 000000 – 3FFFFFh. The SWSM bit of the SWCR defines a multiplication factor of 1 or 2 for the base number of wait states. |

**Figure A.2** Software Wait-Signal Register (SWWSR)

(Courtesy of Texas Instruments Inc.)

| 15–1 | 0 |
|---|---|
| **Reserved** | **SWSM** |
| **R/W-0** | **R/W-0** |

**LEGEND: R = Read, W = Write**

| PIN NO. | PIN NAME | RESET VALUE | FUNCTION |
|---|---|---|---|
| 15–1 | Reserved | 0 | These bits are reserved and are unaffected by writes. |
| 0 | SWSM | 0 | Software wait-state multiplier. Used to multiply the number of wait states defined in the SWWSR by a factor of 1 or 2.<br>• SWSM = 0: wait-state base values are unchanged (multiplied by 1).<br>• SWSM = 1: wait-state base values are multiplied by 2 for a maximum of 14 wait states |

**Figure A.3** Software Wait-State Control Register (SWCR)

(Courtesy of Texas Instruments Inc.)

| 15 | 14 13 | 12 | 11 3 | 2 | 1 | 0 |
|---|---|---|---|---|---|---|
| **CONSEC*** | **DIVFCT** | **IACKOFF** | **Reserved** | **HBH** | **BH** | **Res** |
| **R/W-1** | **R/W-11** | **R/W-1** | **R** | **R/W-0** | **R/W-0** | **R** |

**R = Read, W = Write**

| BIT | NAME | RESET VALUE | FUNCTION |
|---|---|---|---|
| 15 | CONSEC* | 1 | Consecutive bank-switching. Specifies the bank-switching mode.<br>CONSEC* = 0: Bank-switching on 32K bank boundaries only. This bit is cleared if fast access is desired for continuous memory reads (i.e., no starting and trailing cycles between read cycles).<br>CONSEC* = 1: Consecutive bank switches on external memory reads. Each read cycle consists of 3 cycles: starting cycle, read cycle, and trailing cycle. |
| 13–14 | DIVFCT | 11 | CLKOUT output divide factor. The CLKOUT output is driven by an on-chip source having a frequency equal to 1/(DIVFCT + 1) of the DSP clock.<br>DIVFCT = 00: CLKOUT is not divided.<br>DIVFCT = 01: CLKOUT is divided by 2 from the DSP clock.<br>DIVFCT = 10: CLKOUT is divided by 3 from the DSP clock.<br>DIVFCT = 11: CLKOUT is divided by 4 from the DSP clock (default value following reset). |
| 12 | IACKOFF | 1 | IACK* signal output off. Controls the output of the /IACK signal. IACKOFF is set to 1 at reset.<br>IACKOFF = 0: The IACK* signal output off function is disabled.<br>IACKOFF = 1: The IACK* signal output off function is enabled. |
| 11–3 | Rsvd | — | Reserved |
| 2 | HBH | 0 | HPI bus holder. Controls the HPI bus holder. HBH is cleared to 0 at reset.<br>HBH = 0: The bus holder is disabled except when HPI16 = 1.<br>HBH = 1: The bus holder is enabled. When not driven, the HPI data bus, HD[7:0] is held in the previous logic level. |

**Figure A.4** Bank-Switching Control Register (BSCR) *(continued)*

(Courtesy of Texas Instruments Inc.)

| | | | |
|---|---|---|---|
| 1 | BH | 0 | Bus holder. Controls the bus holder. BH is cleared to 0 at reset.<br>BH = 0: The bus holder is disabled.<br>BH = 1: The bus holder is enabled. When not driven, the data bus, D[15:0] is held in the previous logic level. |
| 0 | Rsvd | — | Reserved |

**Figure A.4** Continued

| CLKMD1 | CLKMD2 | CLKMD3 | CLKMD RESET VALUE | CLOCK MODE |
|---|---|---|---|---|
| 0 | 0 | 0 | 0000h | 1/2 (PLL disabled) |
| 0 | 0 | 1 | 9007h | PLL × 10 |
| 0 | 1 | 0 | 4007h | PLL × 5 |
| 1 | 0 | 0 | 1007h | PLL × 2 |
| 1 | 1 | 0 | F007h | PLL × 1 |
| 1 | 1 | 1 | 0000h | 1/2 (PLL disabled) |
| 1 | 0 | 1 | F000h | 1/4 (PLL disabled) |
| 0 | 1 | 1 | — | Reserved (Bypass mode) |

†The external CLKMD1–CLKMD3 pins are sampled to determine the desired clock generation mode while RS is low. Following reset, the clock generation mode can be reconfigured by writing to the internal clock mode register in software.

**Figure A.5** Clock Mode Settings at Reset

(Courtesy of Texas Instruments Inc.)

| | ADDRESS | | |
|---|---|---|---|
| NAME | DEC | HEX | DESCRIPTION |
| IMR | 0 | 0 | Interrupt mask register |
| IFR | 1 | 1 | Interrupt flag register |
| — | 2–5 | 2–5 | Reserved for testing |
| ST0 | 6 | 6 | Status register 0 |
| ST1 | 7 | 7 | Status register 1 |
| AL | 8 | 8 | Accumulator A low word (15–0) |
| AH | 9 | 9 | Accumulator A high word (31–16) |
| AG | 10 | A | Accumulator A guard bits (39–32) |
| BL | 11 | B | Accumulator B low word (15–0) |
| BH | 12 | C | Accumulator B high word (31–16) |
| BG | 13 | D | Accumulator B guard bits (39–32) |
| TREG | 14 | E | Temporary register |
| TRN | 15 | F | Transition register |
| AR0 | 16 | 10 | Auxiliary register 0 |
| AR1 | 17 | 11 | Auxiliary register 1 |
| AR2 | 18 | 12 | Auxiliary register 2 |
| AR3 | 19 | 13 | Auxiliary register 3 |
| AR4 | 20 | 14 | Auxiliary register 4 |
| AR5 | 21 | 15 | Auxiliary register 5 |
| AR6 | 22 | 16 | Auxiliary register 6 |
| AR7 | 23 | 17 | Auxiliary register 7 |
| SP | 24 | 18 | Stack pointer register |
| BK | 25 | 19 | Circular buffer size register |
| BRC | 26 | 1A | Block repeat counter |
| RSA | 27 | 1B | Block repeat start address |
| REA | 28 | 1C | Block repeat end address |
| PMST | 29 | 1D | Processor mode status (PMST) register |
| XPC | 30 | 1E | Extended program page register |
| — | 31 | 1F | Reserved |

**Figure A.6** Memory-Mapped Registers

(Courtesy of Texas Instruments Inc.)

| NAME | ADDRESS | | DESCRIPTION |
|---|---|---|---|
| | DEC | HEX | |
| DRR20 | 32 | 20 | McBSP 0 Data Receive Register 2 |
| DRR10 | 33 | 21 | McBSP 0 Data Receive Register 1 |
| DXR20 | 34 | 22 | McBSP 0 Data Transmit Register 2 |
| DXR10 | 35 | 23 | McBSP 0 Data Transmit Register 1 |
| TIM | 36 | 24 | Timer Register |
| PRD | 37 | 25 | Timer Period Register |
| TCR Timer | 38 | 26 | Control Register |
| — | 39 | 27 | Reserved |
| SWWSR | 40 | 28 | Software Wait-State Register |
| BSCR | 41 | 29 | Bank-Switching Control Register |
| — | 42 | 2A | Reserved |
| SWCR | 43 | 2B | Software Wait-State Control Register |
| HPIC | 44 | 2C | HPI Control Register (HMODE = 0 only) |
| — | 45–47 | 2D–2F | Reserved |
| DRR22 | 48 | 30 | McBSP 2 Data Receive Register 2 |
| DRR12 | 49 | 31 | McBSP 2 Data Receive Register 1 |
| DXR22 | 50 | 32 | McBSP 2 Data Transmit Register 2 |
| DXR12 | 51 | 33 | McBSP 2 Data Transmit Register 1 |
| SPSA2 | 52 | 34 | McBSP 2 Subbank Address Register† |
| SPSD2 | 53 | 35 | McBSP 2 Subbank Data Register† |
| — | 54–55 | 36–37 | Reserved |
| SPSA0 | 56 | 38 | McBSP 0 Subbank Address Register† |
| SPSD0 | 57 | 39 | McBSP 0 Subbank Data Register† |
| — | 58–59 | 3A–3B | Reserved |
| GPIOCR | 60 | 3C | General-Purpose I/O Control Register |
| GPIOSR | 61 | 3D | General-Purpose I/O Status Register |
| CSIDR | 62 | 3E | Device ID Register |
| — | 63 | 3F | Reserved |
| DRR21 | 64 | 40 | McBSP 1 Data Receive Register 2 |
| DRR11 | 65 | 41 | McBSP 1 Data Receive Register 1 |
| DXR21 | 66 | 42 | McBSP 1 Data Transmit Register 2 |
| DXR11 | 67 | 43 | McBSP 1 Data Transmit Register 1 |
| — | 68–71 | 44–47 | Reserved |
| SPSA1 | 72 | 48 | McBSP 1 Subbank Address Register† |
| SPSD1 | 73 | 49 | McBSP 1 Subbank Data Register† |
| — | 74–83 | 4A–53 | Reserved |
| DMPREC | 84 | 54 | DMA Priority and Enable Control Register |

**Figure A.7** Peripheral Memory-Mapped Registers (*continued*)

(Courtesy Texas Instruments Inc.)

| | | | |
|---|---|---|---|
| DMSA DMA | 85 | 55 | Subbank Address Register‡ |
| DMSDI | 86 | 56 | DMA Subbank Data Register with Autoincrement‡ |
| DMSDN | 87 | 57 | DMA Subbank Data Register‡ |
| CLKMD | 88 | 58 | Clock Mode Register (CLKMD) |
| — | 89–95 | 59–5F | Reserved |

†See Table Figure A.8 for a detailed description of the McBSP control registers and their subaddresses.

‡See Table Figure A.9 for a detailed description of the DMA subbank addressed registers.

**Figure A.4** Continued

| McBSP0 | | McBSP1 | | McBSP2 | | SUB | |
|---|---|---|---|---|---|---|---|
| NAME | ADDRESS | NAME | ADDRESS | NAME | ADDRESS | ADDRESS | DESCRIPTION |
| SPCR10 | 39h | SPCR11 | 49h | SPCR12 | 35h | 00h | Serial port control register 1 |
| SPCR20 | 39h | SPCR21 | 49h | SPCR22 | 35h | 01h | Serial port control register 2 |
| RCR10 | 39h | RCR11 | 49h | RCR12 | 35h | 02h | Receive control register 1 |
| RCR20 | 39h | RCR21 | 49h | RCR22 | 35h | 03h | Receive control register 2 |
| XCR10 | 39h | XCR11 | 49h | XCR12 | 35h | 04h | Transmit control register 1 |
| XCR20 | 39h | XCR21 | 49h | XCR22 | 35h | 05h | Transmit control register 2 |
| SRGR10 | 39h | SRGR11 | 49h | SRGR12 | 35h | 06h | Sample rate generator register 1 |
| SRGR20 | 39h | SRGR21 | 49h | SRGR22 | 35h | 07h | Sample rate generator register 2 |
| MCR10 | 39h | MCR11 | 49h | MCR12 | 35h | 08h | Multichannel register 1 |
| MCR20 | 39h | MCR21 | 49h | MCR22 | 35h | 09h | Multichannel register 2 |
| RCERA0 | 39h | RCERA1 | 49h | RCERA2 | 35h | 0Ah | Receive channel enable register partition A |
| RCERB0 | 39h | RCERB1 | 49h | RCERA2 | 35h | 0Bh | Receive channel enable register partition B |
| XCERA0 | 39h | XCERA1 | 49h | XCERA2 | 35h | 0Ch | Transmit channel enable register partition A |
| XCERB0 | 39h | XCERB1 | 49h | XCERA2 | 35h | 0Dh | Transmit channel enable register partition B |
| PCR0 | 39h | PCR1 | 49h | PCR2 | 35h | 0Eh | Pin control register |
| RCERC0 | 39h | RCERC1 | 49h | RCERC2 | 35h | 010h | Additional channel enable register for 128-channel selection |

**Figure A.8** McBSP Control Registers and Subaddresses *(continued)*

(Courtesy of Texas Instruments Inc.)

| | | | | | | | |
|---|---|---|---|---|---|---|---|
| RCERD0 | 39h | RCERD1 | 49h | RCERD2 | 35h | 011h | Additional channel enable register for 128-channel selection |
| XCERC0 | 39h | XCERC1 | 49h | XCERC2 | 35h | 012h | Additional channel enable register for 128-channel selection |
| XCERD0 | 39h | XCERD1 | 49h | XCERD2 | 35h | 013h | Additional channel enable register for 128-channel selection |
| RCERE0 | 39h | RCERE1 | 49h | RCERE2 | 35h | 014h | Additional channel enable register for 128-channel selection |
| RCERF0 | 39h | RCERF | 49h | RCERF2 | 35h | 015h | Additional channel enable register for 128-channel selection |
| XCERE0 | 39h | XCERE1 | 49h | XCERE2 | 35h | 016h | Additional channel enable register for 128-channel selection |
| XCERF0 | 39h | XCERF1 | 49h | XCERF2 | 35h | 017h | Additional channel enable register for 128-channel selection |
| RCERG0 | 39h | RCERG1 | 49h | RCERG2 | 35h | 018h | Additional channel enable register for 128-channel selection |
| RCERH0 | 39h | RCERH | 49h | RCERH2 | 35h | 019h | Additional channel enable register for 128-channel selection |
| XCERG0 | 39h | XCERG1 | 49h | XCERG2 | 35h | 01Ah | Additional channel enable register for 128-channel selection |
| XCERH0 | 39h | XCERH1 | 49h | XCERH2 | 35h | 01Bh | Additional channel enable register for 128-channel selection |

**Figure A.8** Continued

| NAME | ADDRESS | SUB ADDRESS | DESCRIPTION |
|---|---|---|---|
| DMSRC0 | 56h/57h | 00h | DMA channel 0 source address register |
| DMDST0 | 56h/57h | 01h | DMA channel 0 destination address register |
| DMCTR0 | 56h/57h | 02h | DMA channel 0 element count register |
| DMSFC0 | 56h/57h | 03h | DMA channel 0 sync select and frame count register |
| DMMCR0 | 56h/57h | 04h | DMA channel 0 transfer mode control register |
| DMSRC1 | 56h/57h | 05h | DMA channel 1 source address register |
| DMDST1 | 56h/57h | 06h | DMA channel 1 destination address register |
| DMCTR1 | 56h/57h | 07h | DMA channel 1 element count register |
| DMSFC1 | 56h/57h | 08h | DMA channel 1 sync select and frame count register |
| DMMCR1 | 56h/57h | 09h | DMA channel 1 transfer mode control register |
| DMSRC2 | 56h/57h | 0Ah | DMA channel 2 source address register |
| DMDST2 | 56h/57h | 0Bh | DMA channel 2 destination address register |
| DMCTR2 | 56h/57h | 0Ch | DMA channel 2 element count register |
| DMSFC2 | 56h/57h | 0Dh | DMA channel 2 sync select and frame count register |
| DMMCR2 | 56h/57h | 0Eh | DMA channel 2 transfer mode control register |
| DMSRC3 | 56h/57h | 0Fh | DMA channel 3 source address register |
| DMDST3 | 56h/57h | 10h | DMA channel 3 destination address register |
| DMCTR3 | 56h/57h | 11h | DMA channel 3 element count register |
| DMSFC3 | 56h/57h | 12h | DMA channel 3 sync select and frame count register |
| DMMCR3 | 56h/57h | 13h | DMA channel 3 transfer mode control register |
| DMSRC4 | 56h/57h | 14h | DMA channel 4 source address register |
| DMDST4 | 56h/57h | 15h | DMA channel 4 destination address register |
| DMCTR4 | 56h/57h | 16h | DMA channel 4 element count register |
| DMSFC4 | 56h/57h | 17h | DMA channel 4 sync select and frame count register |
| DMMCR4 | 56h/57h | 18h | DMA channel 4 transfer mode control register |
| DMSRC5 | 56h/57h | 19h | DMA channel 5 source address register |
| DMDST5 | 56h/57h | 1Ah | DMA channel 5 destination address register |
| DMCTR5 | 56h/57h | 1Bh | DMA channel 5 element count register |
| DMSFC5 | 56h/57h | 1Ch | DMA channel 5 sync select and frame count register |
| DMMCR5 | 56h/57h | 1Dh | DMA channel 5 transfer mode control register |
| DMSRCP | 56h/57h | 1Eh | DMA source program page address (common channel) |
| DMDSTP | 56h/57h | 1Fh | DMA destination program page address (common channel) |
| DMIDX0 | 56h/57h | 20h | DMA element index address register 0 |
| DMIDX1 | 56h/57h | 21h | DMA element index address register 1 |
| DMFRI0 | 56h/57h | 22h | DMA frame index register 0 |
| DMFRI1 | 56h/57h | 23h | DMA frame index register 1 |
| DMGSA0 | 56h/57h | 24h | DMA global source address reload register, channel 0 |

**Figure A.9** DMA Subbank Addressed Registers *(continued)*

(Courtesy of Texas Instruments Inc.)

| | | | |
|---|---|---|---|
| DMGDA0 | 56h/57h | 25h | DMA global destination address reload register, channel 0 |
| DMGCR0 | 56h/57h | 26h | DMA global count reload register, channel 0 |
| DMGFR0 | 56h/57h | 27h | DMA global frame count reload register, channel 0 |
| XSRCDP | 56h/57h | 28h | DMA extended source data page (currently not supported) |
| XDSTDP | 56h/57h | 29h | DMA extended destination data page (currently not supported) |
| DMGSA1 | 56h/57h | 2Ah | DMA global source address reload register, channel 1 |
| DMGDA1 | 56h/57h | 2Bh | DMA global destination address reload register, channel 1 |
| DMGCR1 | 56h/57h | 2Ch | DMA global count reload register, channel 1 |
| DMGFR1 | 56h/57h | 2Dh | DMA global frame count reload register, channel 1 |
| DMGSA2 | 56h/57h | 2Eh | DMA global source address reload register, channel 2 |
| DMGDA2 | 56h/57h | 2Fh | DMA global destination address reload register, channel 2 |
| DMGCR2 | 56h/57h | 30h | DMA global count reload register, channel 2 |
| DMGFR2 | 56h/57h | 31h | DMA global frame count reload register, channel 2 |
| DMGSA3 | 56h/57h | 32h | DMA global source address reload register, channel 3 |
| DMGDA3 | 56h/57h | 33h | DMA global destination address reload register, channel 3 |
| DMGCR3 | 56h/57h | 34h | DMA global count reload register, channel 3 |
| DMGFR3 | 56h/57h | 35h | DMA global frame count reload register, channel 3 |
| DMGSA4 | 56h/57h | 36h | DMA global source address reload register, channel 4 |
| DMGDA4 | 56h/57h | 37h | DMA global destination address reload register, channel 4 |
| DMGCR4 | 56h/57h | 38h | DMA global count reload register, channel 4 |
| DMGFR4 | 56h/57h | 39h | DMA global frame count reload register, channel 4 |
| DMGSA5 | 56h/57h | 3Ah | DMA global source address reload register, channel 5 |
| DMGDA5 | 56h/57h | 3Bh | DMA global destination address reload register, channel 5 |
| DMGCR5 | 56h/57h | 3Ch | DMA global count reload register, channel 5 |
| DMGFR5 | 56h/57h | 3Dh | DMA global frame count reload register, channel 5 |

**Figure A.9** Continued

| NAME | LOCATION | | PRIORITY | FUNCTION |
|---|---|---|---|---|
| | DECIMA | HEX | | |
| RS, SINTR | 0 | 00 | 1 | Reset (hardware and software reset) |
| NMI, SINT 16 | 4 | 04 | 2 | Nonmaskable interrupt |
| SINT17 | 8 | 08 | — | Software interrupt #17 |
| SINT18 | 12 | 0C | — | Software interrupt #18 |
| SINT19 | 16 | 10 | — | Software interrupt #19 |
| SINT20 | 20 | 14 | — | Software interrupt #20 |
| SINT21 | 24 | 18 | — | Software interrupt #21 |
| SINT22 | 28 | 1C | — | Software interrupt #22 |
| SINT23 | 32 | 20 | — | Software interrupt #23 |
| SINT24 | 36 | 36 | — | Software interrupt #24 |
| SINT25 | 40 | 28 | — | Software interrupt #25 |
| SINT26 | 44 | 2C | — | Software interrupt #26 |
| SINT27 | 48 | 30 | — | Software interrupt #27 |
| SINT28 | 52 | 34 | — | Software interrupt #28 |
| SINT29 | 56 | 38 | — | Software interrupt #29 |
| SINT30 | 60 | 3C | — | Software interrupt #30 |
| INT0, SINT0 | 64 | 40 | 3 | External user interrupt #0 |
| INT1, SINT1 | 68 | 44 | 4 | External user interrupt #1 |
| INT2, SINT2 | 72 | 48 | 5 | External user interrupt #2 |
| TINT, SINT3 | 76 | 4C | 6 | Timer interrupt |
| RINT0, SINT4 | 80 | 50 | 7 | McBSP #0 receive interrupt (default) |
| XINT0, SINT5 | 84 | 54 | 8 | McBSP #0 transmit interrupt (default) |
| RINT2, SINT6 | 88 | 58 | 9 | McBSP #2 receive interrupt (default) |
| XINT2, SINT7 | 92 | 5C | 10 | McBSP #2 transmit interrupt (default) |
| INT3, SINT8 | 96 | 60 | 11 | External user interrupt #3 |
| HINT, SINT9 | 100 | 64 | 12 | HPI interrupt |
| RINT1, SINT10 | 104 | 68 | 13 | McBSP #1 receive interrupt (default) |
| XINT1, SINT11 | 108 | 6C | 14 | McBSP #1 transmit interrupt (default) |
| DMAC4, SINT12 | 112 | 70 | 15 | DMA channel 4 (default) |
| DMAC5, SINT13 | 116 | 74 | 16 | DMA channel 5 (default) |
| Reserved | 120–127 | 78–7F | — | Reserved |

| 15–14 | 13 | 12 | 11 | 10 | 9 | 8 | 7 | 6 | 5 | 4 | 3 | 2 | 1 | 0 |
|---|---|---|---|---|---|---|---|---|---|---|---|---|---|---|
| Resvd | DMAC5 | DMAC4 | XINT1 | RINT1 | HINT | INT3 | XINT2 | RINT2 | XINT0 | RINT0 | TINT | INT2 | INT1 | INT0 |

**Figure A.10** Interrupt Vector Table and Interrupt Mask Register/Interrupt Flag Register (IMR/IFR) (Courtesy of Texas Instruments Inc.)

| 15–12 | 11 | 10 | 9–6 | 5 | 4 | 3–0 |
|---|---|---|---|---|---|---|
| Reserved | Soft | Free | PSC | TRB | TSS | TDDR |

| Bit | Name | Reset Value | Function |
|---|---|---|---|
| 15–12 | Reserved | — | Reserved; always read as 0. |
| 11 | Soft | 0 | Used in conjunction with the Free bit to determine the state of the timer when a breakpoint is encountered in the HLL debugger. When the Free bit is cleared, the Soft bit selects the timer mode.<br>Soft = 0 The timer stops immediately.<br>Soft = 1 The timer stops when the counter decrements to 0. |
| 10 | Free | 0 | Used in conjunction with the Soft bit to determine the state of the timer when a breakpoint is encountered in the HLL debugger. When the Free bit is cleared, the Soft bit selects the timer mode.<br>Free = 0 The Soft bit selects the timer mode.<br>Free = 1 The timer runs free regardless of the Soft bit. |
| 9–6 | PSC | — | Timer prescaler counter. Specifies the count for the on-chip timer. When PSC is decremented past 0 or the timer is reset, PSC is loaded with the contents of TDDR and the TIM is decremented. |
| 5 | TRB | — | Timer reload. Resets the on-chip timer. When TRB is set, the TIM is loaded with the value in the PRD and the PSC is loaded with the value in TDDR. TRB is always read as a 0. |
| 4 | TSS | 0 | Timer stop status. Stops or starts the on-chip timer. At reset, TSS is cleared and the timer immediately starts timing.<br>TSS = 0 The timer is started.<br>TSS = 1 The timer is stopped. |
| 3–0 | TDDR | 0000 | Timer divide-down ratio. Specifies the timer divide-down ratio (period) for the on-chip timer. When PSC is decremented past 0, PSC is loaded with the contents of TDDR. |

**Figure A.11** Timer Control Register (TCR)

(Courtesy of Texas Instruments Inc.)

# Index